# Domain Conditions and Social Rationality

Satish Kumar Jain

# Domain Conditions and Social Rationality

 Springer

Satish Kumar Jain
Formerly Professor
Jawaharlal Nehru University
New Delhi, India

ISBN 978-981-13-9671-7    ISBN 978-981-13-9672-4   (eBook)
https://doi.org/10.1007/978-981-13-9672-4

This Springer imprint is published by the registered company Springer Nature Singapore Pte Ltd.
The registered company address is: 152 Beach Road, #21-01/04 Gateway East, Singapore 189721, Singapore

*For*

*Abha, Avinash, Mayank,*
*Rajendra and Subrata*

# Preface

One implication of the impossibility theorems of social choice theory, Arrow impossibility theorem being the most important of them, is that all 'democratic' methods of arriving at social decisions by combining individual preferences which satisfy Arrow's independence of irrelevant alternatives, a requirement quite crucial for the unambiguity of social choices, fail to generate rational social preferences for some configurations of individual preferences. The problem is exemplified by the famous voting paradox associated with the majority rule. Under majority rule, it is possible to have alternative $x$ defeating alternative $y$ in a majority vote, alternative $y$ defeating alternative $z$ in a majority vote and alternative $z$ defeating alternative $x$ in a majority vote, thereby making it impossible to choose rationally from among these three alternatives. Thus, in the context of any rule that is to be used for arriving at social decisions on the basis of individual preferences, it is important to know the configurations of individual preferences under which it would be possible to choose rationally. This monograph is almost exclusively concerned with the derivation of conditions for various rules and classes of rules under which the social preferences would be rational. To make the monograph essentially self-contained, the basic social choice theoretic concepts, definitions, propositions and theorems needed for the subject matter of the monograph have been given in Chap. 2. Each of the Chaps. 3–10 deals with a particular rule or a class of rules and is essentially independent of other chapters. Chapter 11 depends, in addition to Chap. 2, on Chaps. 3, 6, 9 and 10. The treatment throughout is rigorous. Unlike most of the literature on domain conditions, care is taken in this monograph with respect to the number of individuals in the 'necessity' proofs.

This monograph was written while I was Indian Council of Social Science Research (ICSSR) National Fellow; thanks are due to ICSSR for awarding me the National Fellowship. I wish to thank the Management Development Institute for providing me with affiliation and for providing facilities. I also wish to thank the Economics faculty of the Management Development Institute, particularly Prof. Sunil Ashra and Prof. Rohit Prasad.

    Some of the material included in this book I had published as papers in journals
and in an edited volume; I thank the publishers of these journals and the edited
volume for permission to include material from them in the book. The editorial help
from Springer is gratefully acknowledged; I particularly wish to thank Ms. Nupoor
Singh. Thanks are due to Kaushal Kishore and Amit Kumar for their help in
proofreading.

New Delhi, India                                                    Satish Kumar Jain

# Contents

# About the Author

**Satish Kumar Jain** is an economist who holds a Master's in Economics from Delhi School of Economics and a Doctorate in Economics from the University of Rochester. He was on the faculty of the Centre for Economic Studies and Planning at Jawaharlal Nehru University for three and a half decades. He was Reserve Bank of India Chair Professor from 2011 to 2013, and was an Indian Council of Social Science Research (ICSSR) National Fellow from 2016 to 2018. He has taught at Shri Ram College of Commerce, Delhi; the Indian Statistical Institute, Delhi; and the Indian Institute of Information Technology, Hyderabad. He has authored *Economic Analysis of Liability Rules* (Springer, 2015); edited *Law and Economics* (Oxford University Press, 2010); and co-edited *Economic Growth, Efficiency, and Inequality* (with Anjan Mukherji, Routledge, 2015). His teaching and research interests include social choice theory, and law and economics.

# Abbreviations

| | |
|---|---|
| A | Anonymity |
| AP | Antagonistic preferences |
| AUEV | Absence of unique extremal value |
| CI | Cyclical indifference |
| CP | Conflictive preferences |
| DP | Dichotomous preferences |
| EP | Echoic preferences |
| ER | Extremal restriction |
| EVR | Extremal value restriction |
| I | Independence of irrelevant alternatives |
| LA | Limited agreement |
| LSEVR | Latin Square extremal value restriction |
| LSIRR-Q | Latin Square intransitive relation restriction-Q |
| LSLOR | Latin Square linear ordering restriction |
| LSLOR-Q | Latin Square linear ordering restriction-Q |
| LSPA | Latin Square partial agreement |
| LSPA-Q | Latin Square partial agreement-Q |
| LSUVR | Latin Square unique value restriction |
| $\overline{\text{M}}$ | Strict monotonicity |
| M | Monotonicity |
| MMD | Method of majority decision |
| N | Neutrality |
| $\overline{\text{P}}$ | Pareto-criterion |
| P | Weak Pareto-criterion |
| PA | Partial agreement |
| PI | Pareto-indifference |
| PP | Pareto-preference |
| PQT | Pareto quasi-transitivity |
| PR | Placement restriction |
| SAP(1) | Strongly antagonistic preferences (1) |
| SAP(2) | Strongly antagonistic preferences (2) |
| SC | Single-Cavedness |
| SDR | Social decision rule |

| SG | Separability into two groups |
|---|---|
| SP | Single-Peakedness |
| SPR | Strict placement restriction |
| SWF | Social welfare function |
| TP | Taboo preferences |
| UVR | Unique value restriction |
| VR(1) | Value restriction (1) |
| VR(2) | Value restriction (2) |
| WCP | Weak conflictive preferences |
| WER | Weak extremal restriction |
| WER-Q | Weak extremal restriction-Q |
| WLSEVR | Weak Latin Square extremal value restriction |
| WLSPA | Weak Latin Square partial agreement |
| WM | Weak monotonicity |
| WPQT | Weak Pareto quasi-transitivity |
| WPR | Weak Pareto rule |

# Glossary of Symbols

| | |
|---|---|
| $\sim$ | Not (negation) |
| $\wedge$ | And (conjunction) |
| $\vee$ | Or (disjunction) |
| $\rightarrow$ | If-then (conditional) |
| $\leftrightarrow$ | If and only if (biconditional) |
| iff | If and only if |
| $\forall$ | For all (universal quantifier) |
| $\exists$ | There exists (existential quantifier) |
| $\in$ | Is an element of |
| $\notin$ | Is not an element of |
| $B - A$ | Complement of set $A$ in set $B$ |
| $A \cap B$ | Intersection of sets $A$ and $B = \{x \mid x \in A \text{ and } x \in B\}$ |
| $A \cup B$ | Union of sets $A$ and $B = \{x \mid x \in A \text{ or } x \in B\}$ |
| $A \subseteq B$ | $A$ is a subset of $B$, i.e., $(\forall x)(x \in A \rightarrow x \in B)$ |
| $A = B$ | Sets $A$ and $B$ are equal, i.e., $(A \subseteq B \text{ and } B \subseteq A)$ |
| $A \neq B$ | Sets $A$ and $B$ are not equal, i.e., $\sim(A = B)$ |
| $A \subset B$ | $A$ is a proper subset of $B$, i.e., $(A \subseteq B \text{ and } A \neq B)$ |
| $A \supseteq B$ | $A$ is a superset of $B$, i.e., $(B$ is a subset of $A)$ |
| $A \supset B$ | $A$ is a proper superset of $B$, i.e., $(B$ is a proper subset of $A)$ |
| $A \times B$ | Cartesian product of sets $A$ and $B = \{(x, y) \mid x \in A \text{ and } y \in B\}$ |
| $S$ | The set of social alternatives |
| $s$ | Number of social alternatives |
| $\#X$ | Number of alternatives in the set $X$ |
| $N$ | The set of individuals |
| $n$ | Number of individuals |
| $n(), N()$ | Number of individuals holding the preferences specified in the parentheses |
| $\Phi$ | The set of permutations of alternatives in $S$ |
| $\phi$ | A permutation of alternatives in $S$ |
| $\Theta$ | The set of permutations of individuals in $N$ |
| $\theta$ | A permutation of individuals in $N$ |
| $R$ | Social binary weak preference relation, socially at least as good as |
| $P(R), P$ | Asymmetric part of $R$, socially better than |

| | |
|---|---|
| $I(R), I$ | Symmetric part of $R$, socially indifferent to |
| $C(S, R)$ | The set of best elements in $S$ according to $R$ |
| $R_i$ | Binary weak preference relation of individual $i$, for individual $i$ at least as good as |
| $P(R_i), P_i$ | Asymmetric part of $R_i$, for individual $i$ better than |
| $I(R_i), I_i$ | Symmetric part of $R_i$, for individual $i$ indifferent to |
| $\mathbb{N}$ | The set of positive integers |
| $\mathscr{B}$ | The set of all binary relations on the set of social alternatives $S$ |
| $\mathscr{C}$ | The set of all reflexive and connected binary relations on the set of social alternatives $S$ |
| $\mathscr{A}$ | The set of all reflexive, connected and acyclic binary relations on the set of social alternatives $S$ |
| $\mathscr{Q}$ | The set of all reflexive, connected and quasi-transitive binary relations on the set of social alternatives $S$ |
| $\mathscr{T}$ | The set of all reflexive, connected and transitive binary relations (orderings) on the set of social alternatives $S$ |
| $\mathscr{L}$ | The set of all reflexive, connected, transitive and anti-symmetric binary relations (linear orderings) on the set of social alternatives $S$ |
| $\mathscr{D}$ | A set of binary relations on the set of social alternatives $S$ |
| $\mathscr{T}^n$ | n-fold Cartesian product of $\mathscr{T}$ with itself |
| $\mathscr{Q}^n$ | $n$-fold Cartesian product of $\mathscr{Q}$ with itself |
| $\mathscr{D}^n$ | $n$-fold Cartesian product of $\mathscr{D}$ with itself |
| $(R_1, \ldots, R_n)$ | A profile of individual orderings, an element of $\mathscr{T}^n$; or a profile of reflexive, connected and quasi-transitive individual weak preference relations, an element of $\mathscr{Q}^n$ |
| $\mathbb{D}$ | A set of profiles of individual binary weak preference relations, a subset of $\mathscr{T}^n$ or a subset of $\mathscr{Q}^n$ |
| $R\|A$ | Restriction of $R$ to $A \subseteq S$, i.e., $R\|A = R \cap (A \times A)$ |
| $\mathscr{D}\|A$ | Restriction of $\mathscr{D}$ to $A \subseteq S$, i.e., $\mathscr{D}\|A = \{R\|A \,\|R \in \mathscr{D}\}, \mathscr{D} \subseteq \mathscr{B}$ |
| $D(x, y)$ | Almost decisive for the ordered pair $(x, y)$ |
| $\overline{D}(x, y)$ | Decisive for the ordered pair $(x, y)$ |
| $S(x, y)$ | Almost semidecisive for the ordered pair $(x, y)$ |
| $\overline{S}(x, y)$ | Semidecisive for the ordered pair $(x, y)$ |
| $D_{N-A}(x, y)$ | Almost $(N - A)$—decisive for the ordered pair $(x, y)$ |
| $\overline{D}_{N-A}(x, y)$ | $(N - A)$—decisive for the ordered pair $(x, y)$ |
| $S_{N-A}(x, y)$ | Almost $(N - A)$—semidecisive for the ordered pair $(x, y)$ |
| $\overline{S}_{N-A}(x, y)$ | $(N - A)$—semidecisive for the ordered pair $(x, y)$ |
| $W$ | The set of winning coalitions |
| $W_m$ | The set of minimal winning coalitions |
| $B$ | The set of blocking coalitions |
| $B_s$ | The set of strictly blocking coalitions |
| $WLS(xyzx)$ | Weak Latin Square with the cyclical arrangement $xyzx$ |
| $LS(xyzx)$ | Latin Square with the cyclical arrangement $xyzx$ |

| | |
|---|---|
| $T[WLS(xyzx)]$ | The set of all orderings of the triple $\{x,y,z\}$ that can be part of $WLS(xyzx)$ |
| $T[LS(xyzx)]$ | The set of all orderings of the triple $\{x,y,z\}$ that can be part of $LS(xyzx)$ |
| $Q[WLS(xyzx)]$ | The set of all reflexive, connected and quasi-transitive binary weak preference relations of the triple $\{x,y,z\}$ that can be part of $WLS(xyzx)$ |
| $Q[LS(xyzx)]$ | The set of all reflexive, connected and quasi-transitive binary weak preference relations of the triple $\{x,y,z\}$ that can be part of $LS(xyzx)$ |
| $B[WLS(xyzx)]$ | The set of all binary weak preference relations of the triple $\{x,y,z\}$ that can be part of $WLS(xyzx)$ |
| $B[LS(xyzx)]$ | The set of all binary weak preference relations of the triple $\{x,y,z\}$ that can be part of $LS(xyzx)$ |
| $\lceil t \rceil$ | Smallest integer greater than or equal to $t$ |
| $\lfloor t \rfloor$ | Largest integer smaller than or equal to $t$ |
| $xUy$ | $(\forall i \in N)\,(xR_iy)$ |
| $x\overline{U}y$ | $(\forall i \in N)\,(xP_iy)$ |
| $\mathcal{T}_1^T$ | $\{\mathcal{D} \subseteq \mathcal{T} \mid \text{every } (R_1,\ldots,R_n) \in \mathcal{D}^n \text{ yields transitive } R\}$ |
| $\mathcal{T}_2^T$ | $\{\mathcal{D} \subseteq \mathcal{T} \mid \text{some } (R_1,\ldots,R_n) \in \mathcal{D}^n \text{ yields intransitive } R\}$ |
| $\mathcal{T}_1^Q$ | $\{\mathcal{D} \subseteq \mathcal{T} \mid \text{every } (R_1,\ldots,R_n) \in \mathcal{D}^n \text{ yields quasi-transitive } R\}$ |
| $\mathcal{T}_2^Q$ | $\{\mathcal{D} \subseteq \mathcal{T} \mid \text{some } (R_1,\ldots,R_n) \in \mathcal{D}^n \text{ yields non quasi-transitive } R\}$ |
| $\mathcal{T}_1^A$ | $\{\mathcal{D} \subseteq \mathcal{T} \mid \text{every } (R_1,\ldots,R_n) \in \mathcal{D}^n \text{ yields acyclic } R\}$ |
| $\mathcal{T}_2^A$ | $\{\mathcal{D} \subseteq \mathcal{T} \mid \text{some } (R_1,\ldots,R_n) \in \mathcal{D}^n \text{ yields non-acyclic } R\}$ |

# Introduction

The problem of aggregating individual preferences arises in contexts in which the decision depends, at least partly, on the preferences of more than one individual. Most important of such contexts include elections and decision-making by representative bodies and committees. For combining individual preferences into overall preferences a variety of procedures are used. The most important of these procedures is that of the method of majority decision (majority rule) under which, between any two alternatives, one alternative is better from a collective point of view than the other one if and only if (iff) the number of individuals who consider the former to be better than the latter is greater than the number of individuals who consider the latter to be better than the former. Majority rule is one of the simplest and most commonly used procedures for making decisions at a collective level on the basis of preferences of individual members.

Although the appeal of the majority rule is straightforward, the rule has a serious shortcoming as it can give rise to quite paradoxical social preferences. Consider for instance a society consisting of three individuals, A, B and C, which is to choose one alternative out of the three mutually exclusive alternatives $x$, $y$, and $z$. Let the rankings[1] of these three alternatives in the descending order by the three individuals be:

| A's ranking | B's ranking | C's ranking |
| --- | --- | --- |
| $x$ | $y$ | $z$ |
| $y$ | $z$ | $x$ |
| $z$ | $x$ | $y$ |

---

[1] A preference relation is a ranking or ordering iff it satisfies the properties of reflexivity, connectedness and transitivity. Reflexivity holds iff every alternative is at least as good as itself; connectedness holds iff between any two distinct alternatives $x$ and $y$, $x$ is at least as good as $y$ or $y$ is at least as good as $x$; transitivity holds iff for any three alternatives $x, y, z$, if $x$ is at least as good as $y$, and $y$ is at least as good as $z$, then $x$ is at least as good as $z$.

© Springer Nature Singapore Pte Ltd. 2019
S. K. Jain, *Domain Conditions and Social Rationality*,
https://doi.org/10.1007/978-981-13-9672-4_1

As two out of three individuals prefer $x$ to $y$ and only one individual prefers $y$ to $x$; two out of three individuals prefer $y$ to $z$ and only one individual prefers $z$ to $y$; and two out of three individuals prefer $z$ to $x$ and only one individual prefers $x$ to $z$; we obtain the paradoxical result that $x$ is socially better than $y$, $y$ is socially better than $z$, and $z$ is socially better than $x$. Under this scenario it is impossible for the collective to make a rational choice. No matter which of the three alternatives is chosen, among the rejected alternatives there will be an alternative that will be strictly better than the chosen one. This paradox was discovered by the French philosopher and mathematician Marquis de Condorcet (1785).

In the above example, although each of the three individuals has a ranking (ordering) of alternatives, the social preferences generated by the majority rule fail to be a ranking. At a first glance, transitivity appears to be an essential requirement of rationality. If transitivity is taken to be a basic requirement then the search for a satisfactory rule for aggregating individual preferences into social preferences must for all practical purposes be confined to rules that aggregate individual orderings into a social ordering. This is the framework in which Arrow formulated the problem of social choice in his seminal work Social Choice and Individual Values (1951, 1963). The famous Arrow Impossibility Theorem shows that the following four conditions are logically inconsistent: (i) The rule for aggregating individual orderings into a social ordering, termed social welfare function by Arrow, must work for every logically possible configuration of individual orderings, i.e., the domain of the rule must be unrestricted. (ii) The rule must satisfy the weak Pareto-criterion. The weak Pareto-criterion requires that whenever some alternative $x$ is unanimously preferred by everyone in the society to some other alternative $y$, $x$ must be socially preferred to $y$. (iii) The rule must be non-dictatorial. An individual is a dictator iff it is the case that whenever he/she prefers some alternative $x$ to some other alternative $y$, $x$ is socially preferred to $y$. The rule is non-dictatorial iff no individual is a dictator. (iv) The rule must be such that the social preferences between any two alternatives must depend only on the individual preferences between those two alternatives. In other words, it must not be the case that although no individual changes his/her preferences between a particular pair of alternatives, the social preferences between them change merely because individuals have changed their preferences among some other alternatives. This requirement is known as the condition of independence of irrelevant alternatives.

It is immediate that no rule that violates the weak Pareto-criterion or is dictatorial can be acceptable. The condition of independence of irrelevant alternatives is also quite crucial. In its absence the choice procedure becomes ambiguous. Thus, it follows that any rule for aggregating individual preferences into social preferences, that is non-dictatorial and satisfies the weak Pareto-criterion and independence of irrelevant alternatives, must fail to generate a social ranking corresponding to some configuration(s) of individual orderings. The seriousness of the problem of the failure of the rule to generate a social ordering corresponding to every logically possible configuration of orderings of course will differ from rule to rule. Let a rule for aggregating individual orderings into social preferences that are reflexive and connected be termed a social decision rule. Thus from Arrow's Impossibility Theorem it follows that every non-dictatorial social decision rule, satisfying the weak Pareto-criterion and independence of irrelevant alternatives, must fail to generate transitive social preferences corresponding to some configuration(s) of individual orderings. Given a social decision rule, one can partition the set of configurations of individual orderings into two subsets: (i) The configurations for which the rule generates transitive social preferences. (ii) The configurations for which the rule does not generate transitive social preferences. If transitivity is considered an essential requirement, then the social decision rule can be used in situations where the configuration of individual orderings belongs to subset (i). Thus, it is quite important to know in the context of social decision rules that are to be used for decision-making the precise composition of subset (i). This, however, is unlikely to be an easily solvable problem for most social decision rules. In fact, excepting the method of majority decision, the precise composition of subset (i) is not known for any social decision rule.

A less intractable problem is as follows. Let $f$ be a social decision rule. Let $S$ be the set of social alternatives and let $\mathscr{T}$ be the set of orderings of $S$. The set of all nonempty subsets of $\mathscr{T}$ is $2^{\mathscr{T}} - \{\emptyset\}$. Let the set of individuals be $N$ with cardinality $n$. One can partition the set of all nonempty subsets of $\mathscr{T}$ into two subsets: $(\mathscr{T}_1^T)$ The set of nonempty subsets $\mathscr{D}$ of $\mathscr{T}$ which are such that every profile of individual orderings belonging to $\mathscr{D}^n$ yields transitive social preferences under $f$. $(\mathscr{T}_2^T)$ The set of nonempty subsets $\mathscr{D}$ of $\mathscr{T}$ which are such that at least one profile of individual orderings belonging to $\mathscr{D}^n$ yields intransitive social preferences under $f$. A condition on preferences is called a sufficient condition for transitivity if every nonempty $\mathscr{D}$ satisfying the condition belongs to set $(\mathscr{T}_1^T)$. A condition on preferences is called an Inada-type necessary condition for transitivity if every nonempty $\mathscr{D}$ violating the condition belongs to set $(\mathscr{T}_2^T)$.[2] A condition on preferences is called an Inada-type necessary and sufficient condition for transitivity if every nonempty $\mathscr{D}$ satisfying the condition belongs to set $(\mathscr{T}_1^T)$ and every nonempty $\mathscr{D}$ violating the condition belongs to set $(\mathscr{T}_2^T)$. As an illustration consider the method of majority decision $f$ defined for a set $S = \{x, y, z\}$ of three alternatives and a set $N = \{1, 2, 3\}$ of three individuals. There are 13 logically possible orderings of three alternatives $x, y, z$ given below. Alternatives are written in the descending order of preference, whether vertically or horizontally. Alternatives that are equally good (indifferent) are written together and enclosed in parentheses.

---

[2] An Inada-type necessary condition is not a necessary condition in the sense of logic. The expression 'Inada-type necessary condition for transitivity' will be used only as a shorthand expression for the property mentioned in the text. A similar remark applies to Inada-type necessary conditions for rationality conditions weaker than transitivity. Throughout this text, in the statement of theorems the use of the expression 'Inada-type necessary condition' has been avoided altogether.

| 1. $x$ | 2. $y$ | 3. $z$ | 4. $x$ | 5. $z$ | 6. $y$ |
|---|---|---|---|---|---|
| $y$ | $z$ | $x$ | $z$ | $y$ | $x$ |
| $z$ | $x$ | $y$ | $y$ | $x$ | $z$ |

| 7. $x$ | 8. $y$ | 9. $z$ | 10. $(xy)$ | 11. $(yz)$ | 12. $(zx)$ | 13. $(xyz)$ |
|---|---|---|---|---|---|---|
| $(yz)$ | $(zx)$ | $(xy)$ | $z$ | $x$ | $y$ | |

There are $2^{13} - 1$ nonempty subsets of the set of these 13 orderings. If $\mathscr{D} = \{xyz, (xyz)\}$ then it is immediate that under the method of majority decision every profile of individual orderings belonging to $\mathscr{D}^3$ yields transitive social preferences. Thus, for the method of majority decision defined for $S = \{x, y, z\}$ and $N = \{1, 2, 3\}$, $\{xyz, (xyz)\}$ belongs to set $\mathscr{T}_1^T$. On the other hand we have seen that if $\mathscr{D} = \{xyz, yzx, zxy\}$ and $n = 3$ then there exists a profile of individual orderings that results in intransitive social preferences under the method of majority decision. Thus, for the method of majority decision defined for $S = \{x, y, z\}$ and $N = \{1, 2, 3\}$, $\{xyz, yzx, zxy\}$ belongs to set $\mathscr{T}_2^T$.

In choosing rationally, one chooses an alternative that is best in the sense of being at least as good as every alternative. There can be more than one best alternatives. From the perspective of rational choice it does not matter which of the best alternatives is selected. If the number of alternatives is positive and finite then, in case there is a ranking of all alternatives, a best element will clearly exist. If there is a ranking of all alternatives, one can arrange them from top to bottom in the descending order, and choose the topmost alternative, and in case there are more than one alternatives at the top then choose any one of them. A preference relation is a ranking iff it satisfies the three properties of reflexivity, connectedness, and transitivity. If any one of these three conditions is violated then a best alternative may not exist. We have already seen that if transitivity is violated then it is possible that a best alternative may not exist, as is the case with the Condorcet paradox example. If alternative $x$ is not at least as good as itself, then clearly $x$ cannot be at least as good as every alternative. Thus violation of reflexivity can also result in the non-existence of a best alternative. The reflexivity requirement, although formally needed, is a trivial requirement as the question of some alternative not being at least as good as itself does not arise. Violation of connectedness implies that, for some distinct alternatives $x$ and $y$, it is the case that neither $x$ is at least as good as $y$ nor $y$ is at least as good as $x$, i.e., $x$ and $y$ are non-comparable. It is clear that such non-comparability can lead to non-existence of a best alternative. While violation of any of these three conditions can lead to non-existence of a best alternative, none of these three conditions is a necessary condition for the existence of a best alternative. In fact, it is possible for a preference relation to violate all three conditions and still have a best alternative. Thus, for a preference relation defined over a nonempty finite set, being an ordering is sufficient for the existence of a best alternative, but not necessary. A condition weaker than transitivity, called quasi-transitivity, is also sufficient to ensure the existence of a best element in the case of a reflexive and connected preference relation defined over a nonempty finite set. A preference relation is quasi-transitive iff 'better than' relation is transitive. If 'at least as good as' relation is transitive then both 'better than' relation and 'indifferent to' are transitive. In the case of a quasi-transitive relation the relation 'indifferent to' need not be transitive. Both transitivity and quasi-transitivity are conditions defined over triples of alternatives. If these conditions hold for all triples then they hold over all subsets. There is a condition that is weaker than even quasi-transitivity, called acyclicity, that is also sufficient to ensure the existence of a best element in the case of a reflexive and connected preference relation defined over a nonempty finite set. Acyclicity requires that if there is a chain of alternatives connected by 'better than' relation then the first alternative in the chain must be at least good as the last alternative in the chain. Thus, if acyclicity holds, and we have $x_1$ better than $x_2$, $x_2$ better than $x_3, \ldots, x_{m-1}$ better than $x_m$, then it must be the case that $x_1$ is at least as good as $x_m$. Acyclicity, unlike quasi-transitivity, is not a condition defined over triples only. Acyclicity holding over all triples does not imply that it would hold over all subsets.

It is possible to argue that the transitivity requirement for social preferences is unnecessarily restrictive. What is needed is that the society or the collective should be able to choose an alternative that is best. Thus, in the context of social decision rules which do not invariably yield quasi-transitive (acyclic) social preferences, it is of considerable importance to characterize those nonempty subsets $\mathscr{D}$ of $\mathscr{T}$ which are such that every profile of individual orderings belonging to $\mathscr{D}^n$ gives rise to quasi-transitive (acyclic) social preferences. Let $f$ be a social decision rule which does not invariably yield quasi-transitive social preferences. Consider the following partition of the set of all nonempty subsets of $\mathscr{T}$: $(\mathscr{T}_1^Q)$ The set of nonempty subsets $\mathscr{D}$ of $\mathscr{T}$ which are such that every profile of individual orderings belonging to $\mathscr{D}^n$ yields quasi-transitive social preferences under $f$. $(\mathscr{T}_2^Q)$ The set of nonempty subsets $\mathscr{D}$ of $\mathscr{T}$ which are such that at least one profile of individual orderings belonging to $\mathscr{D}^n$ yields social preferences violating quasi-transitivity under $f$. A condition on preferences is called a sufficient condition for quasi-transitivity under $f$ if every nonempty $\mathscr{D}$ satisfying the condition belongs to set $(\mathscr{T}_1^Q)$. A condition on preferences is called an Inada-type necessary condition for quasi-transitivity under $f$ if every nonempty $\mathscr{D}$ violating the condition belongs to set $(\mathscr{T}_2^Q)$. A condition on preferences is called an Inada-type necessary and sufficient condition for quasi-transitivity under $f$ if every nonempty $\mathscr{D}$ satisfying the condition belongs to set $(\mathscr{T}_1^Q)$ and every nonempty $\mathscr{D}$ violating the condition belongs to set $(\mathscr{T}_2^Q)$.

Next, let $f$ be a social decision rule which does not invariably yield acyclic social preferences. Consider the following partition of the set of all nonempty subsets of $\mathscr{T}$: $(\mathscr{T}_1^A)$ The set of nonempty subsets $\mathscr{D}$ of $\mathscr{T}$ which are such that every profile of individual orderings belonging to $\mathscr{D}^n$ yields acyclic social preferences under $f$. $(\mathscr{T}_2^A)$ The set of nonempty subsets $\mathscr{D}$ of $\mathscr{T}$ which are such that at least one profile of individual orderings belonging to $\mathscr{D}^n$ yields social preferences violating acyclicity under $f$. A condition on preferences is called a sufficient condition for acyclicity under $f$ if every nonempty $\mathscr{D}$ satisfying the condition belongs to set $(\mathscr{T}_1^A)$. A condition on preferences is called an Inada-type necessary condition for acyclicity under $f$ if every nonempty $\mathscr{D}$ violating the condition belongs to set $(\mathscr{T}_2^A)$. A condition on preferences is called an Inada-type necessary and sufficient condition for acyclicity under $f$ if every nonempty $\mathscr{D}$ satisfying the condition belongs to set $(\mathscr{T}_1^A)$ and every nonempty $\mathscr{D}$ violating the condition belongs to set $(\mathscr{T}_2^A)$.

The pioneering contributions on the conditions for transitivity and quasi-transitivity under the majority rule came from Black (1948, 1958), Arrow (1951, 1963), Inada (1964, 1969), and Sen and Pattanaik (1969), among others. Although the majority rule for obvious reasons has attracted the maximum attention,[3] other important rules have also been analysed from the perspective of conditions under which they yield rational social preferences.[4] This monograph is almost exclusively concerned with the Inada-type necessary and sufficient conditions for transitivity and quasi-transitivity under the various social decision rules and classes of social decision rules. The social decision rules and the classes of social decision rules for which the domain conditions for transitivity and quasi-transitivity have been discussed in this text include the method of majority decision, the strict majority rule, the class of semi-strict majority rules, the class of special majority rules, the class of non-minority rules, the class of Pareto-inclusive non-minority rules, the class of simple game social decision rules, and the class of neutral and monotonic binary social decision rules (Chaps. 3–10). Chapter 2 contains the basic social choice theoretic concepts, definitions, propositions and theorems that are required for the subject matter of the monograph. In Chaps. 3–10 domain conditions are discussed under the assumption that every individual has an ordering of social alternatives. Chapter 11 explores how some of the results obtained in Chaps. 3–10 change when it is assumed that every individual has a reflexive, connected and quasi-transitive weak preference relation over the set of social alternatives.

## References

Arrow, Kenneth J. 1951. *Social choice and individual values*, 2nd ed., 1963. New York: Wiley.
Batra, Raveendra, and Prasanta K. Pattanaik. 1971. Transitivity of social decisions under some more general group decision rules than the method of majority voting. *Review of Economic Studies* 38: 295–306.
Batra, Raveendra, and Prasanta K. Pattanaik. 1972. Transitive multi-stage majority decisions with quasi-transitive individual preferences. *Econometrica* 40: 1121–1135.
Black, D. 1948. On the rationale of group decision making. *The Journal of Political Economy* 56: 23–34.

---

[3]The literature on the conditions that ensure rationality of social preferences under the majority rule is vast. See Arrow (1951, 1963), Black (1948, 1958), Fine (1973), Fishburn (1970b, 1972, 1973), Gaertner (1988), Gaertner and Heinecke (1977, 1978), Inada (1964, 1969, 1970), Jain (1985, 2009), Kelly (1974), Nicholson (1965), Pattanaik (1971), Pattanaik and Sengupta (1974), Saposnik (1975), Sen (1966, 1970), Sen and Pattanaik (1969), and Slutsky (1977). The list is by no means exhaustive. For a survey of the literature and an extensive bibliography thereof see Gaertner (2001).
[4]See, among others, Batra and Pattanaik (1971, 1972), Dummett and Farquharson (1961), Fine (1973), Fishburn (1973), Jain (1983, 1984, 1986a,b, 1987, 1989, 1991), Murakami (1968), Pattanaik (1970), and Sen (1970).

Black, Duncan. 1958. *The theory of committees and elections*. London: Cambridge University Press.

Condorcet, Marquis de. 1785. *Essai sur l'application de l'analyse à la probabilité des décisions rendues à la pluralité des voix*. Paris.

Dummett, Michael, and Robin Farquharson. 1961. Stability in voting. *Econometrica* 29: 33–43.

Fine, Kit. 1973. Conditions for the existence of cycles under majority and non-minority rules. *Econometrica* 41: 888–899.

Fishburn, Peter C. 1970b. Conditions for simple majority decision with intransitive individual indifference. *Journal of Economic Theory* 2: 354–367.

Fishburn, Peter C. 1972. Conditions on preferences that guarantee a simple majority winner. *The Journal of Mathematical Sociology* 2: 105–112.

Fishburn, Peter C. 1973. *The theory of social choice*. Princeton: Princeton University Press.

Gaertner, Wulf. 1988. Binary inversions and transitive majorities. In *Measurement in economics*, ed. Wolfgang Eichhorn, 253–267. Berlin: Springer.

Gaertner, Wulf. 2001. *Domain conditions in social choice theory*. London: Cambridge University Press.

Gaertner, Wulf, and Achim Heinecke. 1977. On two sufficient conditions for transitivity of the social preference relation. *Zeitschrift für Nationalökonomie* 37: 61–66.

Gaertner, Wulf, and Achim Heinecke. 1978. Cyclically mixed preferences - A necessary and sufficient condition for transitivity of the social preference relation. In *Decision theory and social ethics*, ed. Hans W. Gottinger, and Werner Leinfellner, 169–186. Dordrecht: D. Reidel.

Inada, Ken-ichi. 1964. A note on the simple majority decision rule. *Econometrica* 32: 525–531.

Inada, Ken-ichi. 1969. The simple majority decision rule. *Econometrica* 37: 490–506.

Inada, Ken-ichi. 1970. Majority rule and rationality. *Journal of Economic Theory* 2: 27–40.

Jain, Satish K. 1983. Necessary and sufficient conditions for quasi-transitivity and transitivity of special majority rules. *Keio Economic Studies* 20: 55–63.

Jain, Satish K. 1984. Non-minority rules: Characterization of configurations with rational social preferences. *Keio Economic Studies* 21: 45–54.

Jain, Satish K. 1985. A direct proof of Inada-Sen-Pattanaik theorem on majority rule. *The Economic Studies Quarterly* 36: 209–215.

Jain, Satish K. 1986a. Special majority rules: Necessary and sufficient condition for quasi-transitivity with quasi-transitive individual preferences. *Social Choice and Welfare* 3: 99–106.

Jain, Satish K. 1986b. Semi-strict majority rules: Necessary and sufficient conditions for quasi-transitivity and transitivity. Unpublished manuscript. Centre for Economic Studies and Planning, Jawaharlal Nehru University.

Jain, Satish K. 1987. Maximal conditions for transitivity under neutral and monotonic binary social decision rules. *The Economic Studies Quarterly* 38: 124–130.

Jain, Satish K. 1989. Characterization theorems for social decision rules which are simple games. Paper presented at the IX World Congress of the International Economic Association, Economics Research Center, Athens School of Economics & Business, Athens, Greece, held on Aug 28 - Sep 1, 1989.

Jain, Satish K. 1991. Non-minority rules: Necessary and sufficient condition for quasi-transitivity with quasi-transitive individual preferences. *Keio Economic Studies* 28: 21–27.

Jain, Satish K. 2009. The method of majority decision and rationality conditions. In *Ethics, welfare, and measurement, Volume 1 of Arguments for a better world: Essays in honor of Amartya Sen*, ed. Kaushik Basu, and Ravi Kanbur, 167–192. New York: Oxford University Press.

Kelly, J.S. 1974. Necessity conditions in voting theory. *Journal of Economic Theory* 8: 149–160.

Murakami, Yasusuke. 1968. *Logic and social choice*. London: Routledge & Kegan Paul.

Nicholson, Michael B. 1965. Conditions for the 'voting paradox' in committee decisions. *Metroeconomica* 7: 29–44.

Pattanaik, Prasanta K. 1970. On social choice with quasitransitive individual preferences. *Journal of Economic Theory* 2: 267–275.

Pattanaik, Prasanta K. 1971. *Voting and collective choice*. Cambridge: Cambridge University Press.

Pattanaik, Prasanta K., and Manimay Sengupta. 1974. Conditions for transitive and quasi-transitive majority decisions. *Economica* 41: 414–423.

Saposnik, Rubin. 1975. On the transitivity of the social preference relation under simple majority rule. *Journal of Economic Theory* 10: 1–7.

Sen, Amartya K. 1966. A possibility theorem on majority decisions. *Econometrica* 34: 491–499.

Sen, Amartya K. 1970. *Collective choice and social welfare*. San Francisco: Holden-Day.

Sen, Amartya K., and Prasanta K. Pattanaik. 1969. Necessary and sufficient conditions for rational choice under majority decision. *Journal of Economic Theory* 1: 178–202.

Slutsky, Steven M. 1977. A characterisation of societies with consistent majority decision. *Review of Economic Studies* 44: 211–225.

# The Preliminaries

This chapter contains the basic social choice theoretic concepts, definitions and propositions that will be needed in the rest of the text. The chapter is divided into three sections, concerned with binary relations, social decision rules and Latin Squares, respectively.[1]

A binary relation is a relation that holds between two things, objects or persons. Thus a binary relation is simply the set of ordered pairs for which the relation holds. 'At least as good as', and 'is brother of' provide examples of binary relations. If the relation holds between $a$ and $b$, but not between $b$ and $a$, then the ordered pair $(a, b)$ belongs to the asymmetric part of the relation. If the relation holds both ways then the ordered pair $(a, b)$ belongs to the symmetric part of the relation. If a binary relation is designated by $R$ then its asymmetric and symmetric parts are designated by $P(R)$ and $I(R)$, respectively. If $R$ is interpreted as 'at least as good as' then $P(R)$ and $I(R)$ will stand for 'better than' and 'indifferent to', respectively; and if $R$ is interpreted as 'is brother of' then $P(R)$ and $I(R)$ will stand for 'brother–sister' and 'brother–brother' relationships, respectively. A binary relation over a set is defined to be reflexive iff if for every element of the set the relation holds with itself; irreflexive iff there is no element of the set for which the relation holds with itself; connected iff the relation holds at least one way between any two distinct elements of the set; symmetric iff between any two elements of the set the relation holds both ways or neither way; asymmetric iff between any two elements of the set the relation holds at most one way; antisymmetric iff between any two distinct elements of the set the relation holds at most one way; transitive iff for any three elements of the set if the relation holds between the first and the second element, and also between the second element and the third element, then it holds between the first element and the third element; quasi-transitive iff the asymmetric part of the relation is transitive; and acyclic iff for any three or more elements of the set, if the relation between the first

---

[1] For the material covered in this chapter see Tarski (1941), Suppes (1957), Arrow (1963), Gibbard (1969), Sen (1970), and Jain (1985), among others.

© Springer Nature Singapore Pte Ltd. 2019
S. K. Jain, *Domain Conditions and Social Rationality*,
https://doi.org/10.1007/978-981-13-9672-4_2

element and the second element belongs to the asymmetric part of the relation, the relation between the second element and the third element belongs to the asymmetric part of the relation, and so on, then the relation must hold between the first element and the last element.

A binary relation is called an ordering iff it is reflexive, connected and transitive; and a linear ordering iff it reflexive, connected, antisymmetric and transitive. The relation 'greater than or equal to' defined over the set of real numbers is an example of a linear ordering. The reflexivity of the relation 'at least as good as' is trivial. For individuals, the relation 'at least as good as' is generally taken to be connected. However, if the set of alternatives contains unfamiliar things, an individual may not be able to compare some alternatives with some other alternatives, in which case the relation will not be connected. For individuals, the relation 'at least as good as' is also generally taken to be transitive. But there are contexts in which an individual's 'at least as good as' relation may not be transitive. If an individual always prefers an alternative of a larger size to an alternative of a smaller size and is indifferent between alternatives of the same size, and three alternatives of decreasing size are such that the difference between the sizes of the first alternative and the second alternative is below the threshold of the individual's perceptive ability, and similarly the difference between the sizes of the second and the third alternatives is below the threshold of the individual's perceptive ability, but the difference between the sizes of the first and the third alternatives is large enough to be perceived by the individual; then the individual will be indifferent between the first alternative and the second alternative, will be indifferent between the second alternative and the third alternative, and prefer the first alternative to the third alternative.

An alternative is best in a set iff it is at least as good as every alternative in the set. If the 'at least as good as' relation defined over a nonempty finite set is reflexive, connected and acyclic then the existence of a best element is guaranteed. If any of these three conditions is violated then it is possible that a best element may not exist. However, these conditions are not necessary. Even if all three conditions are violated, a best element may still exist. Transitivity is strictly stronger than quasi-transitivity, and quasi-transitivity is strictly stronger than acyclicity. Therefore, given the nonemptiness and finiteness of the set of alternatives, reflexivity, connectedness and transitivity suffice for the existence of a best element; as do reflexivity, connectedness and quasi-transitivity.

Social choice theory is concerned with the problem of aggregating individual preferences into social preferences. Arrow Impossibility Theorem, the starting point of the social choice theory, states incompatibility of some extremely reasonable conditions. In the Arrowian framework, the set of social alternatives is assumed to contain at least three alternatives; and the finite set of individuals at least two individuals. Every individual is assumed to have an ordering of the social alternatives. The rule for aggregating individual preferences into social preferences, called social welfare function by Arrow, is assumed to generate a unique social ordering corresponding to every given profile of individual orderings. Arrow showed that there does not exist any social welfare function which can simultaneously satisfy conditions of unrestricted domain, weak Pareto-criterion, independence of irrelevant alternatives, and non-dictatorship. Condition of unrestricted domain requires that the domain of the social welfare function must consist of all logically possible profiles of individual orderings; weak Pareto-criterion requires that if all individuals unanimously prefer one alternative to another then in the social preferences as well the same strict preference must be reflected; independence of irrelevant alternatives requires that the social preferences over any pair of alternatives must be determined solely on the basis of individual preferences over that pair; and non-dictatorship requires that there be no dictator, a dictator being a person whose strict preferences are invariably reflected in social preferences. The proof of Arrow's theorem given in this chapter is essentially that of Arrow (1963), although somewhat simpler.

With respect to a reflexive and connected binary relation, in a triple, i.e., in a set of three distinct alternatives, an alternative is best iff it is at least as good as the other two alternatives; medium iff it is at least as good as one of the two other alternatives, and the other of the two alternatives is at least as good as this one; worst iff if both the other alternatives are at least as good as this one. With respect to a reflexive and connected binary relation, in a triple, (i) an alternative is proper best iff it is better than one of the two other alternatives and at least as good as the other of the two alternatives; (ii) an alternative is proper medium iff it is better than one of the two other alternatives and the other of the two alternatives is at least as good as this alternative; or it is at least as good as one of the two other alternatives and the other of the two alternatives is better than this alternative; (iii) an alternative is proper worst iff one of the two other alternatives is better than this alternative and the other of the two alternatives is at least as good as this alternative. In the context of this text, the notions of weak Latin Square and Latin Square are of fundamental importance. A weak Latin Square forms if there are binary relations that together can create a cyclical arrangement over a triple. In other words, if there is a binary relation in which $a$ is best, $b$ is medium and $c$ worst; and there is a binary relation in which $b$ is best, $c$ is medium and $a$ is worst; and there is a binary relation in which $c$ is best, $a$ is medium and $b$ worst; then together they create the cyclical arrangement $abca$. These binary relations together then form a weak Latin Square, that can be designated as $WLS(abca)$ to indicate the cyclical arrangement involved. The only difference between a weak Latin Square and a Latin Square is that in the definition of Latin Square 'proper medium' replaces 'medium' occurring in the definition of weak Latin Square. Throughout this text, the concepts of weak Latin Square and Latin Square play important roles in the formulation of the relevant conditions for social rationality.

## 2.1 Binary Relations

A binary relation $R$ on a set $S$ is a subset of the Cartesian product of $S$ with itself; $R \subseteq S \times S$. $xRy$ will stand for $(x, y) \in R$; $x, y \in S$. Asymmetric and symmetric parts of a binary relation $R$ on a set $S$, to be denoted by $P(R)$ and $I(R)$, respectively, are defined by: $(\forall x, y \in S)[(x P(R) y \leftrightarrow xRy \wedge \sim yRx) \wedge (x I(R) y \leftrightarrow xRy \wedge yRx)]$.

$P(R)$ and $I(R)$ will be written as $P$ and $I$ in abbreviated form. Asymmetric and symmetric parts of binary relations $R_i$, $R^i$, $R'_i$, $R'$, etc., will be written in abbreviated form as $P_i$, $P^i$, $P'_i$, $P'$, etc., and $I_i$, $I^i$, $I'_i$, $I'$, etc., respectively. If binary relation $R$ stands for 'at least as good as' then $P$ and $I$ stand for 'better than' and 'indifferent to', respectively.

A binary relation $R$ on a set $S$ is defined to be (i) reflexive iff $(\forall x \in S)(xRx)$, (ii) irreflexive iff $(\forall x \in S)(\sim xRx)$, (iii) connected iff $(\forall x, y \in S)(x \neq y \rightarrow xRy \vee yRx)$, (iv) symmetric iff $(\forall x, y \in S)(xRy \rightarrow yRx)$, (v) asymmetric iff $(\forall x, y \in S)(xRy \rightarrow \sim yRx)$, (vi) antisymmetric iff $(\forall x, y \in S)(xRy \wedge yRx \rightarrow x = y)$, (vii) transitive iff $(\forall x, y, z \in S)(xRy \wedge yRz \rightarrow xRz)$, (viii) quasi-transitive iff $(\forall x, y, z \in S)(xPy \wedge yPz \rightarrow xPz)$, (ix) acyclic iff $(\forall x_1, x_2, \ldots, x_m \in S)(x_1 P x_2 \wedge x_2 P x_3 \wedge \ldots \wedge x_{m-1} P x_m \rightarrow x_1 R x_m)$, where $m$ is a positive integer $\geq 3$, (x) an ordering iff it is reflexive, connected and transitive, and (xi) a linear ordering (strong ordering) iff it is reflexive, connected, antisymmetric and transitive.

*Remark 2.1* Expression $(\forall x, y \in S)(xRy \wedge yRx \rightarrow x = y)$ is equivalent to $(\forall x, y \in S)[x \neq y \rightarrow \sim (xRy \wedge yRx)]$, which in turn is equivalent to $(\forall x, y \in S)[x \neq y \rightarrow (xRy \rightarrow \sim yRx)]$. From the last expression it is clear that asymmetry is equivalent to the union of antisymmetry and irreflexivity; and consequently antisymmetry is strictly weaker than asymmetry.                                                                      ◊

**Proposition 2.1** *If binary relation $R$ on set $S$ is transitive then the following hold:*

   *(i)* $(\forall x, y, z \in S)(xPy \wedge yPz \rightarrow xPz)$
  *(ii)* $(\forall x, y, z \in S)(xPy \wedge yIz \rightarrow xPz)$
 *(iii)* $(\forall x, y, z \in S)(xIy \wedge yPz \rightarrow xPz)$
 *(iv)* $(\forall x, y, z \in S)(xIy \wedge yIz \rightarrow xIz)$.

*Proof* (i) Let $xPy \wedge yPz$
$xPy \rightarrow xRy$, by the definition of $P$
$yPz \rightarrow yRz$, by the definition of $P$
$xRy \wedge yRz \rightarrow xRz$, as $R$ is transitive.                                                        (P2.1-1)
Suppose $zRx$
$zRx \wedge xRy \rightarrow zRy$, by transitivity of $R$
$zRy$ contradicts $\sim zRy$ which holds in view of $yPz$
Therefore $zRx$ is not possible
Thus $\sim zRx$ holds.                                                                                       (P2.1-2)
(P2.1-1) and (P2.1-2) establish that $xPz$ holds.

(ii) Let $xPy \wedge yIz$
$xPy \rightarrow xRy$
$yIz \rightarrow yRz$
$xRy \wedge yRz \rightarrow xRz$, as $R$ is transitive.                                                        (P2.1-3)
Suppose $zRx$
$yRz \wedge zRx \rightarrow yRx$, by transitivity of $R$
$yRx$ contradicts $\sim yRx$ which holds in view of $xPy$

Therefore $zRx$ is not possible
Thus $\sim zRx$ holds. (P2.1-4)
(P2.1-3) and (P2.1-4) establish that $xPz$ holds.

(iii) Let $xIy \wedge yPz$
$xIy \rightarrow xRy$
$yPz \rightarrow yRz$
$xRy \wedge yRz \rightarrow xRz$, by $R$-transitivity. (P2.1-5)
Suppose $zRx$
$zRx \wedge xRy \rightarrow zRy$, by transitivity of $R$
$zRy$ contradicts $\sim zRy$ which holds in view of $yPz$
Therefore $zRx$ is not possible
Thus $\sim zRx$ holds. (P2.1-6)
(P2.1-5) and (P2.1-6) establish that $xPz$ holds.

(iv) Let $xIy \wedge yIz$
$xIy \rightarrow xRy \wedge yRx$
$yIz \rightarrow yRz \wedge zRy$
$xRy \wedge yRz \rightarrow xRz$, by $R$-transitivity
$zRy \wedge yRx \rightarrow zRx$, by $R$-transitivity
$xRz \wedge zRx \rightarrow xIz$. □

*Remark 2.2* From Proposition 2.1(i) it follows that transitivity implies quasi-transitivity. Quasi-transitivity is strictly weaker than transitivity as can be seen from the example that follows. ◊

*Example 2.1* Consider the binary relation $R$ on set $S = \{x, y, z\}$ given by:
$\{(x, x), (y, y), (z, z), (x, y), (y, z), (z, y), (x, z), (z, x)\} = xIx, yIy, zIz, xPy,$
$yIz, xIz$.
$R$ violates transitivity as we have: $yRz \wedge zRx \wedge \sim yRx$.
$R$, however, satisfies quasi-transitivity trivially as $\sim (\exists a, b, c \in S)[aPb \wedge bPc]$. ◊

**Proposition 2.2** *Quasi-transitivity implies acyclicity.*

*Proof* Let binary relation $R$ on $S$ be quasi-transitive.
Suppose $x_1Px_2 \wedge x_2Px_3 \wedge \ldots \wedge x_{m-1}Px_m, x_1, x_2, \ldots x_{m-1}, x_m \in S, m \geq 3$.
$x_1Px_2 \wedge x_2Px_3 \rightarrow x_1Px_3$, by quasi-transitivity
$x_1Px_3 \wedge x_3Px_4 \rightarrow x_1Px_4$, by quasi-transitivity
$\vdots$
$x_1Px_{m-1} \wedge x_{m-1}Px_m \rightarrow x_1Px_m$, by quasi-transitivity
$x_1Px_m \rightarrow x_1Rx_m$.
Thus $R$ is acyclic. □

*Remark 2.3* Acyclicity is strictly weaker than quasi-transitivity as can be seen from the example that follows. ◊

*Example 2.2* Consider the binary relation $R$ on set $S = \{x, y, z\}$ given by:
$\{(x, x), (y, y), (z, z), (x, y), (y, z), (x, z), (z, x)\} = xIx, yIy, zIz, xPy, yPz, xIz$.
$R$ violates quasi-transitivity as we have: $xPy \wedge yPz \wedge \sim xPz$.
The only case in which $(a_1Pa_2 \wedge a_2Pa_3 \wedge \ldots \wedge a_{m-1}Pa_m)$, $m \geq 3$, $a_1, \ldots, a_m \in S$, holds is $(xPy \wedge yPz)$; and we have $xRz$. Thus, $(\forall a_1, a_2, \ldots, a_m \in S)[a_1Pa_2 \wedge a_2Pa_3 \wedge \ldots \wedge a_{m-1}Pa_m \to a_1Ra_m]$, $m \geq 3$; and consequently $R$ satisfies acyclicity.                                                                                                   ◊

Let $R$ be a binary relation on a set $S$. $x \in S$ is a best element of $S$ with respect to binary relation $R$ iff $(\forall y \in S)(xRy)$. The set of best elements in $S$ is called its choice set and is denoted by $C(S, R)$.

**Proposition 2.3** *Let $R$ be a binary relation on a finite nonempty set $S$. If $R$ is reflexive, connected and acyclic then $C(S, R) \neq \emptyset$.*

*Proof* Let $S = \{x_1, \ldots, x_s\}$, $s \geq 1$. Let binary relation $R$ on $S$ be reflexive, connected and acyclic.
Suppose $C(S, R) = \emptyset$.
$x_1 \notin C(S, R) \to (\exists y_1 \in S)(\sim x_1Ry_1)$
$y_1 \neq x_1$, as $R$ is reflexive
Therefore, $y_1 \in \{x_2, \ldots, x_s\}$
Without any loss of generality assume: $y_1 = x_2$
$\sim x_1Rx_2 \to x_2Rx_1$, by connectedness
$x_2Rx_1 \wedge \sim x_1Rx_2 \to x_2Px_1$
$x_2 \notin C(S, R) \to (\exists y_2 \in S)(\sim x_2Ry_2)$
$y_2 \neq x_2$, as $R$ is reflexive
Also, $y_2 \neq x_1$, as we have $x_2Rx_1$
Therefore, $y_2 \in \{x_3, \ldots, x_s\}$
Without any loss of generality assume: $y_2 = x_3$
As $\sim x_2Rx_3$ implies $x_3Rx_2$ by connectedness, it follows that $x_3Px_2$ holds
$x_3 \notin C(S, R) \to (\exists y_3 \in S)(\sim x_3Ry_3)$
$y_3 \neq x_3$, as $R$ is reflexive
$y_3 \neq x_2$, as we have $x_3Rx_2$
As $x_3Px_2 \wedge x_2Px_1 \to x_3Rx_1$, by acyclicity; we conclude that $y_3 \neq x_1$
Therefore, $y_3 \in \{x_4, \ldots, x_s\}$
Without any loss of generality assume: $y_3 = x_4$
As $\sim x_3Rx_4$ implies $x_4Rx_3$ by connectedness, it follows that $x_4Px_3$ holds
Continuing this way we conclude: $x_sPx_{s-1} \wedge \ldots \wedge x_3Px_2 \wedge x_2Px_1$.
Thus we obtain:
$x_sRx_t$, $t \in \{1, 2, \ldots, (s-2)\}$, by acyclicity
$x_sRx_{s-1}$, as $x_sPx_{s-1}$ holds
$x_sRx_s$, by reflexivity
Thus $x_s \in C(S, R)$, a contradiction as $C(S, R)$ was assumed to be empty.
This contradiction establishes the proposition.                                              □

The following two corollaries follow immediately from Proposition 2.3, in view of Remark 2.2 and Proposition 2.2.

**Corollary 2.1** *Let R be a binary relation on a finite nonempty set S. If R is reflexive, connected and transitive then $C(S, R) \neq \emptyset$.*

**Corollary 2.2** *Let R be a binary relation on a finite nonempty set S. If R is reflexive, connected and quasi-transitive then $C(S, R) \neq \emptyset$.*

*Remark 2.4* If any of the three conditions of reflexivity, connectedness and acyclicity is violated then a best element may not exist as can be seen from the example given below. ◇

*Example 2.3* (a) Let connected and transitive binary relation $R_1$ on set $S = \{x, y, z\}$ be given by:
$\{(x, y), (y, z), (x, z)\}$.

(b) Let reflexive and transitive binary relation $R_2$ on set $S = \{x, y, z\}$ be given by:
$\{(x, x), (y, y), (z, z), (x, y)\}$.

(c) Let reflexive and connected binary relation $R_3$ on set $S = \{x, y, z\}$ be given by:
$\{(x, x), (y, y), (z, z), (x, y), (y, z), (z, x)\}$.
$R_1$ violates reflexivity, $R_2$ violates connectedness, and $R_3$ violates acyclicity. In each case there is no best element as all three sets $C(S, R_1)$, $C(S, R_2)$, $C(S, R_3)$ are empty. ◇

*Remark 2.5* Proposition 2.3 establishes that reflexivity, connectedness and acyclicity together constitute a set of sufficient conditions for the existence of a best element in a nonempty finite set. These conditions, however, are not necessary. It is possible for a binary relation to violate all three conditions and still have a best element as the example that follows shows. ◇

*Example 2.4* Consider the binary relation $R$ on set $S = \{x, y, z, w\}$ given by:
$\{(x, x), (x, y), (x, z), (x, w), (y, z), (z, w)\}$.
$x$ is best in $S$ as $(\forall t \in S)(x R t)$.
Reflexivity is violated as $\sim wRw$.
Connectedness is violated as $\sim yRw \wedge \sim wRy$.
Acyclicity is violated as $yPz \wedge zPw \wedge \sim yRw$. ◇

*Remark 2.6* If $S$ is infinite then, even when $R$ is an ordering, a best element may not exist, as the example that follows shows. ◇

*Example 2.5* Consider the ordering relation $R$ on set $S = \{x_\nu \mid \nu \in \mathbb{N}\}$ given by:
$(\forall x_i, x_j \in S)[x_i P x_j \leftrightarrow i > j]$, where $\mathbb{N}$ is the set of positive integers.
As for every alternative in $S$ there exists an alternative that is better than it, $S$ has no best element. ◇

## 2.2  Social Decision Rules

The set of social alternatives and the finite set of individuals constituting the society are denoted by $S$ and $N$, respectively. We assume $\#S = s \geq 3$ and $\#N = n \geq 2$, $n \in \mathbb{N}$, where $\mathbb{N}$ denotes the set of positive integers. Each individual $i \in N$ is assumed to have a binary weak preference relation $R_i$ on $S$. Unless otherwise specified, it would be assumed that every individual's $R_i$ is an ordering.

We denote by $\mathscr{B}$ the set of all binary relations on $S$; by $\mathscr{C}$ the set of all reflexive and connected binary relations on $S$; by $\mathscr{A}$ the set of all reflexive, connected and acyclic binary relations on $S$; by $\mathscr{Q}$ the set of all reflexive, connected and quasi-transitive binary relations on $S$; by $\mathscr{T}$ the set of all reflexive, connected and transitive binary relations (orderings) on $S$; and by $\mathscr{L}$ the set of all reflexive, connected, antisymmetric and transitive binary relations (linear orderings) on $S$.

A Social Decision Rule (SDR) is a function $f$ from $\mathbb{D} \subseteq \mathscr{T}^n$ to $\mathscr{C}$; $f : \mathbb{D} \mapsto \mathscr{C}$. A Social Welfare Function (SWF) is a function $f$ from $\mathbb{D} \subseteq \mathscr{T}^n$ to $\mathscr{T}$; $f : \mathbb{D} \mapsto \mathscr{T}$. The social binary weak preference relations corresponding to $(R_1, \ldots, R_n)$, $(R'_1, \ldots, R'_n)$, etc., will be denoted by $R$, $R'$, etc., respectively.

Thus, a social decision rule generates a unique reflexive and connected social weak preference relation for every profile of individual orderings in the domain; and a social welfare function generates a unique social ordering for every profile of individual orderings in the domain.

Let $f : \mathbb{D} \mapsto \mathscr{C}$; $\mathbb{D} \subseteq \mathscr{T}^n$. Let $V \subseteq N$. Let $x, y \in S$, $x \neq y$. We define the set of individuals $V$ to be (i) almost decisive for $(x, y)$ $[D(x, y)]$ iff $(\forall (R_1, \ldots, R_n) \in \mathbb{D})[(\forall i \in V)(x P_i y) \wedge (\forall i \in N - V)(y P_i x) \rightarrow x P y]$, (ii) decisive for $(x, y)$ $[\overline{D}(x, y)]$ iff $(\forall (R_1, \ldots, R_n) \in \mathbb{D})[(\forall i \in V)(x P_i y) \rightarrow x P y]$, (iii) a decisive set iff it is decisive for every $(a, b) \in S \times S$, $a \neq b$, (iv) almost semidecisive for $(x, y)$ $[S(x, y)]$ iff $(\forall (R_1, \ldots, R_n) \in \mathbb{D})[(\forall i \in V)(x P_i y) \wedge (\forall i \in N - V)(y P_i x) \rightarrow x R y]$, (v) semidecisive for $(x, y)$ $[\overline{S}(x, y)]$ iff $(\forall (R_1, \ldots, R_n) \in \mathbb{D})[(\forall i \in V)(x P_i y) \rightarrow x R y]$, (iii) a semidecisive set iff it is semidecisive for every $(a, b) \in S \times S$, $a \neq b$. $V$ is defined to be a minimal decisive set iff it is a decisive set and no proper subset of it is a decisive set.

An individual $j \in N$ is called a dictator iff $(\forall (R_1, \ldots, R_n) \in \mathbb{D})(\forall x, y \in S)[x P_j y \rightarrow x P y]$. Thus individual $j$ is a dictator iff $\{j\}$ is a decisive set. $V$ is an oligarchy iff $V$ is a decisive set and every individual belonging to $V$ is a semidecisive set. $V$ is a strict oligarchy iff $V$ is a decisive set and $(\forall i \in V)(\forall (R_1, \ldots, R_n) \in \mathbb{D})(\forall$ distinct $a, b \in S)(a R_i b \rightarrow a R b)$.

*Remark 2.7* The following implications are immediate from the definitions:
$V$ is $\overline{D}(x, y) \rightarrow V$ is $D(x, y)$
$V$ is $\overline{S}(x, y) \rightarrow V$ is $S(x, y)$
$V$ is $\overline{D}(x, y) \rightarrow V$ is $\overline{S}(x, y)$
$V$ is $D(x, y) \rightarrow V$ is $S(x, y)$.                                                              $\Diamond$

An SDR $f : \mathbb{D} \mapsto \mathscr{C}$, $\mathbb{D} \subseteq \mathscr{T}^n$, (i) satisfies the weak Pareto-criterion (P) iff $(\forall (R_1, \ldots, R_n) \in \mathbb{D})(\forall x, y \in S)[(\forall i \in N)(x P_i y) \rightarrow x P y]$, (ii) satisfies binariness or independence of irrelevant alternatives (I) iff $(\forall(R_1, \ldots, R_n),(R_1', \ldots, R_n') \in \mathbb{D})(\forall x, y \in S)[(\forall i \in N)[(xR_iy \leftrightarrow xR_i'y) \wedge (yR_ix \leftrightarrow yR_i'x)] \rightarrow [(xRy \leftrightarrow xR'y) \wedge (yRx \leftrightarrow yR'x)]]$, (iii) is non-dictatorial iff $\sim (\exists j \in N)(\forall(R_1, \ldots, R_n) \in \mathbb{D})(\forall x, y \in S)[xP_jy \rightarrow xPy]$.

In other words, a social decision rule satisfies condition P iff the set of all individuals $N$ is a decisive set; satisfies the condition of independence of irrelevant alternatives iff the social weak preference relation over every pair of alternatives is determined solely by individual weak preference relations over that pair; and is non-dictatorial iff there is no dictator.

**Lemma 2.1** (Arrow) *Let $\#S \geq 3$; and $\#N = n \geq 2, n \in \mathbb{N}$. Let social welfare function $f : \mathcal{T}^n \mapsto \mathcal{T}$ satisfy weak Pareto-criterion and independence of irrelevant alternatives. Then, if a set of individuals $V \subseteq N$ is almost decisive for some ordered pair of distinct alternatives then it is decisive for every ordered pair of distinct alternatives.*

*Proof* Let $V$ be almost decisive for $(x, y)$, $x \neq y$, $x, y \in S$. Let $z$ be an alternative distinct from $x$ and $y$, and consider the following configuration of individual preferences:

$(\forall i \in V)[xP_iy \wedge yP_iz]$

$(\forall i \in N - V)[yP_ix \wedge yP_iz]$.

In view of the almost decisiveness of $V$ for $(x, y)$ and the fact that $[(\forall i \in V)(xP_iy) \wedge (\forall i \in N - V)(yP_ix)]$, we obtain $xPy$. From $(\forall i \in N)(yP_iz)$ we conclude $yPz$, by condition P. From $xPy$ and $yPz$ we conclude $xPz$, by transitivity of $R$. As $(\forall i \in V)(xP_iz)$, and the preferences of individuals in $N - V$ have not been specified over $\{x, z\}$, it follows, in view of condition I, that $V$ is decisive for $(x, z)$. Similarly, by considering the configuration $[(\forall i \in V)(zP_ix \wedge xP_iy) \wedge (\forall i \in N - V)(zP_ix \wedge yP_ix)]$ we can conclude $[D(x, y) \rightarrow \overline{D}(z, y)]$. From the above demonstration, it is clear that if $a, b, c$ are three distinct alternatives then from $D(a, b)$ we can conclude both $\overline{D}(a, c)$ and $\overline{D}(c, b)$.

Now,

$D(x, y) \rightarrow \overline{D}(x, z)$

$\rightarrow D(x, z)$

$\rightarrow \overline{D}(y, z) \wedge \overline{D}(x, y)$

$D(x, y) \rightarrow \overline{D}(z, y)$

$\rightarrow D(z, y)$

$\rightarrow \overline{D}(z, x)$

$\rightarrow D(z, x)$

$\rightarrow \overline{D}(y, x)$

It has been shown that $D(x, y) \rightarrow \overline{D}(a, b)$, for all $(a, b) \in \{x, y, z\} \times \{x, y, z\}$, where $a \neq b$. To prove the assertion for any $(a, b) \in S \times S, a \neq b$, first we note that if $\#[\{x, y\} \cap \{a, b\}] = 2$ then $\overline{D}(a, b)$ has already been deduced. If $\#[\{x, y\} \cap \{a, b\}] = 1$ then $\overline{D}(a, b)$ can be deduced by considering the triple which includes all of $x, y, a, b$. If $\{x, y\} \cap \{a, b\} = \emptyset$, then one first considers the triple $\{x, y, a\}$ and deduces $\overline{D}(x, a)$ and hence $D(x, a)$, and then considers the triple $\{x, a, b\}$ and obtains $\overline{D}(a, b)$.                                                                                               □

**Lemma 2.2** *Let* $\#S \geq 3$; *and* $\#N = n \geq 2, n \in \mathbb{N}$. *Let social welfare function* $f : \mathscr{T}^n \mapsto \mathscr{T}$ *satisfy weak Pareto-criterion and independence of irrelevant alternatives. Then, if a set of individuals* $V \subseteq N$ *is almost semidecisive for some ordered pair of distinct alternatives then it is semidecisive for every ordered pair of distinct alternatives.*

*Proof* Let $V$ be almost semidecisive for $(x, y)$, $x \neq y$, $x, y \in S$. Let $z$ be an alternative distinct from $x$ and $y$, and consider the following configuration of individual preferences:
$(\forall i \in V)[x P_i y \wedge y P_i z]$
$(\forall i \in N - V)[y P_i x \wedge y P_i z]$.
In view of the almost semidecisiveness of $V$ for $(x, y)$ and the fact that $[(\forall i \in V)(x P_i y) \wedge (\forall i \in N - V)(y P_i x)]$, we obtain $x R y$. From $(\forall i \in N)(y P_i z)$ we conclude $y P z$, by condition P. As $z P x$, in view of $y P z$ and transitivity of social $R$ would imply $y P x$, contradicting $x R y$; we conclude that it must be the case that $x R z$ holds. As $(\forall i \in V)(x P_i z)$, and the preferences of individuals in $N - V$ have not been specified over $\{x, z\}$, it follows, in view of condition I, that $V$ is semidecisive for $(x, z)$. Similarly, by considering the configuration $[(\forall i \in V)(z P_i x \wedge x P_i y) \wedge (\forall i \in N - V)(z P_i x \wedge y P_i x)]$ we can conclude $[S(x, y) \rightarrow \overline{S}(z, y)]$. From the above demonstration, it is clear that if $a$, $b$, $c$ are three distinct alternatives then from $S(a, b)$ we can conclude both $\overline{S}(a, c)$ and $\overline{S}(c, b)$. By appropriate interchanges of alternatives, as in the proof of Arrow's Lemma, it follows that $S(x, y) \rightarrow \overline{S}(a, b)$, for all $(a, b) \in \{x, y, z\} \times \{x, y, z\}$, where $a \neq b$. For establishing the assertion for any $(a, b) \in S \times S, a \neq b$, one employs the same argument as in the proof of Arrow's Lemma.                                                                                                □

*Remark 2.8* In the proof of Arrow's Lemma transitivity of social $R$ was used only to deduce $(x P y \wedge y P z \rightarrow x P z)$ and $(z P x \wedge x P y \rightarrow z P y)$. In the proof of Lemma 2.2 transitivity of social $R$ was used only to deduce $(y P z \wedge z P x \rightarrow y P x)$. These deductions can still be made if transitivity is replaced by quasi-transitivity. Thus both Arrow's Lemma and Lemma 2.2 can be generalized as given below.                                   ◇

**Lemma 2.3** *Let* $\#S \geq 3$; *and* $\#N = n \geq 2, n \in \mathbb{N}$. *Let social decision rule* $f : \mathscr{T}^n \mapsto \mathscr{Q}$ *satisfy weak Pareto-criterion and independence of irrelevant alternatives. Then, if a set of individuals* $V \subseteq N$ *is almost decisive for some ordered pair of distinct alternatives then it is decisive for every ordered pair of distinct alternatives.*[2]

**Lemma 2.4** *Let* $\#S \geq 3$; *and* $\#N = n \geq 2, n \in \mathbb{N}$. *Let social decision rule* $f : \mathscr{T}^n \mapsto \mathscr{Q}$ *satisfy weak Pareto-criterion and independence of irrelevant alternatives. Then, if a set of individuals* $V \subseteq N$ *is almost semidecisive for some ordered*

---

[2]Gibbard (1969).

*pair of distinct alternatives then it is semidecisive for every ordered pair of distinct alternatives.*[3]

**Proposition 2.4** *Let $\#S \geq 3$; and $\#N = n \geq 2, n \in \mathbb{N}$. Let social decision rule $f$ : $\mathscr{T}^n \mapsto \mathscr{Q}$ satisfy weak Pareto-criterion and independence of irrelevant alternatives. Then there is a unique minimal decisive set.*[4]

*Proof* The set of all individuals is decisive, by condition P. Therefore, in view of condition P and the finiteness of $N$, a nonempty minimal decisive set exists. Suppose there exist more than one minimal decisive sets. Let $V_1$ and $V_2$ be two distinct minimal decisive sets. Now $V_1 \cap V_2 \neq \emptyset$; otherwise we will get a contradiction as follows. Suppose $V_1 \cap V_2 = \emptyset$. Assume the following configuration of preferences: $\forall i \in V_1 : x P_i y, \forall i \in V_2 : y P_i x$. By the definition of a decisive set $x P y$ and $y P x$, which is impossible. Therefore $V_1 \cap V_2 \neq \emptyset$. As $V_1$ and $V_2$ are distinct minimal decisive sets, it follows that both $V_1 - V_2$ and $V_2 - V_1$ must be nonempty.

Let $x, y, z \in S$ be distinct alternatives. Consider the following configuration of preferences:

$(\forall i \in V_1 \cap V_2)(x P_i y P_i z)$
$(\forall i \in V_1 - V_2)(z P_i x P_i y)$
$(\forall i \in V_2 - V_1)(y P_i z P_i x)$
$(\forall i \in N - (V_1 \cup V_2))(z P_i y P_i x).$

Now,

$(\forall i \in V_1)(x P_i y) \rightarrow x P y$
$(\forall i \in V_2)(y P_i z) \rightarrow y P z$
$x P y \wedge y P z \rightarrow x P z$, by quasi-transitivity of $R$.

We have $(\forall i \in V_1 \cap V_2)(x P_i z) \wedge (\forall i \in N - (V_1 \cap V_2))(z P_i x) \wedge x P z$. By condition I then it follows that $V_1 \cap V_2$ is almost decisive for $(x, z)$. By Lemma 2.3 then it follows that $V_1 \cap V_2$ is a decisive set. However, $V_1 \cap V_2$ is a proper subset of both $V_1$ and $V_2$ which were assumed to be minimal decisive sets. This contradiction establishes the proposition.                                                                    □

The following corollary follows immediately from Proposition 2.4.

**Corollary 2.3** *Let $\#S \geq 3$; and $\#N = n \geq 2, n \in \mathbb{N}$. Let social welfare function $f$ : $\mathscr{T}^n \mapsto \mathscr{T}$ satisfy weak Pareto-criterion and independence of irrelevant alternatives. Then there is a unique minimal decisive set.*

**Proposition 2.5** *Let $\#S \geq 3$; and $\#N = n \geq 2, n \in \mathbb{N}$. Let social welfare function $f : \mathscr{T}^n \mapsto \mathscr{T}$ satisfy weak Pareto-criterion and independence of irrelevant alternatives. Then the unique minimal decisive set consists of a single individual.*

---

[3]Sen (1970).
[4]Gibbard (1969).

*Proof* Let $V$ be the unique minimal decisive set; and let $\#V \geq 2$. Let $(V_1, V_2)$ be a partition[5] of $V$ [i.e., $V_1 \neq \emptyset$, $V_2 \neq \emptyset$, $V_1 \cap V_2 = \emptyset$, $V_1 \cup V_2 = V$]. Let $x, y, z \in S$ be distinct alternatives. Consider the following configuration of individual preferences over $\{x, y, z\} \subseteq S$:

$(\forall i \in V_1)(x P_i y P_i z)$

$(\forall i \in V_2)(y P_i z P_i x)$

$(\forall i \in N - V)(z P_i x P_i y)$.

From $(\forall i \in V)(y P_i z)$, we obtain $y P z$, by the decisiveness of $V$.

$y P x \vee x R y$, as $R$ is connected.

$y P x \to V_2$ is almost decisive for $(y, x)$, by condition I

$\to V_2$ is a decisive set, by Arrow's Lemma

This contradicts the minimality of $V$.

$x R y \to x P z$, by transitivity of $R$ in view of $y P z$

$\to V_1$ is almost decisive for $(x, z)$, by condition I

$\to V_1$ is a decisive set, by Arrow's Lemma

This contradicts the minimality of $V$.

The proposition is established as both $y P x$ and $x R y$ lead to contradiction. □

**Theorem 2.1** (Arrow Impossibility Theorem) *Let $\#S \geq 3$; and $\#N = n \geq 2$, $n \in$ $\mathbb{N}$. There exists no social welfare function $f : \mathscr{T}^n \mapsto \mathscr{T}$ satisfying weak Pareto-criterion and independence of irrelevant alternatives which is non-dictatorial.*

*Proof* Given $\#S \geq 3$; and $\#N = n \geq 2$, $n \in \mathbb{N}$; by Proposition 2.5 it follows that if $f : \mathscr{T}^n \mapsto \mathscr{T}$ satisfies weak Pareto-criterion and independence of irrelevant alternatives then there exists an individual who is a dictator. Therefore there cannot possibly exist a social welfare function $f : \mathscr{T}^n \mapsto \mathscr{T}$ satisfying weak Pareto-criterion and independence of irrelevant alternatives which is non-dictatorial, given $\#S \geq 3$; and $\#N = n \geq 2$, $n \in \mathbb{N}$. □

From Arrow Impossibility Theorem it follows that if $f$ is a social decision rule $f : \mathscr{T}^n \mapsto \mathscr{C}$, and $\#S \geq 3$, $\#N = n \geq 2$, $n \in \mathbb{N}$, then it must be the case that at least one of the following is true: (i) $f$ does not yield transitive social $R$ for every $(R_1, \ldots, R_n) \in \mathscr{T}^n$, (ii) $f$ violates weak Pareto-criterion, (iii) $f$ violates independence of irrelevant alternatives, (iv) $f$ is dictatorial. The example that follows illustrates it.

*Example 2.6* (i) $f : \mathscr{T}^n \mapsto \mathscr{C}$ is the Method of Majority Decision (MMD) iff $(\forall (R_1, \ldots, R_n) \in \mathscr{T}^n)(\forall x, y \in S)[x R y \leftrightarrow n(x P_i y) \geq n(y P_i x)]$, where $n()$ denotes[6] the number of individuals having the preferences specified in the parentheses.

From the definition of MMD it follows that $[x P y \leftrightarrow n(x P_i y) > n(y P_i x)] \wedge [x I y \leftrightarrow n(x P_i y) = n(y P_i x)]$.

Let $j, k \in N$; $x, y, z \in S$.

---

[5]Let $A$ be a nonempty set. $(A_1, A_2, \ldots, A_m)$ is a partition of $A$ iff $(\forall i, j \in \{1, 2, \ldots, m\})[(A_i \neq \emptyset) \wedge (i \neq j \to A_i \cap A_j = \emptyset)] \wedge \cup_{i=1}^{m} A_i = A$.

[6]Throughout this text $n()$ and $N()$ will denote the number of individuals having the preferences specified in the parentheses.

Suppose $x P_j y P_j z \wedge y P_k z P_k x \wedge (\forall i \in N - \{j, k\})(x I_i y I_i z)$
$\rightarrow [n(x P_i y) = 1 \wedge n(y P_i x) = 1] \wedge [n(y P_i z) = 2 \wedge n(z P_i y) = 0] \wedge [n(x P_i z) = 1 \wedge n(z P_i x) = 1]$
$\rightarrow x I y \wedge y P z \wedge x I z$, which violates transitivity.
Therefore MMD does not yield transitive social $R$ for every $(R_1, \ldots, R_n) \in \mathcal{T}^n$.
Suppose $(\forall i \in N)(x P_i y)$
$\rightarrow n(x P_i y) = n \wedge n(y P_i x) = 0$
$\rightarrow x P y$.
Thus MMD satisfies condition P.
Let $(\forall i \in N)[(x R_i y \leftrightarrow x R_i' y) \wedge (y R_i x \leftrightarrow y R_i' x)], x, y \in S$
$\rightarrow n(x P_i y) = n(x P_i' y) \wedge n(y P_i x) = n(y P_i' x)$
$\rightarrow (x R y \leftrightarrow x R' y) \wedge (y R x \leftrightarrow y R' x)$.
Therefore MMD satisfies condition I.
Let $j \in N; x, y \in S$.
Suppose $x P_j y \wedge (\forall i \in N - \{j\})(y P_i x)$
$\rightarrow n(x P_i y) = 1 \wedge n(y P_i x) = n - 1$
$\rightarrow y R x$, as $n \geq 2$.
This shows that MMD is non-dictatorial.

(ii) Let $\overline{R}$ be a given ordering of $S$. Let $f : \mathcal{T}^n \mapsto \mathcal{C}$ be defined by:
$(\forall (R_1, \ldots, R_n) \in \mathcal{T}^n)[f(R_1, \ldots, R_n) = \overline{R}]$.
As social $R$ is always $\overline{R}$, it follows that $f$ yields transitive social $R$ for every $(R_1, \ldots, R_n) \in \mathcal{T}^n$.
Suppose $x \overline{R} y; x, y \in S$.
Consider $(R_1, \ldots, R_n) \in \mathcal{T}^n$ such that: $(\forall i \in N)(y P_i x)$.
$(\forall i \in N)(y P_i x) \wedge \sim y \overline{P} x$ is a violation of condition P.
As social $R$ is the same for all $(R_1, \ldots, R_n) \in \mathcal{T}^n$, the consequent of condition I is satisfied. So condition I holds trivially.
As condition P is violated, it follows that $f$ is non-dictatorial.

(iii) Let $S$ be finite; $S = \{x, y, z_1, \ldots, z_{s-2}\}$.
Let $B_i(\alpha) = \{\beta \in S \mid \beta P_i \alpha\}$, $W_i(\alpha) = \{\beta \in S \mid \alpha P_i \beta\}$, $b_i(\alpha) = \# B_i(\alpha)$, $w_i(\alpha) = \# W_i(\alpha), \alpha \in S, i \in N$.
$B_i(\alpha)$ is the set of alternatives which individual $i$ considers to be better than $\alpha$; and $W_i(\alpha)$ is the set of alternatives which individual $i$ considers to be worse than $\alpha$.
Let $m_i(\alpha) = \frac{1}{s - b_i(\alpha) - w_i(\alpha)} \sum_{t=w_i(\alpha)+1}^{s - b_i(\alpha)} t, t \in \mathbb{N}$; and $M(\alpha) = \sum_{i \in N} m_i(\alpha)$.
Let $f : \mathcal{T}^n \mapsto \mathcal{C}$ be the rank-order method defined by:
$(\forall (R_1, \ldots, R_n) \in \mathcal{T}^n)(\forall a, b \in S)[a R b \leftrightarrow M(a) \geq M(b)]$.
Let $a R b \wedge b R c; a, b, c \in S$
$\rightarrow M(a) \geq M(b) \wedge M(b) \geq M(c)$
$\rightarrow M(a) \geq M(c)$
$\rightarrow a R c$.
Thus $f$ yields transitive $R$ for every $(R_1, \ldots, R_n) \in \mathcal{T}^n$.
Let $(\forall i \in N)(a P_i b)$
$\rightarrow M(a) > M(b)$
$\rightarrow a P b$.

Thus condition P holds.

Let $(R_1, \ldots, R_n) = (x P_1 y P_1 z_1 I_1 \ldots I_1 z_{s-2}, y P_2 z_1 I_2 \ldots I_2 z_{s-2} P_2 x, (\forall i \in N - \{1, 2\})(x I_i y I_i z_1 I_i \ldots I_i z_{s-2}))$ $\wedge (R_1', \ldots, R_n') = (x P_1' z_1 I_1' \ldots I_1' z_{s-2} P_1' y, y P_2' x P_2' z_1 I_2' \ldots I_2' z_{s-2}, (\forall i \in N - \{1, 2\})(x I_i' y I_i' z_1 I_i' \ldots I_i' z_{s-2}))$.

For $(R_1, \ldots, R_n)$ we have:

$m_1(x) = s, m_2(x) = 1, m_i(x) = \frac{1}{s} \sum_{t=1}^{s} t = \frac{s+1}{2}, i \in N - \{1, 2\}$; and so $M(x) = s + 1 + (n - 2)\frac{s+1}{2}$

$m_1(y) = s - 1, m_2(y) = s, m_i(y) = \frac{s+1}{2}, i \in N - \{1, 2\}$; and so $M(y) = 2s - 1 + (n - 2)\frac{s+1}{2}$

$m_1(z_k) = \frac{1}{s-2} \sum_{t=1}^{s-2} t = \frac{s-1}{2}, m_2(z_k) = \frac{1}{s-2} \sum_{t=2}^{s-1} t = \frac{s+1}{2}, m_i(z_k) = \frac{s+1}{2}, i \in N - \{1, 2\}, k \in \{1, \ldots, s - 2\}$; and so $M(z_k) = s + (n - 2)\frac{s+1}{2}$.

For $(R_1', \ldots, R_n')$ we have:

$m_1(x) = s, m_2(x) = s - 1, m_i(x) = \frac{s+1}{2}, i \in N - \{1, 2\}$; and so $M(x) = 2s - 1 + (n - 2)\frac{s+1}{2}$

$m_1(y) = 1, m_2(y) = s, m_i(y) = \frac{s+1}{2}, i \in N - \{1, 2\}$; and so $M(y) = s + 1 + (n - 2)\frac{s+1}{2}$

$m_1(z_k) = \frac{1}{s-2} \sum_{t=2}^{s-1} t = \frac{s+1}{2}, m_2(z_k) = \frac{1}{s-2} \sum_{t=1}^{s-2} t = \frac{s-1}{2}, m_i(z_k) = \frac{s+1}{2}, i \in N - \{1, 2\}, k \in \{1, \ldots, s - 2\}$; and so $M(z_k) = s + (n - 2)\frac{s+1}{2}$.

As $\#S \geq 3$, we obtain:

$y P x P z_1 I \ldots I z_{s-2}$ and $x P' y P' z_1 I' \ldots I' z_{s-2}$.

$(\forall i \in N)[(x R_i y \leftrightarrow x R_i' y) \wedge (y R_i x \leftrightarrow y R_i' x)] \wedge y P x \wedge x P' y$ is a violation of condition I.

Take any individual $j \in N$.

Consider the profile: $[x P_j y P_j z_1 I_j \ldots I_j z_{s-2}, (\forall i \in N - \{j\})(y P_i z_1 I_i \ldots I_i z_{s-2} P_i x)]$.

We obtain:

$m_j(x) = s, m_i(x) = 1, i \in N - \{j\}$; and so $M(x) = s + n - 1$

$m_j(y) = s - 1, m_i(y) = s, i \in N - \{j\}$; and so $M(y) = ns - 1$.

As $\#S \geq 3$ and $\#N \geq 2$, we obtain $y P x$ which establishes that $j$ is not a dictator. Thus $f$ is non-dictatorial.

(iv) Let $f : \mathscr{T}^n \mapsto \mathscr{C}$ be defined by:

$(\forall (R_1, \ldots, R_n) \in \mathscr{T}^n)[f(R_1, \ldots, R_n) = R_1]$.

As social $R$ is identical to $R_1$ for every profile in the domain, it follows that $f$ yields transitive social $R$ for every $(R_1, \ldots, R_n) \in \mathscr{T}^n$.

Suppose $(\forall i \in N)(x P_i y); x, y \in S$

$\rightarrow x P_1 y$

$\rightarrow x P y$.

Thus condition P holds.

Let $(\forall i \in N)[(x R_i y \leftrightarrow x R_i' y) \wedge (y R_i x \leftrightarrow y R_i' x)], x, y \in S$

$\rightarrow [(x R_1 y \leftrightarrow x R_1' y) \wedge (y R_1 x \leftrightarrow y R_1' x)]$

$\rightarrow [(x R y \leftrightarrow x R' y) \wedge (y R x \leftrightarrow y R' x)]$.

Therefore condition I holds.

As social $R$ for every $(R_1, \ldots, R_n) \in \mathscr{T}^n$ is identical to $R_1$, $f$ is dictatorial.    $\Diamond$

We now define some more conditions on social decision rules.

A social decision rule $f : \mathbb{D} \mapsto \mathscr{C}, \mathbb{D} \subseteq \mathscr{T}^n$, satisfies (i) Pareto-Preference (PP) iff $(\forall (R_1, \ldots, R_n) \in \mathbb{D})(\forall x, y \in S)[(\forall i \in N)(x R_i y) \wedge (\exists i \in N)(x P_i y) \rightarrow x P y]$, (ii) Pareto-Indifference (PI) iff $(\forall (R_1, \ldots, R_n) \in \mathbb{D})(\forall x, y \in S)[(\forall i \in N)(x I_i y) \rightarrow x I y]$, (iii) the Pareto-criterion $(\overline{P})$ iff Pareto-preference and Pareto-indifference hold.

Let $\Phi$ be the set of all permutations of alternatives in $S$. Let $\phi \in \Phi$. Corresponding to a binary relation $R$ on set $S$, we define the binary relation $\phi(R)$ on $S$ by: $(\forall x, y \in S)[\phi(x) \phi(R) \phi(y) \leftrightarrow x R y]$. A social decision rule $f : \mathbb{D} \mapsto \mathscr{C}, \mathbb{D} \subseteq \mathscr{T}^n$, satisfies neutrality (N) iff $(\forall (R_1, \ldots, R_n), (R'_1, \ldots, R'_n) \in \mathbb{D})(\forall \phi \in \Phi)[(\forall i \in N)[R'_i = \phi(R_i)] \rightarrow R' = \phi(R)]$.

Let $\Theta$ be the set of all permutations of individuals in $N$, i.e., the set of all permutations of positive integers $1, 2, \ldots, n$. Let $\theta \in \Theta$. A social decision rule $f : \mathbb{D} \mapsto \mathscr{C}, \mathbb{D} \subseteq \mathscr{T}^n$, satisfies anonymity (A) iff $(\forall (R_1, \ldots, R_n), (R'_1, \ldots, R'_n) \in \mathbb{D})[(\exists \theta \in \Theta)(\forall i \in N)(R_i = R'_{\theta(i)}) \rightarrow R' = R]$.

Neutrality requires that all alternatives are to be treated equally; and anonymity that all individuals are to be treated equally.

*Remark 2.9*  In the social choice literature, at times the anonymity condition is defined as follows:

A social decision rule $f : \mathbb{D} \mapsto \mathscr{C}, \mathbb{D} \subseteq \mathscr{T}^n$, satisfies anonymity iff $(\forall (R_1, \ldots, R_n), (R'_1, \ldots, R'_n) \in \mathbb{D})(\forall j, k \in N)[(\forall i \in N - \{j, k\})(R_i = R'_i) \wedge (R_j = R'_k) \wedge (R_k = R'_j) \rightarrow R' = R]$.

If $\mathbb{D} = \mathscr{T}^n$ then it is clear that this definition is equivalent to the definition of anonymity given above. However, if $\mathbb{D} \subset \mathscr{T}^n$ then it is possible that the SDR is anonymous in the sense of this definition but not in the sense of the definition given earlier, as can be seen from the example given below:

*Example 2.7*  Let:

$S = \{x, y, z\}$, $N = \{1, 2, 3\}$.

$\mathbb{D} = \{(x P_1 y P_1 z, z P_2 y P_2 x, x I_3 y I_3 z), (x I_1 y I_1 z, x P_2 y P_2 z, z P_3 y P_3 x)\}$.

Let SDR $f : \mathbb{D} \mapsto \mathscr{C}$ be defined as follows:

$f(x P_1 y P_1 z, z P_2 y P_2 x, x I_3 y I_3 z) = x P y P z$ and $f(x I_1 y I_1 z, x P_2 y P_2 z, z P_3 y P_3 x) = x I y I z$. As the set of orderings in the two profiles is the same but the social preferences corresponding to them are different, it follows that anonymity condition is violated as defined earlier. However, anonymity in the sense of the latter definition is trivially satisfied as the antecedent of the condition does not hold.    $\Diamond$

Thus, if the definition is to formalize the idea of anonymity correctly regardless of the domain of the rule then it must be defined as was done in the earlier definition. $\Diamond$

A social decision rule $f : \mathbb{D} \mapsto \mathscr{C}, \mathbb{D} \subseteq \mathscr{T}^n$, satisfies monotonicity (M) iff $(\forall (R_1, \ldots, R_n), (R'_1, \ldots, R'_n) \in \mathbb{D})(\forall x \in S)[(\forall i \in N)[(\forall a, b \in S - \{x\})(a R_i b \leftrightarrow a R'_i b) \wedge (\forall y \in S - \{x\})[(x P_i y \rightarrow x P'_i y) \wedge (x I_i y \rightarrow x R'_i y)]] \rightarrow (\forall y \in S - \{x\})[(x P y \rightarrow x P' y) \wedge (x I y \rightarrow x R' y)]]$. Monotonicity requires that if an alternative goes up or stays at the same level in everyone's esteem, there being no other change, then it must not go down in the social preferences.

It is clear from the definitions of conditions I, N and M that a social decision rule $f$ : $T^n \mapsto C$ satisfying condition I satisfies (i) neutrality iff $(\forall (R_1, \ldots, R_n), (R'_1, \ldots, R'_n) \in T^n)(\forall x, y, z, w \in S)[(\forall i \in N)[(x R_i y \leftrightarrow z R'_i w) \wedge (y R_i x \leftrightarrow w R'_i z)] \rightarrow [(x R y \leftrightarrow z R' w) \wedge (y R x \leftrightarrow w R' z)]]$, and (ii) monotonicity iff $(\forall (R_1, \ldots, R_n), (R'_1, \ldots, R'_n) \in T^n)(\forall x, y \in S)[(\forall i \in N)[(x P_i y \rightarrow x P'_i y) \wedge (x I_i y \rightarrow x R'_i y)] \rightarrow [(x P y \rightarrow x P' y) \wedge (x I y \rightarrow x R' y)]]$.

## 2.3 Latin Squares

Let $A \subseteq S$ and let $R$ be a binary relation on $S$. We define restriction of $R$ to $A$, denoted by $R|A$, by $R|A = R \cap (A \times A)$. Let $\mathscr{D} \subseteq \mathscr{B}$. We define restriction of $\mathscr{D}$ to $A$, denoted by $\mathscr{D}|A$, by $\mathscr{D}|A = \{R|A \mid R \in \mathscr{D}\}$.

A set of three distinct alternatives will be called a triple of alternatives. Let $R$ be a binary relation on $S$ and let $A = \{x, y, z\} \subseteq S$ be a triple of alternatives. We define $R$ to be unconcerned over $A$ iff $(\forall a, b \in A)(a I b)$. $R$ is defined to be concerned over $A$ iff it is not unconcerned over $A$. We define in $A$, according to $R$, $x$ to be best iff $(x R y \wedge x R z)$; to be medium iff $[(y R x \wedge x R z) \vee (z R x \wedge x R y)]$; to be worst iff $(y R x \wedge z R x)$; to be proper best iff $[(x P y \wedge x R z) \vee (x R y \wedge x P z)]$; to be proper medium iff $[(y P x \wedge x R z) \vee (y R x \wedge x P z) \vee (z P x \wedge x R y) \vee (z R x \wedge x P y)]$; and to be proper worst iff $[(y P x \wedge z R x) \vee (y R x \wedge z P x)]$.

Weak Latin Square (WLS): Let $A = \{x, y, z\} \subseteq S$ be a triple of alternatives and let $R^s$, $R^t$, $R^u$ be binary relations on $S$. The set $\{R^s|A, R^t|A, R^u|A\}$ forms a weak Latin Square over $A$ iff $(\exists$ distinct $a, b, c \in A)$ [(in $R^s|A$ $a$ is best and $b$ is medium and $c$ is worst) $\wedge$ (in $R^t|A$ $b$ is best and $c$ is medium and $a$ is worst) $\wedge$ (in $R^u|A$ $c$ is best and $a$ is medium and $b$ is worst)]. The above weak Latin Square will be denoted by $WLS(abca)$.

*Remark 2.10* $R^s|A$, $R^t|A$, $R^u|A$ in the definition of weak Latin Square need not be distinct. $\{x I y I z\}$ forms a weak Latin Square over the triple $\{x, y, z\}$.          $\Diamond$

*Remark 2.11* If $R^s|A$, $R^t|A$, $R^u|A$ are orderings over $A$, then the set $\{R^s|A, R^t|A, R^u|A\}$ forms a weak Latin Square over A iff $(\exists$ distinct $a, b, c \in A)[a R^s b R^s c \wedge b R^t c R^t a \wedge c R^u a R^u b]$.          $\Diamond$

Latin Square (LS): Let $A = \{x, y, z\} \subseteq S$ be a triple of alternatives and let $R^s$, $R^t$, $R^u$ be binary relations on $S$. The set $\{R^s|A, R^t|A, R^u|A\}$ forms a Latin Square over $A$ iff $(\exists$ distinct $a, b, c \in A)$ [(in $R^s|A$ $a$ is best and $b$ is proper medium and $c$ is worst) $\wedge$ (in $R^t|A$ $b$ is best and $c$ is proper medium and $a$ is worst) $\wedge$ (in $R^u|A$ $c$ is best and $a$ is proper medium and $b$ is worst)]. The above Latin Square will be denoted by $LS(abca)$.

*Remark 2.12* If $R^s|A$, $R^t|A$, $R^u|A$ are orderings over $A$, then the set $\{R^s|A, R^t|A, R^u|A\}$ forms a Latin Square over A iff $R^s|A$, $R^t|A$, $R^u|A$ are concerned over $A$ and $(\exists$ distinct $a, b, c \in A)[aR^sbR^sc \wedge bR^tcR^ta \wedge cR^uaR^ub]$.                    $\Diamond$

*Remark 2.13* From the definitions of weak Latin Square and Latin Square it is clear that if $R^s|A$, $R^t|A$, $R^u|A$ are orderings and concerned over $A$ then they form a Latin Square iff they form a weak Latin Square.                    $\Diamond$

Let $A = \{x, y, z\} \subseteq S$ be a triple of alternatives. For any distinct $a, b, c \in A$, we define:
$B[WLS(abca)] = \{R \in \mathscr{B}|A \mid (a$ is best and $b$ is medium and $c$ is worst in $R) \vee (b$ is best and $c$ is medium and $a$ is worst in $R) \vee (c$ is best and $a$ is medium and $b$ is worst in $R)\}$
$B[LS(abca)] = \{R \in \mathscr{B}|A \mid (a$ is best and $b$ is proper medium and $c$ is worst in $R) \vee (b$ is best and $c$ is proper medium and $a$ is worst in $R) \vee (c$ is best and $a$ is proper medium and $b$ is worst in $R)\}$
$T[WLS(abca)] = \{R \in \mathscr{T}|A \mid (aRbRc \vee bRcRa \vee cRaRb)\}$
$T[LS(abca)] \quad = \quad \{R \in \mathscr{T}|A \mid R$ is concerned over $A \wedge (aRbRc \vee bRcRa \vee cRaRb)\}$.
Thus we have:
$T[WLS(xyzx)] = T[WLS(yzxy)] = T[WLS(zxyz)] = \{xPyPz, xPyIz, xIyPz,$
$yPzPx, yPzIx, yIzPx, zPxPy, zPxIy, zIxPy, xIyIz\}$
$T[WLS(xzyx)] = T[WLS(zyxz)] = T[WLS(yxzy)] = \{xPzPy, xPzIy, xIzPy,$
$zPyPx, zPyIx, zIyPx, yPxPz, yPxIz, yIxPz, xIyIz\}$
$T[LS(xyzx)] = T[LS(yzxy)] = T[LS(zxyz)] = T[WLS(xyzx)] - \{xIyIz\}$
$T[LS(xzyx)] = T[LS(zyxz)] = T[LS(yxzy)] = T[WLS(xzyx)] - \{xIyIz\}$.

## References

Arrow, Kenneth J. 1951. *Social choice and individual values*, 2nd ed., 1963. New York: Wiley.
Gibbard, Allan. 1969. *Social choice and the Arrow conditions*. Discussion Paper: Department of Philosophy, University of Michigan.
Jain, Satish K. 1985. A direct proof of Inada-Sen-Pattanaik theorem on majority rule. *The Economic Studies Quarterly* 36: 209–215.
Sen, Amartya K. 1970. *Collective choice and social welfare*. San Francisco: Holden-Day.
Suppes, Patrick. 1957. *Introduction to logic*. New York: Van Nostrand.
Tarski, Alfred. 1941. *Introduction to logic and to the methodology of deductive sciences*. New York: Oxford University Press.

# The Method of Majority Decision

The Method of Majority Decision (MMD) is the most important rule for collective decision-making and is widely used in varied contexts. Under the method of majority decision, an alternative $x$ is socially better than another alternative $y$ iff the number of individuals preferring $x$ over $y$ is greater than the number of individuals preferring $y$ over $x$; and alternatives $x$ any $y$ are equally good iff the number of individuals preferring $x$ over $y$ is equal to the number of individuals preferring $y$ over $x$. MMD is simple to use; in order to determine social preferences over any pair of alternatives, one has to consider only individual preferences over that pair, and not over other alternatives. Thus, MMD satisfies the condition of independence of irrelevant alternatives. In addition to simplicity, it has several other attractive features. It does not discriminate among individuals; all individuals are treated equally. A permutation of orderings among individuals leaves the social preferences unchanged. It also does not discriminate among candidates. The rule for deciding social preferences over any particular pair of alternatives is the same as that for any other pair of alternatives. In other words, MMD satisfies the properties of anonymity and neutrality. It satisfies the property of strict monotonicity also. Strict monotonicity is a stronger requirement than monotonicity. Strict monotonicity requires, in addition to what is required by the monotonicity, that if between two alternatives $x$ and $y$, every individual either changes preferences in favour of $x$ or does not change preferences between them and some individuals do change their preferences in favour of alternative $x$, then in case there is social indifference between $x$ and $y$ to begin with, then after the change $x$ must be strictly preferred over $y$ in social preferences. Thus, not only do the social preferences generated by the majority rule never move perversely in relation to individual preferences, they in fact move positively with them. It is also the case that the MMD is the only social decision rule that satisfies all the four properties of independence of irrelevant alternatives, neutrality, anonymity and strict monotonicity. The characterization of the MMD as a social decision rule satisfying

© Springer Nature Singapore Pte Ltd. 2019
S. K. Jain, *Domain Conditions and Social Rationality*,
https://doi.org/10.1007/978-981-13-9672-4_3

the properties of independence of irrelevant alternatives, neutrality, anonymity and strict monotonicity was obtained by May (1952).[1]

As under the method of majority decision it is possible to have intransitive, even cyclical, social preferences, an important problem is that of characterizing sets of orderings defined over the set of alternatives which are such that if in a profile every individual's ordering belongs to one of these sets then the social weak preference relation generated by the social decision rule is invariably rational (transitive, quasi-transitive or acyclic). This chapter is concerned with the derivation of conditions that completely characterize such sets for transitivity and quasi-transitivity under the MMD.

Let $\mathscr{D}$ be a set of orderings of the set of alternatives $S$. For the transitivity and quasi-transitivity under the MMD the following results hold. (i) If the number of individuals $n$ is even and greater than or equal to 2, then every profile of individual orderings belonging to $\mathscr{D}^n$ yields transitive social $R$ under the MMD iff $\mathscr{D}$ satisfies the condition of extremal restriction. (ii) If the number of individuals $n$ is odd and greater than or equal to 3, then every profile of individual orderings belonging to $\mathscr{D}^n$ yields transitive social $R$ under the MMD iff $\mathscr{D}$ satisfies the condition of Weak Latin Square partial agreement. (iii) If the number of individuals $n$ is greater than or equal to 5, then every profile of individual orderings belonging to $\mathscr{D}^n$ yields quasi-transitive social $R$ under the MMD iff $\mathscr{D}$ satisfies the condition of Latin Square partial agreement. (iv) If the number of individuals $n$ is four, then every profile of individual orderings belonging to $\mathscr{D}^n$ yields quasi-transitive social $R$ under the MMD iff $\mathscr{D}$ satisfies the condition of weak extremal restriction. (v) If the number of individuals $n$ is three, then every profile of individual orderings belonging to $\mathscr{D}^n$ yields quasi-transitive social $R$ under the MMD iff $\mathscr{D}$ satisfies the condition of Latin Square linear ordering restriction. The logical relationships among these five conditions are given by: extremal restriction implies Weak Latin Square partial agreement; Weak Latin Square partial agreement implies Latin Square partial agreement; Latin Square partial agreement implies weak extremal restriction; and weak extremal restriction implies Latin Square linear ordering restriction.

Extremal restriction requires that if a set of orderings of a triple contains a linear ordering of the triple, then in any ordering belonging to the set which is of the same Latin Square as the one associated with the linear ordering, the alternative which is the best in the linear ordering must be considered to be at least as good as the alternative which is the worst in the linear ordering. Weak Latin Square partial agreement requires fulfilment of what is required by extremal restriction if a weak Latin Square involving a linear ordering exists. Latin Square partial agreement requires fulfilment of what is required by extremal restriction in case a Latin Square involving a linear ordering exists. Weak extremal restriction requires that in case there is a linear ordering of the triple in question, then it must not be the case that there is an ordering of the triple in which the worst alternative of the linear ordering is uniquely best,

---

[1] For alternative characterizations of majority rule, see Asan and Sanver (2002), Campbell and Kelly (2000), and Woeginger (2003), among others.

there is an ordering of the triple in which the best alternative of the linear ordering is uniquely worst, and both these orderings belong to the same Latin Square which is associated with the linear ordering. Latin Square linear ordering restriction merely requires that there be no Latin Square involving more than one linear ordering.

The chapter is divided into six sections. The first section contains the statement and the proof of May's characterization theorem. The second section formally defines the five conditions that turn out to be relevant for transitivity and quasi-transitivity under the MMD. The third and fourth sections contain the transitivity and quasi-transitivity theorems, respectively, mentioned above. In the context of the method of majority decision, there are a large number of conditions relevant for transitivity or quasi-transitivity that have emerged since the pioneering work of Black (1948) and Arrow (1951). Section five contains a discussion of most of these conditions and their interrelationships. The last section of the paper contains a brief survey of the literature pertaining to transitivity and quasi-transitivity under the method of majority decision. The last two sections have been put in the appendix to the chapter.

## 3.1 Characterization of the Method of Majority Decision

The Method of Majority Decision (MMD): MMD $f : \mathscr{T}^n \mapsto \mathscr{C}$ is defined by: $(\forall (R_1, \ldots, R_n) \in \mathscr{T}^n)(\forall x, y \in S)[xRy \leftrightarrow n(xP_iy) \geq n(yP_ix)]$, where $n()$ denotes the number of individuals having the preferences specified in the parentheses.

From the definition of MMD, it follows that $(\forall (R_1, \ldots, R_n) \in \mathscr{T}^n)(\forall x, y \in S)[[xPy \leftrightarrow n(xP_iy) > n(yP_ix)] \wedge [xIy \leftrightarrow n(xP_iy) = n(yP_ix)]]$. Thus under the method of majority decision, an alternative $x$ is socially preferred to another alternative $y$ iff the number of people who prefer $x$ to $y$ is greater than the number of people who prefer $y$ to $x$; and $x$ is socially indifferent to $y$ iff the number of people who prefer $x$ to $y$ is equal to the number of people who prefer $y$ to $x$.

Let social decision rule $f : \mathscr{T}^n \mapsto \mathscr{C}$ satisfy condition I. $f$ is strictly monotonic $(\overline{M})$ iff $(\forall (R_1, \ldots, R_n), (R'_1, \ldots, R'_n) \in \mathscr{T}^n)(\forall x, y \in S)[(\forall i \in N)[(xP_iy \rightarrow xP'_iy) \wedge (xI_iy \rightarrow xR'_iy)] \wedge (\exists i \in N)[(yP_ix \wedge xR'_iy) \vee (xI_iy \wedge xP'_iy)] \rightarrow [xRy \rightarrow xP'y]]$. Strict monotonicity in the context of rules satisfying condition I thus requires that: if between-alternatives $x$ and $y$ every individual's preferences either remain unchanged or change in favour of $x$, and at least one individual changes preferences in favour of $x$, then if $x$ is socially preferred to $y$ or socially indifferent to $y$ before the change then after the change $x$ must be socially preferred to $y$. Strict monotonicity is a stronger requirement than monotonicity.

**Theorem 3.1** *(May's Theorem) A social decision rule $f : \mathscr{T}^n \mapsto \mathscr{C}$ is the method of majority decision iff it satisfies the conditions of independence of irrelevant alternatives, neutrality, anonymity and strict monotonicity.*

*Proof* Let $f : \mathscr{T}^n \mapsto \mathscr{C}$ be the method of majority decision.

Let $(\forall i \in N)[(xR_i y \leftrightarrow xR'_i y) \wedge (yR_i x \leftrightarrow yR'_i x)], x, y \in S$

$\rightarrow n(xP_i y) = n(xP'_i y) \wedge n(yP_i x) = n(yP'_i x)$

$\rightarrow (xRy \leftrightarrow xR'y) \wedge (yRx \leftrightarrow yR'x)$.

Therefore, MMD satisfies condition I.

Let $(\forall i \in N)[(xR_i y \leftrightarrow zR'_i w) \wedge (yR_i x \leftrightarrow wR'_i z)], x, y, z, w \in S$

$\rightarrow n(xP_i y) = n(zP'_i w) \wedge n(yP_i x) = n(wP'_i z)$

$\rightarrow (xRy \leftrightarrow zR'w) \wedge (yRx \leftrightarrow wR'z)$.

Therefore, MMD satisfies neutrality.

Let $(R'_1, \ldots, R'_n)$ be a permutation of $(R_1, \ldots, R_n)$.

Then we have: $(\forall x, y \in S)[n(xP'_i y) = n(xP_i y)]$

$\rightarrow R' = R$.

Thus, MMD satisfies anonymity.

Let $\quad (\forall i \in N)[(xP_i y \rightarrow xP'_i y) \wedge (xI_i y \rightarrow xR'_i y)] \wedge (\exists i \in N)[(yP_i x \wedge xR'_i y) \vee (xI_i y \wedge xP'_i y)]$.

Suppose $xRy$.

$xRy \rightarrow n(xP_i y) \geq n(yP_i x)$                                       (T3.1-1)

$(\forall i \in N)[(xP_i y \rightarrow xP'_i y) \wedge (xI_i y \rightarrow xR'_i y)]$

$\rightarrow n(xP'_i y) \geq n(xP_i y) \wedge n(yP'_i x) \leq n(yP_i x)$                  (T3.1-2)

(T3.1-1) $\wedge$ (T3.1-2) $\rightarrow n(xP'_i y) \geq n(xP_i y) \geq n(yP_i x) \geq n(yP'_i x)$    (T3.1-3)

$(\exists i \in N)(yP_i x \wedge xR'_i y) \wedge$ (T3.1-3) $\rightarrow n(xP'_i y) \geq n(xP_i y) \geq n(yP_i x) > n(yP'_i x)$

(T3.1-4)

$(\exists i \in N)(xI_i y \wedge xP'_i y) \wedge \quad$ (T3.1-3) $\quad \rightarrow n(xP'_i y) > n(xP_i y) \geq n(yP_i x) \geq$
$n(yP'_i x)$

(T3.1-5)

(T3.1-4) $\wedge$ (T3.1-5) $\rightarrow n(xP'_i y) > n(yP'_i x)$

$\rightarrow xP'y$.

Thus, MMD satisfies strict monotonicity.

Let $f : \mathscr{T}^n \mapsto \mathscr{C}$ satisfy conditions I, N, A and $\overline{M}$.

By condition I, the social $R$ over any pair of alternatives is entirely determined by individual preferences over that pair. By neutrality, the rule for determining social $R$ is the same for every pair of alternatives. By anonymity then it follows that social $R \mid \{x, y\}; x, y \in S$ is entirely determined by what numbers $n(xP_i y), n(xI_i y)$ and $n(yP_i x)$ are.

Consider any profile $(R_1, \ldots, R_n)$ such that: $n(xP_i y) = n(yP_i x)$. Consider the permutation $\phi$ of $S$ given by: $[(\forall z \in S - \{x, y\})(\phi(z) = z) \wedge \phi(x) = y \wedge \phi(y) = x]$. By neutrality, if individual preferences become $(\phi(R_1), \ldots, \phi(R_n))$, then social preferences would become $\phi(R)$. But in view of $n(xP_i y) = n(yP_i x)$, it follows that interchange of $x$ and $y$ would leave the numbers $n(x P_i y), n(x I_i y), n(y P_i x)$ unchanged. So by anonymity social preferences over $\{x, y\}$ must remain unchanged. Thus, it must be the case that $R \mid \{x, y\} = \phi(R) \mid \{x, y\}$. From this, it follows that if $n(x P_i y) = n(y P_i x)$ then $x I y$ must hold.

Next consider any profile $(R_1, \ldots, R_n)$ such that: $n(x P_i y) > n(y P_i x)$. Let $N_{xy} = \{i \in N \mid x P_i y\}$, $N_{(xy)} = \{i \in N \mid x I_i y\}$, $N_{yx} = \{i \in N \mid y P_i x\}$. Let $(R'_1, \ldots, R'_n)$ be such that: $(\forall i \in N'_{xy})(x P'_i y) \wedge (\forall i \in (N_{xy} - N'_{xy}) \cup N_{(xy)})(x I'_i y) \wedge (\forall i \in N_{yx})$ $(y P'_i x)$; where $N'_{xy} \subset N_{xy}$ and $\#N'_{xy} = \#N_{yx}$. By construction we have: $n(x P'_i y) = n(y P'_i x)$. Consequently, $x I' y$ holds. We now have: $(\forall i \in N)[(x P'_i y \rightarrow x P_i y) \wedge (x I'_i y \rightarrow x R_i y)] \wedge (\exists i \in N)(x I'_i y \wedge x P_i y)$. By strict monotonicity we obtain $x P y$ in view of $x I' y$.

Thus, we have shown that $(\forall (R_1, \ldots, R_n) \in \mathscr{T}^n)(\forall x, y \in S)[(n(x P_i y) = n(y P_i x) \rightarrow x I y) \wedge (n(x P_i y) > n(y P_i x) \rightarrow x P y)]$. Thus, $f$ is the MMD. $\qquad \square$

## 3.2 Restrictions on Preferences

We define and discuss several restrictions on sets of orderings.

Extremal Restriction (ER): $\mathscr{D} \subseteq \mathscr{T}$ satisfies ER over the triple $A \subseteq S$ iff ($\forall$ distinct $a, b, c \in A)[(\exists R \in \mathscr{D}|A)(a P b P c) \rightarrow (\forall R \in \mathscr{D}|A \cap T[LS(abca)])(a R c)]$. $\mathscr{D}$ satisfies ER iff it satisfies ER over every triple contained in $S$.

Thus, the satisfaction of extremal restriction by $\mathscr{D}$ over the triple $A$ requires that in case $\mathscr{D}|A$ contains a linear ordering of $A$, then in every ordering in $\mathscr{D}|A$ which belongs to the same Latin Square as the linear ordering, the alternative which is the best in the linear ordering must be at least as good as the alternative which is the worst in the linear ordering.

*Remark 3.1* Extremal restriction is usually defined as follows: $\mathscr{D} \subseteq \mathscr{T}$ satisfies ER over the triple $A \subseteq S$ iff ($\forall$ distinct $a, b, c \in A)[(\exists R \in \mathscr{D}|A)(a P b P c) \rightarrow (\forall R \in \mathscr{D}|A)(\sim c P a \vee c P b P a)]$. $\mathscr{D}$ satisfies ER iff it satisfies ER over every triple contained in $S$. It is clear that the two definitions are equivalent to each other. $\qquad \Diamond$

Weak Latin Square Partial Agreement (WLSPA): $\mathscr{D} \subseteq \mathscr{T}$ satisfies WLSPA over the triple $A \subseteq S$ iff ($\forall$ distinct $a, b, c \in A)[(\exists R^s, R^t, R^u \in \mathscr{D}|A)(a P^s b P^s c \wedge b R^t c R^t a \wedge c R^u a R^u b) \rightarrow (\forall R \in \mathscr{D}|A \cap T[LS(abca)])(a R c)]$. $\mathscr{D}$ satisfies WLSPA iff it satisfies WLSPA over every triple contained in $S$.

The satisfaction of Weak Latin Square partial agreement by $\mathscr{D}$ over the triple $A$ requires that in case $\mathscr{D}|A$ contains a weak Latin Square involving a linear ordering of $A$ then in every ordering in $\mathscr{D}|A$ which belongs to the same weak Latin Square, the alternative which is the best in the linear ordering must be at least as good as the alternative which is the worst in the linear ordering.

Latin Square Partial Agreement (LSPA): $\mathscr{D} \subseteq \mathscr{T}$ satisfies LSPA over the triple $A \subseteq S$ iff ($\forall$ distinct $a, b, c \in A)[(\exists R^s, R^t, R^u \in \mathscr{D}|A)(R^s, R^t, R^u$ are concerned over $A \wedge a P^s b P^s c \wedge b R^t c R^t a \wedge c R^u a R^u b) \rightarrow (\forall R \in \mathscr{D}|A \cap T[LS(abca)])(a R c)]$. $\mathscr{D}$ satisfies LSPA iff it satisfies LSPA over every triple contained in $S$.

The satisfaction of Latin Square partial agreement by $\mathscr{D}$ over the triple $A$ requires that in case $\mathscr{D}|A$ contains a Latin Square involving a linear ordering of $A$ then in every ordering in $\mathscr{D}|A$ which belongs to the same Latin Square, the alternative which is the best in the linear ordering must be at least as good as the alternative which is the worst in the linear ordering.

Weak Extremal Restriction (WER): $\mathscr{D} \subseteq \mathscr{T}$ satisfies WER over the triple $A \subseteq S$ iff $\sim (\exists$ distinct $a, b, c \in A)(\exists R^s, R^t, R^u \in \mathscr{D}|A)(aP^sbP^sc \wedge bR^tcP^ta \wedge cP^u aR^ub)$. $\mathscr{D}$ satisfies WER iff it satisfies WER over every triple contained in $S$.

The satisfaction of weak extremal restriction by $\mathscr{D}$ over the triple $A$ requires that $\mathscr{D}|A$ must not contain a Latin Square which is such that one of the orderings involved in the formation of the Latin Square is a linear ordering and in the other two orderings the alternative which is the worst in the linear ordering is preferred to the alternative which is the best in the linear ordering.

Latin Square Linear Ordering Restriction (LSLOR): $\mathscr{D} \subseteq \mathscr{T}$ satisfies LSLOR over the triple $A \subseteq S$ iff $\sim (\exists$ distinct $a, b, c \in A)(\exists R^s, R^t, R^u \in \mathscr{D}|A)(R^s, R^t, R^u$ are concerned over $A \wedge aP^sbP^sc \wedge bP^tcP^ta \wedge cR^uaR^ub)$. $\mathscr{D}$ satisfies LSLOR iff it satisfies LSLOR over every triple contained in $S$.

The satisfaction of Latin Square linear ordering restriction by $\mathscr{D}$ over the triple $A$ requires that $\mathscr{D}|A$ must not contain a Latin Square involving more than one linear ordering.

*Remark 3.2* From the definitions of the five conditions defined above, it is clear that ER implies WLSPA; WLSPA implies LSPA; LSPA implies WER; and WER implies LSLOR.                                                                            ◇

---

## 3.3  Conditions for Transitivity

This section is concerned with characterizing sets of orderings $\mathscr{D} \in 2^{\mathscr{T}} - \{\emptyset\}$ which are such that every logically possible $(R_1, \ldots, R_n) \in \mathscr{D}^n$ gives rise to transitive social binary weak preference relation under the MMD. If the number of individuals is even and greater than or equal to 2, then extremal restriction constitutes a characterizing condition (Theorem 3.2); and when the number of individuals is odd and greater than or equal to 3, then Weak Latin Square partial agreement constitutes a characterizing condition (Theorem 3.3). Both the results follow directly from two elementary observations about the MMD elaborated in Lemmas 3.1 and 3.2.

**Lemma 3.1** *Let* $f : \mathscr{T}^n \mapsto \mathscr{C}$ *be the method of majority decision. Let* $(R_1, \ldots, R_n) \in \mathscr{T}^n$ *and* $R = f(R_1, \ldots, R_n)$. *Then we have:*

(i) $(\forall$ distinct $x, y \in S)[xRy \rightarrow n(xR_iy) \geq \frac{n}{2}]$

(ii) $(\forall$ distinct $x, y \in S)[xPy \rightarrow n(xR_iy) > \frac{n}{2}]$.

*Proof* As each $R_i$, $i \in N$, is connected, it follows that for any distinct $x, y \in S$, we have: $n(xR_iy) + n(yR_ix) \geq n$. (L3.1-1)

Now, $xRy \rightarrow n(xR_iy) \geq n(yR_ix)$, and (L3.1-2)

$xPy \rightarrow n(xR_iy) > n(yR_ix)$. (L3.1-3)

(i) follows from (L3.1-1) and (L3.1-2), and (ii) follows from (L3.1-1) and (L3.1-3). $\qquad\square$

**Lemma 3.2** *Let $f : \mathscr{T}^n \mapsto \mathscr{C}$ be the method of majority decision. Let $(R_1, \ldots, R_n) \in \mathscr{T}^n$ and $R = f(R_1, \ldots, R_n)$. Let $A = \{x, y, z\} \subseteq S$ be a triple of alternatives and suppose $xPy$, $yRz$ and $zRx$. Then we have:*

*(i)  $(\exists i \in N)[xP_iyP_iz \vee yP_izP_ix \vee zP_ixP_iy]$*
*(ii)  $(\exists i, j, k \in N)[R_i|A, R_j|A, R_k|A \in T[WLS(xyzx)] \wedge xP_iy \wedge yP_jz \wedge zP_kx]$.*

*Proof*  $xPy \rightarrow n(xP_iy) > n(yP_ix)$ (L3.2-1)

$yRz \rightarrow n(yP_iz) \geq n(zP_iy)$ (L3.2-2)

$zRx \rightarrow n(zP_ix) \geq n(xP_iz)$ (L3.2-3)

(L3.2-1), (L3.2-2), and (L3.2-3) imply, respectively:

$n(zP_ixP_iy) + n(zI_ixP_iy) + n(xP_izP_iy) + n(xP_izI_iy) + n(xP_iyP_iz) > n(zP_iy$
$P_ix) + n(zI_iyP_ix) + n(yP_izP_ix) + n(yP_izI_ix) + n(yP_ixP_iz)$ (L3.2-4)

$n(xP_iyP_iz) + n(xI_iyP_iz) + n(yP_ixP_iz) + n(yP_ixI_iz) + n(yP_izP_ix) \geq n(xP_iz$
$P_iy) + n(xI_izP_iy) + n(zP_ixP_iy) + n(zP_ixI_iy) + n(zP_iyP_ix)$ (L3.2-5)

$n(yP_izP_ix) + n(yI_izP_ix) + n(zP_iyP_ix) + n(zP_iyI_ix) + n(zP_ixP_iy) \geq n(yP_ix$
$P_iz) + n(yI_ixP_iz) + n(xP_iyP_iz) + n(xP_iyI_iz) + n(xP_izP_iy)$ (L3.2-6)

Adding (L3.2-4), (L3.2-5) and (L3.2-6), we obtain:

$n(xP_iyP_iz) + n(yP_izP_ix) + n(zP_ixP_iy) > n(xP_izP_iy) + n(zP_iyP_ix) +$
$n(yP_ixP_iz)$ (L3.2-7)

Adding (L3.2-7) to (L3.2-4), (L3.2-5) and (L3.2-6) we obtain, respectively:

$2n(zP_ixP_iy) + 2n(xP_iyP_iz) + n(zI_ixP_iy) + n(xP_iyI_iz) > 2n(zP_iyP_ix) +$
$2n(yP_ixP_iz) + n(zI_iyP_ix) + n(yP_ixI_iz)$ (L3.2-8)

$2n(xP_iyP_iz) + 2n(yP_izP_ix) + n(xI_iyP_iz) + n(yP_izI_ix) > 2n(xP_izP_iy) +$
$2n(zP_iyP_ix) + n(xI_izP_iy) + n(zP_iyI_ix)$ (L3.2-9)

$2n(yP_izP_ix) + 2n(zP_ixP_iy) + n(yI_izP_ix) + n(zP_ixI_iy) > 2n(yP_ixP_iz) +$
$2n(xP_izP_iy) + n(yI_ixP_iz) + n(xP_izI_iy)$ (L3.2-10)

(L3.2-7)-(L3.2-10) imply, respectively:

$n(xP_iyP_iz) + n(yP_izP_ix) + n(zP_ixP_iy) > 0$ (L3.2-11)

$$n(zP_ixP_iy) + n(xP_iyP_iz) + n(zI_ixP_iy) + n(xP_iyI_iz) > 0 \qquad \text{(L3.2-12)}$$

$$n(xP_iyP_iz) + n(yP_izP_ix) + n(xI_iyP_iz) + n(yP_izI_ix) > 0 \qquad \text{(L3.2-13)}$$

$$n(yP_izP_ix) + n(zP_ixP_iy) + n(yI_izP_ix) + n(zP_ixI_iy) > 0 \qquad \text{(L3.2-14)}$$

$$\text{(L3.2-11)} \rightarrow (\exists i \in N)(xP_iyP_iz \lor yP_izP_ix \lor zP_ixP_iy) \qquad \text{(L3.2-15)}$$

$$\text{(L3.2-12)} \rightarrow (\exists i \in N)(R_i|A \in T[WLS(xyzx)] \land xP_iy) \qquad \text{(L3.2-16)}$$

$$\text{(L3.2-13)} \rightarrow (\exists j \in N)(R_j|A \in T[WLS(xyzx)] \land yP_jz) \qquad \text{(L3.2-17)}$$

$$\text{(L3.2-14)} \rightarrow (\exists k \in N)(R_k|A \in T[WLS(xyzx)] \land zP_kx) \qquad \text{(L3.2-18)}$$

(L3.2-15)–(L3.2-18) establish the lemma. $\qquad\qquad\qquad\qquad\qquad\qquad\qquad\qquad$ $\square$

**Theorem 3.2** *Let $\#S \geq 3$ and $\#N = n = 2k, k \geq 1$. Let $\mathscr{D} \subseteq \mathscr{T}$. Then the method of majority decision $f$ yields transitive social $R$, $R = f(R_1, \ldots, R_n)$, for every $(R_1, \ldots, R_n) \in \mathscr{D}^n$ iff $\mathscr{D}$ satisfies the condition of extremal restriction.*

*Proof* Suppose $f$ does not yield transitive social $R$ for every $(R_1, \ldots, R_n) \in \mathscr{D}^n$. Then $(\exists (R_1, \ldots, R_n) \in \mathscr{D}^n)(\exists x, y, z \in S)(xPy \land yRz \land zRx)$. This, by Lemma 3.2, implies that:
$(\exists i \in N)(xP_iyP_iz \lor yP_izP_ix \lor zP_ixP_iy)$, and $\qquad\qquad\qquad\qquad$ (T3.2-1)
$(\exists i, j, k \in N)[R_i|\{x, y, z\}, R_j|\{x, y, z\}, R_k|\{x, y, z\} \in T[WLS(xyzx)] \land xP_iy \land yP_jz \land zP_kx]$. $\qquad\qquad\qquad\qquad\qquad\qquad\qquad\qquad\qquad\qquad\qquad$ (T3.2-2)

(T3.2-1) and (T3.2-2) imply that $\mathscr{D}$ violates ER, which establishes the sufficiency of ER.

Let $\mathscr{D} \subseteq \mathscr{T}$ violate ER. This implies that $(\exists x, y, z \in S)(\exists R^s, R^t \in \mathscr{D})[xP^syP^sz \land (yR^tzP^tx \lor zP^txR^ty)]$. Now consider any $(R_1, \ldots, R_n) \in \mathscr{D}^n$ such that $\#\{i \in N \mid R_i = R^s\} = \#\{i \in N \mid R_i = R^t\} = k = \frac{n}{2}$. The MMD then yields $(xIy \land yPz \land xIz)$ or $(xPy \land yIz \land xIz)$ depending on whether $R^t|\{x, y, z\}$ is $yR^tzP^tx$ or $zP^txR^ty$. In either case, transitivity is violated, which establishes the theorem. $\quad\square$

**Theorem 3.3** *Let $\#S \geq 3$ and $\#N = n = 2k + 1, k \geq 1$. Let $\mathscr{D} \subseteq \mathscr{T}$. Then the method of majority decision $f$ yields transitive social $R$, $R = f(R_1, \ldots, R_n)$, for every $(R_1, \ldots, R_n) \in \mathscr{D}^n$ iff $\mathscr{D}$ satisfies the condition of Weak Latin Square partial agreement.*

*Proof* Suppose $f$ does not yield transitive social $R$ for every $(R_1, \ldots, R_n) \in \mathscr{D}^n$. Then $(\exists (R_1, \ldots, R_n) \in \mathscr{D}^n)(\exists x, y, z \in S)(xPy \land yRz \land zRx)$. $xPy, yRz$ and $zRx$ imply, by Lemma 3.1, respectively:

$$N(xR_iy) > \frac{n}{2} \qquad\qquad\qquad\qquad\qquad\qquad\qquad\qquad\qquad\qquad \text{(T3.3-1)}$$

$$N(yR_iz) \geq \frac{n}{2} \qquad\qquad\qquad\qquad\qquad\qquad\qquad\qquad\qquad\qquad \text{(T3.3-2)}$$

$$N(zR_ix) \geq \frac{n}{2}. \qquad\qquad\qquad\qquad\qquad\qquad\qquad\qquad\qquad\qquad \text{(T3.3-3)}$$

As $n$ is odd, (T3.3-2) and (T3.3-3) imply, respectively:

$$N(y R_i z) > \frac{n}{2} \tag{T3.3-4}$$

$$N(z R_i x) > \frac{n}{2}. \tag{T3.3-5}$$

$$(\text{T3.3-1}) \wedge (\text{T3.3-4}) \rightarrow (\exists i \in N)(x R_i y R_i z) \tag{T3.3-6}$$

$$(\text{T3.3-4}) \wedge (\text{T3.3-5}) \rightarrow (\exists i \in N)(y R_i z R_i x) \tag{T3.3-7}$$

$$(\text{T3.3-5}) \wedge (\text{T3.3-1}) \rightarrow (\exists i \in N)(z R_i x R_i y) \tag{T3.3-8}$$

By Lemma 3.2, we have:

$$(\exists i \in N)(x P_i y P_i z \vee y P_i z P_i x \vee z P_i x P_i y), \text{ and} \tag{T3.3-9}$$

$$(\exists i, j, k \in N)[R_i|\{x, y, z\}, R_j|\{x, y, z\}, R_k|\{x, y, z\} \in T[WLS(xyzx)] \wedge x P_i y \wedge y P_j z \wedge z P_k x]. \tag{T3.3-10}$$

(T3.3-6)-(T3.3-10) imply that WLSPA is violated, which establishes the sufficiency of WLSPA.

Now, let $\mathscr{D} \subseteq \mathscr{T}$ violate WLSPA. This implies that $(\exists x, y, z \in S)(\exists R^s, R^t, R^u \in \mathscr{D})[(x P^s y P^s z \wedge y R^t z P^t x \wedge z R^u x R^u y) \vee (x P^s y P^s z \wedge y R^u z R^u x \wedge z P^t x R^t y)]$. Consider any $(R_1, \ldots, R_n) \in \mathscr{D}^n$ such that $\#\{i \in N \mid R_i = R^s\} = \#\{i \in N \mid R_i = R^t\} = k = \frac{n-1}{2}$, and $\#\{i \in N \mid R_i = R^u\} = 1$. The MMD then yields $[x R y \wedge y R z \wedge z R x \wedge \sim (y R x \wedge x R z \wedge z R y)]$ violating transitivity, which establishes the theorem. $\qquad \square$

## 3.4  Conditions for Quasi-transitivity

This section is concerned with characterizing sets of orderings $\mathscr{D} \in 2^{\mathscr{T}} - \{\emptyset\}$ which are such that every logically possible $(R_1, \ldots, R_n) \in \mathscr{D}^n$ gives rise to quasi-transitive social binary weak preference relation under the MMD. If the number of individuals is greater than or equal to 5, then Latin Square partial agreement constitutes a characterizing condition (Theorem 3.4); if the number of individuals is 4 then weak extremal restriction constitutes a characterizing condition (Theorem 3.5); and if the number of individuals is 3, then Latin Square linear ordering restriction constitutes a characterizing condition (Theorem 3.6). All the three theorems follow directly from Lemmas 3.2 and 3.4. Lemma 3.4 is a simple consequence of Lemma 3.3 which is similar to Lemma 3.1.

Let $A \subseteq S$ be a triple of alternatives. We denote by $n_A$ the number of individuals who are concerned over $A$.

**Lemma 3.3** *Let $f : \mathscr{T}^n \mapsto \mathscr{C}$ be the method of majority decision. Let $(R_1, \ldots, R_n) \in \mathscr{T}^n$ and $R = f(R_1, \ldots, R_n)$. Then we have:*

(i) *($\forall$ distinct $x, y, z \in S)[x R y \rightarrow n(R_i$ concerned over $A = \{x, y, z\} \wedge x R_i y) \geq \frac{n_A}{2}]$*

(ii) *($\forall$ distinct $x, y, z \in S)[x P y \rightarrow n(R_i$ concerned over $A = \{x, y, z\} \wedge x R_i y) > \frac{n_A}{2}]$.*

*Proof* Let $A = \{x, y, z\} \subseteq S$ be a triple of alternatives.

$xRy \rightarrow n(xP_iy) \geq n(yP_ix)$, and                                     (L3.3-1)

$xPy \rightarrow n(xP_iy) > n(yP_ix)$                                              (L3.3-2)

Adding $n(R_i$ concerned over $A \wedge xI_iy)$ to both sides of inequalities of (L3.3-1) and (L3.3-2) we obtain:

$xRy \rightarrow n(R_i$ concerned over $A \wedge xR_iy) \geq n(R_i$ concerned over $A \wedge yR_ix)$, and                                                                       (L3.3-3)

$xPy \rightarrow n(R_i$ concerned over $A \wedge xR_iy) > n(R_i$ concerned over $A \wedge yR_ix)$

                                                                                  (L3.3-4)

As each $R_i, i \in N$, is connected, it follows that:

$n(R_i$ concerned over $A \wedge xR_iy) + n(R_i$ concerned over $A \wedge yR_ix) \geq n_A$

                                                                                  (L3.3-5)

(i) follows from (L3.3-3) and (L3.3-5); and (ii) follows from (L3.3-4) and (L3.3-5).                                                                           $\square$

**Lemma 3.4** *Let $f : \mathscr{T}^n \mapsto \mathscr{C}$ be the method of majority decision. Let $(R_1, \ldots, R_n) \in \mathscr{T}^n$ and $R = f(R_1, \ldots, R_n)$. Let $A = \{x, y, z\} \subseteq S$ be a triple of alternatives and suppose $xPy$, $yPz$ and $zRx$. Then we must have: $(\exists i, j, k \in N)[R_i, R_j, R_k$ are concerned over $\{x, y, z\} \wedge xR_iyR_iz \wedge yR_jzR_jx \wedge zR_kxR_ky]$.*

*Proof* By Lemma 3.3, $xPy$, $yPz$ and $zRx$ imply, respectively:

$n(R_i$ concerned over $A \wedge xR_iy) > \frac{n_A}{2}$                          (L3.4-1)

$n(R_i$ concerned over $A \wedge yR_iz) > \frac{n_A}{2}$                          (L3.4-2)

$n(R_i$ concerned over $A \wedge zR_ix) \geq \frac{n_A}{2}$                        (L3.4-3)

(L3.4-1) $\wedge$ (L3.4-2) $\rightarrow$ $(\exists i \in N)(R_i$ concerned over $A \wedge xR_iyR_iz)$     (L3.4-4)

(L3.4-2) $\wedge$ (L3.4-3) $\rightarrow$ $(\exists j \in N)(R_j$ concerned over $A \wedge yR_jzR_jx)$     (L3.4-5)

(L3.4-3) $\wedge$ (L3.4-1) $\rightarrow$ $(\exists k \in N)(R_k$ concerned over $A \wedge zR_kxR_ky)$     (L3.4-6)

(L3.4-4)-(L3.4-6) establish the lemma.                                          $\square$

**Theorem 3.4** *Let $\#S \geq 3$ and $\#N = n \geq 5$. Let $\mathscr{D} \subseteq \mathscr{T}$. Then the method of majority decision $f$ yields quasi-transitive social $R$, $R = f(R_1, \ldots, R_n)$, for every $(R_1, \ldots, R_n) \in \mathscr{D}^n$ iff D satisfies the condition of Latin Square partial agreement.*

*Proof* Suppose $f$ does not yield quasi-transitive social $R$ for every $(R_1, \ldots, R_n) \in \mathscr{D}^n$. Then $(\exists(R_1, \ldots, R_n) \in \mathscr{D}^n)(\exists x, y, z \in S)(xPy \wedge yPz \wedge zRx)$. By Lemmas 3.4 and 3.2, we obtain:

$(\exists i, j, k \in N)(R_i, R_j, R_k$ are concerned over $\{x, y, z\} \wedge xR_iyR_iz \wedge yR_jzR_jx$

$\wedge zR_kxR_ky)$                                                              (T3.4-1)

$(\exists i \in N)(xP_iyP_iz \vee yP_izP_ix \vee zP_ixP_iy)$                       (T3.4-2)

$(\exists i, j, k \in N)[R_i|\{x, y, z\}, \; R_j|\{x, y, z\}, \; R_k|\{x, y, z\} \in T[LS(xyzx)] \wedge x P_i y \wedge$
$y P_j z \wedge z P_k x].$ 

(T3.4-3)

(T3.4-1)-(T3.4-3) imply that LSPA is violated, which establishes the sufficiency of LSPA.

Now, let $\mathscr{D} \subseteq \mathscr{T}$ violate LSPA. This implies that $(\exists x, y, z \in S)(\exists R^s, R^t, R^u \in \mathscr{D})[(x P^s y P^s z \wedge y P^t z P^t x \wedge z P^u x P^u y) \vee (x P^s y P^s z \wedge y P^t z P^t x \wedge z P^u x I^u y) \vee (x P^t y P^t z \wedge y P^s z P^s x \wedge z I^u x P^u y) \vee (x P^t y P^t z \wedge y P^u z I^u x \wedge z P^s x I^s y) \vee (x P^t y P^t z \wedge y I^s z P^s x \wedge z I^u x P^u y) \vee (x P^s y P^s z \wedge y I^t z P^t x \wedge z P^u x I^u y)].$ If $n = 3k$, $k \geq 2$, consider any $(R_1, \ldots, R_n) \in \mathscr{D}^n$ such that $\#\{i \in N \mid R_i = R^s\} = k + 1$, $\#\{i \in N \mid R_i = R^t\} = k$ and $\#\{i \in N \mid R_i = R^u\} = k - 1$; if $n = 3k + 1, k \geq 2$, consider any $(R_1, \ldots, R_n) \in \mathscr{D}^n$ such that $\#\{i \in N \mid R_i = R^s\} = k + 1, \#\{i \in N \mid R_i = R^t\} = k$ and $\#\{i \in N \mid R_i = R^u\} = k$; and if $n = 3k + 2, k \geq 1$, consider any $(R_1, \ldots, R_n) \in \mathscr{D}^n$ such that $\#\{i \in N \mid R_i = R^s\} = k + 1, \#\{i \in N \mid R_i = R^t\} = k + 1$ and $\#\{i \in N \mid R_i = R^u\} = k$. In each case, MMD yields social $R$ violating quasi-transitivity over $\{x, y, z\}$, which establishes the theorem. $\qquad\square$

**Theorem 3.5** *Let $\#S \geq 3$ and $\#N = n = 4$. Let $\mathscr{D} \subseteq \mathscr{T}$. Then the method of majority decision $f$ yields quasi-transitive social $R, R = f(R_1, \ldots, R_4)$, for every $(R_1, \ldots, R_4) \in \mathscr{D}^4$ iff $\mathscr{D}$ satisfies the condition of weak extremal restriction.*

*Proof* Suppose $f$ does not yield quasi-transitive social $R$ for every $(R_1, \ldots, R_4) \in \mathscr{D}^4$. Then $(\exists (R_1, \ldots, R_4) \in \mathscr{D}^4)(\exists x, y, z \in S)(x P y \wedge y P z \wedge z R x)$. By Lemmas 3.4 and 3.2, we obtain:

$(\exists i, j, k \in N)(R_i, R_j, R_k$ are concerned over $\{x, y, z\} \wedge x R_i y R_i z \wedge y R_j z R_j x$
$\wedge z R_k x R_k y)$ 

(T3.5-1)

$(\exists i \in N)(x P_i y P_i z \vee y P_i z P_i x \vee z P_i x P_i y)$ 

(T3.5-2)

$(T3.5-1) \rightarrow (\exists i, j, k \in N)(x P_i z \wedge y P_j x \wedge z P_k y)$ 

(T3.5-3)

$(\exists j \in N)(y P_j x) \wedge x P y \rightarrow n(x P_i y) \geq 2 \wedge n(x R_i y) = 3 \wedge n(y P_i x) = 1$ 

(T3.5-4)

$(T3.5-4) \rightarrow n(R_i$ concerned over $\{x, y, z\} \wedge y R_i z R_i x) = 1 \wedge n(z P_i y P_i x) = 0$
$\wedge n(y P_i x P_i z) = 0$ 

(T3.5-5)

$(\exists k \in N)(z P_k y) \wedge y P z \rightarrow N(y P_i z) \geq 2 \wedge n(y R_i z) = 3 \wedge n(z P_i y) = 1$ 

(T3.5-6)

$(T3.5-6) \rightarrow n(R_i$ concerned over $\{x, y, z\} \wedge z R_i x R_i y) = 1 \wedge n(x P_i z P_i y) = 0$

(T3.5-7)

$(T3.5-1) \wedge (T3.5-5) \wedge (T3.5-7) \rightarrow n(x R_i y R_i z) = 2 \wedge n(R_i$ concerned over $\{x, y, z\} \wedge x R_i y R_i z) \geq 1 \wedge n(R_i$ concerned over $\{x, y, z\} \wedge y R_i z R_i x) = 1 \wedge n(R_i$ concerned over $\{x, y, z\} \wedge z R_i x R_i y) = 1$ 

(T3.5-8)

$z R x \wedge N(x P_i z) = 1 \wedge (T3.5-4) \wedge (T3.5-6) \wedge (T3.5-8) \rightarrow (\exists i, j, k \in N)[(x P_i y P_i z \wedge y P_j z P_j x \wedge z R_i x P_i y) \vee (x P_i y P_i z \wedge y P_j z R_j x \wedge z P_i x P_i y)]$
$\rightarrow$ WER is violated. 

(T3.5-9)

$zRx \wedge n(xP_iz) = 2 \wedge$(T3.5-4) $\wedge$ (T3.5-6) $\wedge$(T3.5-8) $\rightarrow (\exists i, j, k \in N)(xP_iyP_iz \wedge yR_jzP_jx \wedge zP_kxR_ky) \vee (\exists i, j, k, l \in L)(xP_iyI_iz \wedge xI_jyP_jz \wedge yP_k zP_kx \wedge zP_lxP_ly)$

$\rightarrow$ WER is violated.                                                                                    (T3.5-10)

(T3.5-9) and (T3.5-10) establish that WER is violated, which proves the sufficiency of WER.

Suppose $\mathscr{D} \subseteq \mathscr{T}$ violates WER. This implies that $(\exists x, y, z \in S)(\exists R^s, R^t, R^u \in \mathscr{D})(xP^syP^sz \wedge yR^tzP^tx \wedge zP^uxR^uy)$. Consider any $(R_1, \ldots, R_4) \in \mathscr{D}^4$ such that $\#\{i \in N \mid R_i = R^s\} = 2, \quad \#\{i \in N \mid R_i = R^t\} = 1$ and $\#\{i \in N \mid R_i = R^u\} = 1$. MMD then yields $(xPy \wedge yPz \wedge zIx)$, which violates quasi-transitivity. This establishes the theorem. $\qquad\square$

**Theorem 3.6** *Let $\#S \geq 3$ and $\#N = n = 3$. Let $\mathscr{D} \subseteq \mathscr{T}$. Then the method of majority decision $f$ yields quasi-transitive social $R$, $R = f(R_1, R_2, R_3)$, for every $(R_1, R_2, R_3) \in \mathscr{D}^3$ iff $\mathscr{D}$ satisfies the condition of Latin Square linear ordering restriction.*

*Proof* Suppose $f$ does not yield quasi-transitive social $R$ for every $(R_1, R_2, R_3) \in \mathscr{D}^3$. Then $(\exists(R_1, R_2, R_3) \in \mathscr{D}^3)(\exists x, y, z \in S)(xPy \wedge yPz \wedge zRx)$. By Lemma 3.4, we have:

$(\exists i, j, k \in N)(R_i, R_j, R_k$ are concerned over $\{x, y, z\} \wedge xR_iyR_iz \wedge yR_jzR_jx$

$\wedge zR_kxR_ky)$                                                                                            (T3.6-1)

$\rightarrow (\exists i, j, k \in N)(xP_iz \wedge yP_jx \wedge zP_ky)$                                          (T3.6-2)

$xPy \wedge (\exists j \in N)(yP_jx) \wedge$ (T3.6-1) $\rightarrow (\exists i, j, k \in N)(R_i, R_j, R_k$ are concerned over $\{x, y, z\} \wedge xP_iyR_iz \wedge yR_jzR_jx \wedge zR_kxP_ky)$                                     (T3.6-3)

$yPz \wedge (\exists k \in N)(zP_ky) \wedge$ (T3.6-3) $\rightarrow (\exists i, j, k \in N)(xP_iyP_iz \wedge yP_jzR_jx \wedge zR_k xP_ky)$                                                                              (T3.6-4)

$zRx \wedge (\exists i \in N)(xP_iz) \wedge$ (T3.6-4) $\rightarrow (\exists i, j, k \in N)[(xP_iyP_iz \wedge yP_jzP_jx \wedge zR_k xP_ky) \vee (xP_iyP_iz \wedge yP_jzR_jx \wedge zP_kxP_ky)]$.                     (T3.6-5)

(T3.6-5) implies that LSLOR is violated, which establishes sufficiency of LSLOR.

Suppose $\mathscr{D} \subseteq \mathscr{T}$ violates LSLOR. This implies that $(\exists x, y, z \in S)(\exists R^s, R^t, R^u \in \mathscr{D})[xP^syP^sz \wedge yP^tzP^tx \wedge (zP^uxR^uy \vee zR^uxP^uy)]$. Consider any $(R_1, R_2, R_3) \in \mathscr{D}^3$ such that $\#\{i \in N \mid R_i = R^s\} = \#\{i \in N \mid R_i = R^t\} = \#\{i \in N \mid R_i = R^u\} = 1$. MMD then yields $(xRy \wedge yPz \wedge zPx)$ or $(xPy \wedge yPz \wedge zRx)$ depending on whether $R^u$ over $\{x, y, z\}$ is $zP^uxR^uy$ or $zR^uxP^uy$. As quasi-transitivity is violated in either case, the theorem is established. $\qquad\square$

# Appendix

## 3.5 Relationships Among Conditions on Preferences

Let $\mathscr{D}$ be a set of orderings of set $S$; and let $A = \{x, y, z\} \subseteq S$ be a triple of alternatives. Let $L$ be a linear ordering of the triple $A$. We define $x$ to be between $y$ and $z$, denoted by $B_L(y, x, z)$, iff $[(yLx \wedge xLz) \vee (zLx \wedge xLy)]$.

### 3.5.1 Single-Peakedness

Single-Peakedness (SP): $\mathscr{D} \subseteq \mathscr{T}$ satisfies SP over the triple $A \subseteq S$ iff ($\exists$ a linear ordering $L$ of $A$)($\forall a, b, c \in A$)($\forall R \in \mathscr{D}|A$)$[aRb \wedge B_L(a, b, c) \rightarrow bPc]$. $\mathscr{D}$ satisfies SP iff it satisfies SP over every triple contained in $S$.

Thus, the satisfaction of single-peakedness by $\mathscr{D}$ over the triple $A$ requires that there be a way of arranging the three alternatives of the triple such that the following holds for every $R \in \mathscr{D}|A$: If an alternative not lying in between (non-between-alternative) is at least as good as the one which lies in between (the between-alternative), then the between-alternative must be strictly preferred to the other non-between-alternative.

Let orderings $\mathscr{D}|A$ satisfy single-peakedness with $x$ between $y$ and $z$. Then we have: ($\forall R \in \mathscr{D}|A$)$[xPy \vee xPz]$; consequently, $x$ is not worst in any $R \in \mathscr{D}|A$. Similarly, if single-peakedness holds with $y$ between $x$ and $z$ then $y$ is not worst in any $R \in \mathscr{D}|A$; and if single-peakedness holds with $z$ between $x$ and $y$ then $z$ is not worst in any $R \in \mathscr{D}|A$. Thus, satisfaction of single-peakedness by a set of orderings of a triple implies that there is an alternative in the triple which is not worst in any of the orderings of the set.

The set of all orderings of the triple $A$ in which $x$ is not worst is given by: $\{xPyPz, zPxPy, xPzPy, yPxPz, xPyIz, xIyPz, zIxPy\}$. For every ordering in the set, the implication ($yRx \rightarrow xPz$) holds. Therefore, the set of all orderings of $A$ in which $x$ is not worst satisfies the single-peakedness condition with $x$ between $y$ and $z$. Similarly, it can be checked that both the set of orderings of $A$ in which $y$ is not worst and the set of orderings of $A$ in which $z$ is not worst satisfy the condition of single-peakedness. This, together with the demonstration of the preceding paragraph, establishes that a set of orderings of a triple satisfies the condition of single-peakedness iff there is an alternative in the triple such that it is not worst in any of the orderings of the set.

### 3.5.2 Single-Cavedness

Single-Cavedness (SC): $\mathscr{D} \subseteq \mathscr{T}$ satisfies SC over the triple $A \subseteq S$ iff ($\exists$ a linear ordering $L$ of $A$)($\forall a, b, c \in A$)($\forall R \in \mathscr{D}|A$)$[bRa \wedge B_L(a, b, c) \rightarrow cPb]$. $\mathscr{D}$ satisfies SC iff it satisfies SC over every triple contained in $S$.

Thus, the satisfaction of single-cavedness by $\mathscr{D}$ over the triple $A$ requires that there be a way of arranging the three alternatives of the triple such that the following holds for every $R \in \mathscr{D}|A$: If the between-alternative is at least as good as one of the non-between-alternatives, then the other non-between-alternative must be strictly preferred to the between-alternative.

Let orderings $\mathscr{D}|A$ satisfy single-cavedness with $x$ between $y$ and $z$. Then we have: $(\forall R \in \mathscr{D}|A)[yPx \vee zPx]$; consequently, $x$ is not best in any $R \in \mathscr{D}|A$. Similarly, if single-cavedness holds with $y$ between $x$ and $z$ then $y$ is not best in any $R \in \mathscr{D}|A$; and if single-cavedness holds with $z$ between $x$ and $y$ then $z$ is not best in any $R \in \mathscr{D}|A$. Thus satisfaction of single-cavedness by a set of orderings of a triple implies that there is an alternative in the triple which is not best in any of the orderings of the set.

The set of all orderings of the triple $A$ in which $x$ is not best is given by: $\{yPzPx, zPxPy, zPyPx, yPxPz, yPzIx, zPxIy, yIzPx\}$. For every ordering in the set, the implication $(xRy \rightarrow zPx)$ holds. Therefore, the set of all orderings of $A$ in which $x$ is not best satisfies the single-cavedness condition with $x$ between $y$ and $z$. Similarly, it can be checked that both the set of orderings of $A$ in which $y$ is not best and the set of orderings of $A$ in which $z$ is not best satisfy the condition of single-cavedness. This, together with the demonstration of the preceding paragraph, establishes that a set of orderings of a triple satisfies the condition of single-cavedness iff there is an alternative in the triple such that it is not best in any of the orderings of the set.

### 3.5.3  Separability into Two Groups

Separability into Two Groups (SG): $\mathscr{D} \subseteq \mathscr{T}$ satisfies SG over the triple $A \subseteq S$ iff $(\exists A_1, A_2 \subset A)[[A_1 \neq \emptyset \wedge A_2 \neq \emptyset \wedge A_1 \cap A_2 = \emptyset \wedge A_1 \cup A_2 = A] \wedge (\forall R \in \mathscr{D}|A)[(\forall a \in A_1)(\forall b \in A_2)(aPb) \vee (\forall a \in A_1)(\forall b \in A_2)(bPa)]]$. $\mathscr{D}$ satisfies SG iff it satisfies SG over every triple contained in $S$.

Thus, the satisfaction of separability into two groups by $\mathscr{D}$ over the triple $A$ requires that there be a way of partitioning the triple into two subsets such that in every $R \in \mathscr{D}|A$ either alternatives belonging to one of the subsets are strictly preferred to alternatives in the other subset or alternatives belonging to the latter subset are strictly preferred to alternatives in the former subset.

Let orderings $\mathscr{D}|A$ satisfy separability into two groups with the two subsets being $\{x\}$ and $\{y, z\}$. Then we have: $(\forall R \in \mathscr{D}|A)[(xPy \wedge xPz) \vee (yPx \wedge zPx)]$; consequently, $x$ is not medium in any $R \in \mathscr{D}|A$. Similarly, if separability into two groups holds with the two subsets being $\{y\}$ and $\{x, z\}$ then $y$ is not medium in any $R \in \mathscr{D}|A$; and if separability into two groups holds with the two subsets being $\{z\}$ and $\{x, y\}$ then $z$ is not medium in any $R \in \mathscr{D}|A$. Thus satisfaction of separability into two groups by a set of orderings of a triple implies that there is an alternative in the triple which is not medium in any of the orderings of the set.

The set of all orderings of the triple $A$ in which $x$ is not medium is given by: $\{xPyPz, yPzPx, xPzPy, zPyPx, xPyIz, yIzPx\}$. For every ordering in the set, $[(xPy \wedge xPz) \vee (yPx \wedge zPx)]$ holds. Therefore, the separability into two groups holds with the two subsets being $\{x\}$ and $\{y, z\}$. Similarly, it can be checked that both the set of orderings of $A$ in which $y$ is not medium and the set of orderings of $A$ in which $z$ is not medium satisfy the condition of separability into two groups. This, together with the demonstration of the preceding paragraph, establishes that a set of orderings of a triple satisfies the condition of separability into two groups iff there is an alternative in the triple such that it is not medium in any of the orderings of the set.

### 3.5.4  Value Restriction

There are two versions of value restriction.

First Version of Value Restriction (VR(1)): $\mathscr{D} \subseteq \mathscr{T}$ satisfies VR(1) over the triple $A \subseteq S$ iff $(\exists$ distinct $a, b, c \in A)[(\forall R \in \mathscr{D}|A)[bPa \vee cPa] \vee (\forall R \in \mathscr{D}|A)[(aPb \wedge aPc) \vee (bPa \wedge cPa)] \vee (\forall R \in \mathscr{D}|A)[aPb \vee aPc]]$. $\mathscr{D}$ satisfies VR(1) iff it satisfies VR(1) over every triple contained in $S$.

Thus, the satisfaction of VR(1) by $\mathscr{D}$ over the triple $A$ requires that there be an alternative in $A$ such that it is not best in any $R \in \mathscr{D}|A$, or that it is not medium in any $R \in \mathscr{D}|A$, or that it is not worst in any $R \in \mathscr{D}|A$.

The following proposition is immediate.

**Proposition 3.1** *A set of orderings of a triple satisfies value restriction (1) iff it satisfies at least one of the three conditions of single-peakedness, single-cavedness and separability into two groups.*

**Proposition 3.2** *A set of orderings of a triple violates value restriction (1) iff there is a weak Latin Square.*

*Proof* Let the set of orderings $\mathscr{D}|A$ over the triple $A = \{x, y, z\}$ contain a weak Latin Square. Without any loss of generality, assume that $\mathscr{D}|A$ contains $WLS(xyzx)$. Then we have: $(\exists R^s, R^t, R^u \in \mathscr{D}|A)[xR^s yR^s z \wedge yR^t zR^t x \wedge zR^u xR^u y]$. $x$ is best in $R^s$, $y$ is best in $R^t$, $z$ is best in $R^u$; $x$ is medium in $R^u$, $y$ is medium in $R^s$, $z$ is medium in $R^t$; $x$ is worst in $R^t$, $y$ is worst in $R^u$, $z$ is worst in $R^s$. Therefore VR(1) is violated. This establishes that if a set of orderings over a triple contains a weak Latin Square then it violates VR(1).  (P3.2-1)

Let $\mathscr{D}|A$ be a set of orderings of the triple $A$. $\mathscr{D}|A$ would violate VR(1) only if it contains an ordering in which $x$ is best, contains an ordering in which $y$ is best, and contains an ordering in which $z$ is best, i.e., only if $\mathscr{D}|A$ contains $[(xR^1 yR^1 z \vee xR^1 zR^1 y) \wedge (yR^2 zR^2 x \vee yR^2 xR^2 z) \wedge (zR^3 xR^3 y \vee zR^3 yR^3 x)]$. Thus, if $\mathscr{D}|A$ is to violate VR(1), then it must contain at least one of the following eight sets:
$(i)$ $\{xR^1 yR^1 z, yR^2 zR^2 x, zR^3 xR^3 y\}$,   $(ii)$ $\{xR^1 yR^1 z, yR^2 zR^2 x, zR^3 yR^3 x\}$,
$(iii)$ $\{xR^1 yR^1 z, yR^2 xR^2 z, zR^3 xR^3 y\}$,   $(iv)$ $\{xR^1 yR^1 z, yR^2 xR^2 z, zR^3 yR^3 x\}$,
$(v)$ $\{xR^1 zR^1 y, yR^2 zR^2 x, zR^3 xR^3 y\}$,   $(vi)$ $\{xR^1 zR^1 y, yR^2 zR^2 x, zR^3 yR^3 x\}$,
$(vii)$ $\{xR^1 zR^1 y, yR^2 xR^2 z, zR^3 xR^3 y\}$, $(viii)$ $\{xR^1 zR^1 y, yR^2 xR^2 z, zR^3 yR^3 x\}$.
Set $(i)$ forms $WLS(xyzx)$ and set $(viii)$ $WLS(xzyx)$.

Let $\mathscr{D}|A$ contain set (ii). $\mathscr{D}|A$ violates VR(1) only if in addition to the orderings in set (ii) it contains an ordering in which $x$ is medium, i.e., only if it contains $(zR^4xR^4y \vee yR^4xR^4z)$. If $\mathscr{D}|A$ contains $zR^4xR^4y$, then it contains $WLS(xyzx)$. If $\mathscr{D}|A$ contains set (ii) and $yR^4xR^4z$, then VR(1) would be violated only if it contains an ordering in which $y$ is worst, i.e., only if it contains $(zR^5xR^5y \vee xR^5zR^5y)$. If $\mathscr{D}|A$ contains set (ii), $yR^4xR^4z$ and $zR^5xR^5y$, then $WLS(xyzx)$ is contained in it; and if it contains set (ii), $yR^4xR^4z$, and $xR^5zR^5y$ then $WLS(xzyx)$ is contained in it.

Let $\mathscr{D}|A$ contain set (iii). $\mathscr{D}|A$ violates VR(1) only if in addition to the orderings in set (iii) it contains an ordering in which $z$ is medium, i.e., only if it contains $(yR^4zR^4x \vee xR^4zR^4y)$. If $\mathscr{D}|A$ contains $yR^4zR^4x$, then it contains $WLS(xyzx)$. If $\mathscr{D}|A$ contains set (iii) and $xR^4zR^4y$, then VR(1) would be violated only if it contains an ordering in which $x$ is worst, i.e., only if it contains $(yR^5zR^5x \vee zR^5yR^5x)$. If $\mathscr{D}|A$ contains set (iii), $xR^4zR^4y$, and $yR^5zR^5x$ then $WLS(xyzx)$ is contained in it; and if it contains set (iii), $xR^4zR^4y$, and $zR^5yR^5x$ then $WLS(xzyx)$ is contained in it.

Let $\mathscr{D}|A$ contain set (iv). $\mathscr{D}|A$ violates VR(1) only if in addition to the orderings in set (iv) it contains an ordering in which $z$ is medium, i.e., only if it contains $(yR^4zR^4x \vee xR^4zR^4y)$. If $\mathscr{D}|A$ contains $xR^4zR^4y$, then it contains $WLS(xzyx)$. If $\mathscr{D}|A$ contains set (iv) and $yR^4zR^4x$, then VR(1) would be violated only if it contains an ordering in which $y$ is worst, i.e., only if it contains $(zR^5xR^5y \vee xR^5zR^5y)$. If $\mathscr{D}|A$ contains set (iv), $yR^4zR^4x$, and $zR^5xR^5y$ then $WLS(xyzx)$ is contained in it; and if it contains set (iv), $yR^4zR^4x$, and $xR^5zR^5y$ then $WLS(xzyx)$ is contained in it.

Let $\mathscr{D}|A$ contain set (v). $\mathscr{D}|A$ violates VR(1) only if in addition to the orderings in set (v) it contains an ordering in which $y$ is medium, i.e., only if it contains $(xR^4yR^4z \vee zR^4yR^4x)$. If $\mathscr{D}|A$ contains $xR^4yR^4z$, then it contains $WLS(xyzx)$. If $\mathscr{D}|A$ contains set (v) and $zR^4yR^4x$, then VR(1) would be violated only if it contains an ordering in which $z$ is worst, i.e., only if it contains $(xR^5yR^5z \vee yR^5xR^5z)$. If $\mathscr{D}|A$ contains set (v), $zR^4yR^4x$, and $xR^5yR^5z$ then $WLS(xyzx)$ is contained in it; and if it contains set (v), $zR^4yR^4x$, and $yR^5xR^5z$ then $WLS(xzyx)$ is contained in it.

Next, let $\mathscr{D}|A$ contain set (vi). $\mathscr{D}|A$ violates VR(1) only if in addition to the orderings in set (vi) it contains an ordering in which $x$ is medium, i.e., only if it contains $(zR^4xR^4y \vee yR^4xR^4z)$. If $\mathscr{D}|A$ contains $yR^4xR^4z$, then it contains $WLS(xzyx)$. If $\mathscr{D}|A$ contains set (vi) and $zR^4xR^4y$, then VR(1) would be violated only if it contains an ordering in which $z$ is worst, i.e., only if it contains $(xR^5yR^5z \vee yR^5xR^5z)$. If $\mathscr{D}|A$ contains set (vi), $zR^4xR^4y$, and $xR^5yR^5z$ then $WLS(xyzx)$ is contained in it; and if it contains set (vi), $zR^4xR^4y$, and $yR^5xR^5z$ then $WLS(xzyx)$ is contained in it.

Finally, let $\mathscr{D}|A$ contain set (vii). $\mathscr{D}|A$ violates VR(1) only if in addition to the orderings in set (vii) it contains an ordering in which $y$ is medium, i.e., only if it contains $(xR^4yR^4z \vee zR^4yR^4x)$. If $\mathscr{D}|A$ contains $zR^4yR^4x$, then it contains $WLS(xzyx)$. If $\mathscr{D}|A$ contains set (vii) and $xR^4yR^4z$, then VR(1) would be violated only if it contains an ordering in which $x$ is worst, i.e., only if it contains $(yR^5zR^5x \vee zR^5yR^5x)$. If

$\mathscr{D}|A$ contains set (vii), $xR^4yR^4z$, and $yR^5zR^5x$ then $WLS(xyzx)$ is contained in it; and if it contains set (vii), $xR^4yR^4z$, and $zR^5yR^5x$ then $WLS(xzyx)$ is contained in it.

This establishes that if a set of orderings over a triple violates VR(1) then it must contain a weak Latin Square. (P3.2-2)

(P3.2-1) and (P3.2-2) establish the proposition. $\square$

Second Version of Value Restriction (VR(2)): $\mathscr{D} \subseteq \mathscr{T}$ satisfies VR(2) over the triple $A \subseteq S$ iff ($\exists$ distinct $a, b, c \in A$)[($\forall$ concerned $R \in \mathscr{D}|A$)[$bPa \vee cPa$] $\vee$ ($\forall$ concerned $R \in \mathscr{D}|A$)[$(aPb \wedge aPc) \vee (bPa \wedge cPa)$] $\vee$ ($\forall$ concerned $R \in \mathscr{D}|A$)[$aPb \vee aPc$]]. $\mathscr{D}$ satisfies VR(2) iff it satisfies VR(2) over every triple contained in $S$.

Thus, the satisfaction of VR(2) by $\mathscr{D}$ over the triple $A$ requires that there be an alternative in $A$ such that it is not best in any concerned $R \in \mathscr{D}|A$, or that it is not medium in any concerned $R \in \mathscr{D}|A$, or that it is not worst in any concerned $R \in \mathscr{D}|A$.

**Proposition 3.3** *A set of orderings of a triple violates value restriction (2) iff there is a Latin Square.*

The proof of this proposition is omitted as it is similar to the proof of Proposition 3.2.

### 3.5.5 Dichotomous Preferences

Dichotomous Preferences (DP): $\mathscr{D} \subseteq \mathscr{T}$ satisfies DP over the triple $A \subseteq S$ iff $\sim$ ($\exists$ distinct $a, b, c \in A$)($\exists R \in \mathscr{D}|A$)[$aPbPc$]. $\mathscr{D}$ satisfies DP iff it satisfies DP over every triple contained in $S$.

Thus, a set of orderings over a triple satisfies DP iff the set does not contain a linear ordering of the triple.

### 3.5.6 Echoic Preferences

Echoic Preferences (EP): $\mathscr{D} \subseteq \mathscr{T}$ satisfies EP over the triple $A \subseteq S$ iff ($\forall$ distinct $a, b, c \in A$)[$aPbPc \in \mathscr{D}|A \rightarrow (\forall R \in \mathscr{D}|A)(aRc)$]. $\mathscr{D}$ satisfies EP iff it satisfies EP over every triple contained in $S$.

Thus, a set of orderings of a triple satisfies EP iff it does not contain an ordering in which the worst alternative of some linear ordering contained in the set is preferred to the best alternative in that linear ordering.[2]

---

[2]The definition of echoic preferences given here subsumes dichotomous preferences. Thus, the echoic preferences condition given here is the union of Inada's echoic preferences and dichotomous preferences. The advantage of defining echoic preferences like this is that if a set of orderings of a triple satisfies echoic preferences in the sense used here, then every nonempty subset of it would do likewise, a property not satisfied by Inada's echoic preferences.

### 3.5.7   Antagonistic Preferences

Antagonistic Preferences (AP): $\mathcal{D} \subseteq \mathcal{T}$ satisfies AP over the triple $A \subseteq S$ iff ($\forall$ distinct $a, b, c \in A)[aPbPc \in \mathcal{D}|A \rightarrow (\forall R \in \mathcal{D}|A)(aPbPc \vee cPbPa \vee aIc)]$. $\mathcal{D}$ satisfies AP iff it satisfies AP over every triple contained in $S$.

Thus, a set of orderings of a triple satisfies AP iff in case there is a linear ordering, say $xPyPz$, then the set is a subset of $\{xPyPz, zPyPx, zIxPy, yPxIz, xIyIz\}$.[3]

### 3.5.8   Extremal Restriction

**Proposition 3.4** *A set of orderings of a triple satisfies extremal restriction iff it satisfies at least one of the conditions of dichotomous preferences, echoic preferences and antagonistic preferences.*

*Proof* Let the set of orderings $\mathcal{D}|A$ over the triple $A = \{x, y, z\}$ satisfy ER. The cases of satisfaction of ER over $\mathcal{D}|A$ can be divided into the following mutually exclusive and exhaustive cases:

(i) ER is trivially satisfied if $\mathcal{D}|A$ does not contain any strong ordering. If $\mathcal{D}|A$ does not contain any strong ordering then all three conditions of DP, EP and AP are satisfied.

(ii) There is only one strong ordering, say $xPyPz$. As all non-strong concerned orderings of a triple belong to both the Latin Squares and ER holds, it must be the case that $xRz$ holds for all non-strong orderings. Therefore it follows that in this case EP holds.

(iii) There are more than one strong orderings. It can be checked that if there are two strong orderings belonging to the same Latin Square, then ER would be violated. From this, it follows that satisfaction of ER implies that there can be at most two strong orderings; and in case the set contains two strong orderings, then one of them would belong to $LS(xyzx)$ and the other to $LS(xzyx)$. Without any loss of generality assume that $\mathcal{D}|A$ contains $xPyPz$ and one strong ordering of $LS(xzyx)$. First consider the subcase when the ordering of $LS(xzyx)$ in $\mathcal{D}|A$ is $xPzPy$. As all non-strong concerned orderings over a triple belong to both Latin Squares and ER holds, it follows that for all non-strong orderings $R \in \mathcal{D}|A$ we must have both $xRz$ and $xRy$. Thus in this case, $\mathcal{D}|A$ must be a subset of $\{xPyPz, xPzPy, xPyIz, xIyPz, zIxPy, xIyIz\}$. In all orderings of this set, the best alternative of each of the two strong orderings is at least as good as the worst

---

[3]The definition of antagonistic preferences given here subsumes dichotomous preferences. Thus, the antagonistic preferences condition given here is the union of Inada's antagonistic preferences and dichotomous preferences. As in the case of echoic preferences, the advantage of defining antagonistic preferences like this is that if a set of orderings of a triple satisfies antagonistic preferences in the sense used here, then every nonempty subset of it would do likewise, a property not satisfied by Inada's antagonistic preferences.

alternative; consequently, EP is satisfied. Next consider the subcase when the ordering belonging to $LS(xzyx)$ contained in $\mathscr{D}|A$ is $yPxPz$. As ER is satisfied and all concerned non-strong orderings belong to both Latin Squares, for all non-strong orderings $R \in \mathscr{D}|A$ we must have both $xRz$ and $yRz$. Thus in this case, $\mathscr{D}|A$ must be a subset of $\{xPyPz, yPxPz, xPyIz, yPxIz, xIyPz, xIyIz\}$. In all orderings of this set, the best alternative of each of the two strong orderings is at least as good as the worst alternative; consequently, EP is satisfied. Finally consider the subcase when the ordering belonging to $LS(xzyx)$ contained in $\mathscr{D}|A$ is $zPyPx$. As ER is satisfied and all concerned non-strong orderings belong to both Latin Squares, for all non-strong orderings $R \in \mathscr{D}|A$, we must have both $xRz$ and $zRz$. Thus in this case, $\mathscr{D}|A$ must be a subset of $\{xPyPz, zPyPx, yPzIx, zIxPy, xIyIz\}$. This set satisfies AP. The foregoing establishes that satisfaction of ER implies that at least one of the three conditions of DP, EP and AP holds. (P3.4-1)

Let $\mathscr{D}|A$ violate ER. Then $\mathscr{D}|A$ contains a strong ordering of the triple, and an ordering belonging to the same Latin Square as the one to which the strong ordering belongs and in which the alternative worst in the strong ordering is strictly preferred to the alternative best in the strong ordering. Without any loss of generality assume that $\mathscr{D}|A$ contains $xPyPz$ and an ordering belonging to $LS(xyzx)$ with $zPx$, i.e., $yRzPx$ or $zPxRy$. Each of $\{xPyPz, yRzPx\}$ and $\{xPyPz, zPxRy\}$ violates DP as there is a strong ordering; violates EP as there is a strong ordering and an ordering in which the alternative best in the strong ordering is not at least as good as the alternative worst in the strong ordering; and violates AP as there is a strong ordering, namely $xPyPz$, but we do not have for every $R \in \mathscr{D}|A$, $(xPyPz \lor zPyPx \lor xIz)$. Thus, violation of ER implies that all three conditions of DP, EP and AP are violated. (P3.4-2)

(P3.4-1) and (P3.4-2) establish the proposition. □

The way EP and AP have been defined here, they trivially hold over a triple if the set of orderings over the triple does not contain a strong ordering of the triple. Thus, DP implies EP as well as AP. Consequently, it follows that, in view of Proposition 3.4, the following proposition also holds.

**Proposition 3.5** *A set of orderings of a triple satisfies extremal restriction iff it satisfies echoic preferences or antagonistic preferences.*

### 3.5.9 Taboo Preferences

Taboo Preferences (TP): $\mathscr{D} \subseteq \mathscr{T}$ satisfies TP over the triple $A \subseteq S$ iff $xIyIz \notin \mathscr{D}|A \land (\exists$ distinct $a, b \in A)(\forall R \in \mathscr{D}|A)(aRb)$. $\mathscr{D}$ satisfies TP iff it satisfies TP over every triple contained in $S$.

Thus, a set of orderings of a triple satisfies TP iff the set does not contain the unconcerned ordering and there exists a pair of distinct alternatives $a, b \in A$ such that in every ordering of the set $a$ is at least as good as $b$.

### 3.5.10   Weak Latin Square Partial Agreement

**Proposition 3.6** *A set of orderings of a triple satisfies Weak Latin Square partial agreement iff it satisfies at least one of the three conditions of value restriction (1), extremal restriction and taboo preferences.*

*Proof* Let WLSPA hold over the set of orderings $\mathscr{D}|A$ of triple $A = \{x, y, z\}$.
(i) If $\mathscr{D}|A$ does not contain a weak Latin Square then VR(1) is satisfied; as a set of orderings of a triple violates VR(1) iff the set contains a weak Latin Square.
(ii) Suppose $\mathscr{D}|A$ contains one of the two weak Latin Squares, say $LS(xyzx)$, and does not contain the other one. This implies that:
$\mathscr{D}|A$ does not contain the ordering $xIyIz$.                                    (P3.6-1)
As $\mathscr{D}|A$ does not contain $WLS(xzyx)$ we have: $\sim [(\exists R^s, R^t, R^u \in \mathscr{D}|A)(xR^s z\ R^s y \wedge zR^t yR^t x \wedge yR^u xR^u z)]$. Without any loss of generality assume that $\mathscr{D}|A$ does not contain any ordering $R$ such that $xRzRy$. This implies that we must have:
$(\forall R \in \mathscr{D}|A)[(xRz \to yPz) \wedge (zRy \to zPx)]$.                        (P3.6-2)
Consequently, we have:
$(\forall R \in \mathscr{D}|A)[(xRyRz \to xRyPz) \wedge (zRxRy \to zPxRy)]$.              (P3.6-3)
If no $R \in \mathscr{D}|A \cap T[LS(xyzx)]$ is strong then (P3.6-3) implies that $(\forall R \in \mathscr{D}|A)(yRx)$, and therefore, in view of (P3.6-1), TP is satisfied.
Next consider the case when some ordering belonging to $\mathscr{D}|A \cap T[LS(xyzx)]$ is strong. This strong ordering cannot be $xPyPz$ in view of $zPxRy$ belonging to $\mathscr{D}|A$, otherwise WLSPA would be violated. This strong ordering cannot be $zPxPy$ either in view of $xRyPz$ belonging to $\mathscr{D}|A$, otherwise WLSPA would be violated. Thus, if an $R \in \mathscr{D}|A \cap T[LS(xyzx)]$ is strong, it has to be $yPzPx$. Once again, it follows that: $(\forall R \in \mathscr{D}|A)(yRx)$, and therefore, in view of (1), TP is satisfied.
(iii) Finally consider the case when $\mathscr{D}|A$ contains both the weak Latin Squares. If no ordering belonging to $\mathscr{D}|A$ is strong, then DP holds, and consequently, ER is satisfied.
Suppose $\mathscr{D}|A$ contains a strong ordering, say, $xPyPz$.
WLSPA holds over $\mathscr{D}|A \wedge \mathscr{D}|A$ contains both the weak Latin Squares $\wedge \mathscr{D}|A$ contains
$xPyPz \to (\forall R \in \mathscr{D}|A)[(yRzRx \to yRzIx) \wedge (zRxRy \to zIxRy)]$   (P3.6-4)
(P3.6-4) implies that $\mathscr{D}|A$ does not contain any of $yPzPx, yIzPx, zPxPy, zPxIy$
(P3.6-5)
If $xPyPz$ is the only strong ordering in $\mathscr{D}|A$ then (P3.6-5) implies that $(\forall R \in \mathscr{D}|A)(xRz)$. Therefore, EP holds, and consequently, ER is satisfied.
Next suppose that, in addition to $xPyPz$, $\mathscr{D}|A$ contains another strong ordering. In view of (P3.6-5), it follows that this strong ordering must belong to $WLS(xzyx)$.
First consider the case when this other strong ordering contained in $\mathscr{D}|A$ is $xPzPy$.

WLSPA holds over $\mathscr{D}|A \wedge \mathscr{D}|A$ contains both the weak Latin Squares $\wedge\ \mathscr{D}|A$ contains $xPzPy \rightarrow (\forall R \in \mathscr{D}|A)[(yRxRz \rightarrow yIxRz) \wedge (zRyRx \rightarrow zRyIx)]$

(P3.6-6)

(P3.6-6) implies that $\mathscr{D}|A$ does not contain any of $yPxPz, yPxIz, zPyPx, zIyPx$

(P3.6-7)

(P3.6-5) and (P3.6-7) imply that $\mathscr{D}|A \subseteq \{xPyPz, xPzPy, xPyIz, xIyPz, xIzPy, xIyIz\}$. Therefore, EP holds, and consequently, ER is satisfied.

Next consider the case when this other strong ordering contained in $\mathscr{D}|A$ is $yPxPz$. WLSPA holds over $\mathscr{D}|A \wedge \mathscr{D}|A$ contains both the weak Latin Squares $\wedge\ \mathscr{D}|A$ contains $yPxPz \rightarrow (\forall R \in \mathscr{D}|A)[(xRzRy \rightarrow xRzIy) \wedge (zRyRx \rightarrow zIyRx)]$ (P3.6-8)

(P3.6-8) implies that $\mathscr{D}|A$ does not contain any of $xPzPy, xIzPy, zPyPx, zPyIx$

(P3.6-9)

(P3.6-5) and (P3.6-9) imply that $\mathscr{D}|A \subseteq \{xPyPz, yPxPz, xPyIz, xIyPz, yPxIz, xIyIz\}$. Therefore, EP holds, and consequently, ER is satisfied.

Finally consider the case when this other strong ordering contained in $\mathscr{D}|A$ is $zPyPx$. WLSPA holds over $\mathscr{D}|A \wedge \mathscr{D}|A$ contains both the weak Latin Squares $\wedge\ \mathscr{D}|A$ contains $zPyPx \rightarrow (\forall R \in \mathscr{D}|A)[(xRzRy \rightarrow xIzRy) \wedge (yRxRz \rightarrow yRxIz)]$ (P3.6-10)

(P3.6-10) implies that $\mathscr{D}|A$ does not contain any of $xPzPy, xPzIy, yPxPz, yIxPz$

(P3.6-11)

(P3.6-5) and (P3.6-11) imply that $\mathscr{D}|A \subseteq \{xPyPz, zPyPx, yPzIx, zIxPy, xIyIz\}$. Therefore, AP holds, and consequently, ER is satisfied.

This completes the proof of the assertion that satisfaction of WLSPA implies satisfaction of at least one of the three conditions of value restriction (1), taboo preferences and extremal restriction.

Let $\mathscr{D}|A$ violate WLSPA. Without any loss of generality, assume that $\mathscr{D}|A$ contains $WLS(xyzx)$ involving a strong ordering, say $xPyPz$, and also contains an ordering belonging to $LS(xyzx)$ with $zPx$. That is to say, $(\exists R^s, R^t, R^u \in \mathscr{D}|A)[(xP^syP^sz \wedge yR^tzP^tx \wedge zR^uxR^uy) \vee (xP^syP^sz \wedge yR^tzR^tx \wedge zP^uxR^uy)]$. (P3.6-12)

$\mathscr{D}|A$ contains $R^s, R^t, R^u$ such that $(xP^syP^sz \wedge yR^tzR^tx \wedge zR^uxR^uy)$ implies that VR(1) is violated. (P3.6-13)

$xP^syP^sz \in \mathscr{D}|A \rightarrow (\exists R \in \mathscr{D}|A)(xPy) \wedge (\exists R \in \mathscr{D}|A)(yPz) \wedge (\exists R \in \mathscr{D}|A)(xPz)$

(P3.6-14)

$yR^tzR^tx \in \mathscr{D}|A \rightarrow (\exists R \in \mathscr{D}|A)(yPx \vee xIyIz)$ (P3.6-15)

$zR^uxR^uy \in \mathscr{D}|A \rightarrow (\exists R \in \mathscr{D}|A)(zPy \vee xIyIz)$ (P3.6-16)

(P3.6-12) $\rightarrow$ $(\exists R \in \mathscr{D}|A)(zPx)$                                               (P3.6-17)

$xIyIz \in \mathscr{D}|A$ implies that TP is violated.                       (P3.6-18)

From (P3.6-14)-(P3.6-17) it follows that if $xIyIz \notin \mathscr{D}|A$ then also TP is violated.

                                                      (P3.6-19)

(P3.6-18) and (P3.6-19) establish that TP is violated.                 (P3.6-20)

From (P3.6-12) it follows that: $(xP^syP^sz, yR^tzP^tx \in \mathscr{D}|A) \vee (xP^syP^sz, zP^ux$ $R^uy \in \mathscr{D}|A)$                                                 (P3.6-21)

(P3.6-21) implies that ER is violated.                                       (P3.6-22)

(P3.6-13), (P3.6-20), and (P3.6-22) establish the assertion that if a set of orderings of a triple satisfies at least one of the three conditions of value restriction (1), taboo preferences and extremal restriction, then it satisfies WLSPA.   □

### 3.5.11 Limited Agreement

Limited Agreement (LA): $\mathscr{D} \subseteq \mathscr{T}$ satisfies LA over the triple $A \subseteq S$ iff ($\exists$ distinct $a, b \in A)(\forall R \in \mathscr{D}|A)(aRb)$. $\mathscr{D}$ satisfies LA iff it satisfies LA over every triple contained in $S$.

Thus, a set of orderings of a triple satisfies LA iff there exists a pair of distinct alternatives $a, b \in A$ such that in every ordering of the set $a$ is at least as good as $b$.

### 3.5.12 Latin Square Partial Agreement

**Proposition 3.7** *A set of orderings of a triple satisfies Latin Square partial agreement iff it satisfies at least one of the three conditions of value restriction (2), extremal restriction and limited agreement.*

*Proof* Let LSPA hold over the set of orderings $\mathscr{D}|A$ of triple $A = \{x, y, z\}$.

(i) If $\mathscr{D}|A$ does not contain a Latin Square then VR(2) is satisfied; as a set of orderings of a triple violates VR(2) iff the set contains a Latin Square.

(ii) Suppose $\mathscr{D}|A$ contains one of the two Latin Squares, say $LS(xyzx)$, and does not contain the other one. As $\mathscr{D}|A$ does not contain $LS(xzyx)$ we have: $\sim [(\exists$ concerned $R^s, R^t, R^u \in \mathscr{D}|A)(xR^szR^sy \wedge zR^tyR^tx \wedge yR^uxR^uz)]$. Without any loss of generality assume that $\mathscr{D}|A$ does not contain any concerned ordering $R$ such that $xRzRy$. This implies that we must have:

$(\forall$ concerned $R \in \mathscr{D}|A)[(xRz \rightarrow yPz) \wedge (zRy \rightarrow zPx)]$.       (P3.7-1)

Consequently we have:

$(\forall$ concerned $R \in \mathscr{D}|A)[(xRyRz \rightarrow xRyPz) \wedge (zRxRy \rightarrow zPxRy)]$.   (P3.7-2)

If no $R \in \mathscr{D}|A \cap T[LS(xyzx)]$ is strong then (P3.7-2) implies that $(\forall R \in \mathscr{D}|A)(yRx)$, and therefore LA is satisfied.

Next consider the case when some ordering belonging to $\mathcal{D}|A \cap T[LS(xyzx)]$ is strong. This strong ordering cannot be $xPyPz$ in view of $zPxRy$ belonging to $\mathcal{D}|A$, otherwise LSPA would be violated. This strong ordering cannot be $zPxPy$ either in view of $xRyPz$ belonging to $\mathcal{D}|A$, otherwise LSPA would be violated. Thus, if an $R \in \mathcal{D}|A \cap T[LS(xyzx)]$ is strong, it has to be $yPzPx$. Once again, it follows that: $(\forall R \in \mathcal{D}|A)(yRx)$, and therefore LA is satisfied.

(iii) Finally consider the case when $\mathcal{D}|A$ contains both the Latin Squares. If no ordering belonging to $\mathcal{D}|A$ is strong then ER is satisfied.

Suppose $\mathcal{D}|A$ contains a strong ordering, say, $xPyPz$.

LSPA holds over $\mathcal{D}|A \wedge \mathcal{D}|A$ contains both the Latin Squares $\wedge \mathcal{D}|A$ contains

$$xPyPz \to (\forall \text{ concerned } R \in \mathcal{D}|A)[(yRzRx \to yPzIx) \wedge (zRxRy \to zIxPy)]$$
(P3.7-3)

(P3.7-3) implies that $\mathcal{D}|A$ does not contain any of $yPzPx, yIzPx, zPxPy, zPxIy$
(P3.7-4)

$\mathcal{D}|A$ contains both the Latin Squares $\to (\exists \text{ concerned } R \in \mathcal{D}|A)(zRyRx)$
(P3.7-5)

(P3.7-4) $\wedge$ (P3.7-5) $\to (\exists R \in \mathcal{D}|A)(zPyPx)$                      (P3.7-6)

(P3.7-6) $\to (\forall \text{ concerned } R \in \mathcal{D}|A)[(yRxRz \to yPxIz) \wedge (xRzRy \to xIzPy)]$
(P3.7-7)

(P3.7-7) implies that $\mathcal{D}|A$ does not contain any of $yPxPz, yIxPz, xPzPy, xPzIy$
(P3.7-8)

(P3.7-4) $\wedge$ (P3.7-8) $\to \mathcal{D}|A \subseteq \{xPyPz, zPyPx, yPzIx, zIxPy, xIyIz\}$.
(P3.7-9)

(P3.7-9) implies that ER is satisfied. This completes the proof of the assertion that satisfaction of LSPA implies satisfaction of at least one of the three conditions of value restriction (2), limited agreement and extremal restriction.

Let $\mathcal{D}|A$ violate LSPA. Without any loss of generality, assume that $\mathcal{D}|A$ contains $LS(xyzx)$ involving a strong ordering, say $xPyPz$, and also contains an ordering belonging to $LS(xyzx)$ with $zPx$. That is to say, $(\exists \text{ concerned } R^s, R^t, R^u \in \mathcal{D}|A)[(xP^syP^sz \wedge yR^tzP^tx \wedge zR^uxR^uy) \vee (xP^syP^sz \wedge yR^tzR^tx \wedge zP^uxR^uy)]$.
(P3.7-10)

$\mathcal{D}|A$ contains concerned $R^s, R^t, R^u$ such that $(xP^syP^sz \wedge yR^tzR^tx \wedge zR^uxR^uy)$ implies that VR(2) is violated.
(P3.7-11)

$xP^syP^sz \in \mathcal{D}|A \to (\exists R \in \mathcal{D}|A)(xPy) \wedge (\exists R \in \mathcal{D}|A)(yPz) \wedge (\exists R \in \mathcal{D}|A)(xPz)$
(P3.7-12)

concerned $yR^tzR^tx \in \mathcal{D}|A \to (\exists R \in \mathcal{D}|A)(yPx)$                 (P3.7-13)

concerned $zR^uxR^uy \in \mathcal{D}|A \to (\exists R \in \mathcal{D}|A)(zPy)$                 (P3.7-14)

(P3.7-10) $\to (\exists R \in \mathcal{D}|A)(zPx)$                                            (P3.7-15)

(P3.7-12)-(P3.7-15) imply that LA is violated.                                    (P3.7-16)

From (P3.7-10) it follows that: $(x P^s y P^s z, y R^t z P^t x \in \mathscr{D}|A) \vee (x P^s y P^s z, z P^u x$

$R^u y \in \mathscr{D}|A)$                                                         (P3.7-17)

(P3.7-17) implies that ER is violated.                                            (P3.7-18)

(P3.7-11), (P3.7-16), and (P3.7-18) establish the assertion that if a set of orderings
of a triple satisfies at least one of the three conditions of value restriction (2), limited
agreement and extremal restriction, then it satisfies LSPA.                        □

## 3.6  Notes on Literature

The first result providing a condition under which an alternative will necessarily exist
that will be able to defeat every other alternative in a majority vote, i.e., a Condorcet
winner, is due to Black (1948). Black showed that if there exists a way of arranging
alternatives from left to right such that every individual's preferences are single-
peaked and the number of individuals is odd then there will exist an alternative that
will defeat every other alternative in a majority vote. Arrow (1951) formalized and
slightly generalized Black's condition of single-peakedness. Arrow showed that if
there exists a linear (strong) ordering of all alternatives such that for each individual
$i$, $x R_i y$ and $B(x, y, z)$ together imply $y P_i z$, and if the number of individuals is odd,
then the social $R$ generated by the MMD is transitive.

   If there exists a strong ordering of all alternatives such that, for each individual
$i$, $x R_i y$ and $B(x, y, z)$ together imply $y P_i z$, it is immediate that for every triple of
alternatives there would exist a strong ordering such that, for each individual $i$, $x R_i y$
and $B(x, y, z)$ together would imply $y P_i z$. But the converse is not true. That is, even
if there exists a strong ordering over every triple of alternatives such that, for each
individual $i$, $x R_i y$ and $B(x, y, z)$ together imply $y P_i z$, there may not exist a strong
ordering of all alternatives such that, for each individual $i$, $x R_i y$ and $B(x, y, z)$
together imply $y P_i z$. This can be seen by the following example.

*Example 3.1* Let $S = \{x, y, z, u\}$; $N = \{1, 2, 3, 4\}$.
$R_1 : x P_1 y P_1 z P_1 u$
$R_2 : z P_2 x P_2 y P_2 u$
$R_3 : x P_3 z P_3 u P_3 y$
$R_4 : u P_4 x P_4 z P_4 y$
In this case, there exists no strong ordering of four alternatives for which all four
orderings can be expressed as single-peaked preferences. But it can easily be veri-
fied that the assumption of single-peaked preferences is satisfied for every triple of
alternatives.                                                                      ◊

   Although Arrow assumed single-peakedness of individual preferences over all
alternatives, his proof used only single-peakedness over every triple of alternatives.
Thus, Arrow's proof is valid for the more general theorem. This was first pointed

out by Inada (1964). In this paper, Inada also formulated single-cavedness and separability conditions and showed that, given that the number of individuals is odd, each of them is sufficient to guarantee transitivity under the MMD. He also introduced the condition of dichotomous preferences and showed that it was sufficient for transitivity under the MMD regardless of the number of individuals.

The three conditions of single-peakedness, single-cavedness and separability were combined into a single condition of value restriction [referred to as the first version of value restriction in this text] by Sen. He also pointed out that single-peakedness, single-cavedness and separability conditions over a triple were equivalent, respectively, to requiring that there is some alternative that is not worst in any ordering over the triple; that there is some alternative that is not best in any ordering over the triple; that there is some alternative that is not medium in any ordering over the triple.

The most important contribution in the context of conditions for transitivity under the MMD is that of Inada (1969). In this paper, Inada introduced three more conditions of echoic preferences, antagonistic preferences and taboo preferences. He showed that echoic preferences and antagonistic preferences were sufficient conditions for transitivity under the MMD regardless of the number of individuals. He also showed that, if there were no restrictions on the number of individuals, then if a set of orderings of a triple violated all three conditions of dichotomous preferences, echoic preferences and antagonistic preferences, then for some assignment of individuals over these orderings the social $R$ generated by the MMD would be intransitive. In other words, if there were no restrictions on the number of individuals, then apart from dichotomous preferences, echoic preferences and antagonistic preferences, no other conditions were to be found which could ensure transitivity under the MMD. He also showed, given the number of individuals to be odd, sufficiency of taboo preferences for transitivity under the MMD. For the case of odd number of individuals, Inada showed that if any set of orderings violated all of the conditions of VR(1), taboo preferences, dichotomous preferences, echoic preferences and antagonistic preferences, then for some assignment of individuals over these orderings the social $R$ generated by the MMD would be intransitive. Thus, Inada's 1969 paper established (i) conditions which completely characterize all sets of orderings of a triple which invariably give rise to transitive social $R$ under the MMD, when there are no restrictions on the number of individuals; and (ii) conditions which completely characterize all sets of orderings of a triple which invariably give rise to transitive social $R$ under the MMD, when the number of individuals is odd.

The three conditions of dichotomous preferences, echoic preferences and antagonistic preferences were combined into a single condition of extremal restriction by Sen and Pattanaik (1969). They also showed that the union of the second version of value restriction, limited agreement and extremal restriction completely characterizes the sets of orderings of a triple that invariably give rise to quasi-transitive social preferences under the MMD.

The three conditions of second version of value restriction, limited agreement and extremal restriction were combined into a single condition of Latin Square partial agreement in Jain (1985). The conditions of the first version of value restriction, taboo preferences, dichotomous preferences and antagonistic preferences were combined into a single condition of Weak Latin Square partial agreement in Jain (2009).

As the notion of a social decision rule is defined for a specified set of individuals, strictly speaking, one should make a distinction among MMDs defined for different numbers of individuals. Kelly (1974) pointed out that the characterization of sets of orderings of a triple which guarantee quasi-transitivity or acyclicity depends on the number of individuals. From this perspective, the union of three conditions of second version of value restriction, limited agreement and extremal restriction characterizes the sets of orderings of a triple that invariably give rise to quasi-transitive social preferences under the MMD only for five or more individuals, and not for the cases of three and four individuals. The conditions for quasi-transitivity for the cases of three and four individuals were established in Jain (2009).

# References

Arrow, Kenneth J. 1951. *Social choice and individual values*, 2nd ed., 1963. New York: Wiley.

Aşan, Göksel, and M. Remzi Sanver. 2002. Another characterization of the majority rule. *Economics Letters* 75: 409–413.

Black, D. 1948. On the rationale of group decision making. *The Journal of Political Economy* 56: 23–34.

Campbell, Donald E., and Jerry S. Kelly. 2000. A simple characterization of majority rule. *Economic Theory* 15: 689–700.

Inada, Ken-ichi. 1964. A note on the simple majority decision rule. *Econometrica* 32: 525–531.

Inada, Ken-ichi. 1969. The simple majority decision rule. *Econometrica* 37: 490–506.

Jain, Satish K. 1985. A direct proof of Inada-Sen-Pattanaik theorem on majority rule. *The Economic Studies Quarterly* 36: 209–215.

Jain, Satish K. 2009. The method of majority decision and rationality conditions. In *Ethics, welfare, and measurement, Volume 1 of Arguments for a better world: Essays in honor of Amartya Sen*, ed. Kaushik Basu, and Ravi Kanbur, 167–192. New York: Oxford University Press.

Kelly, J.S. 1974. Necessity conditions in voting theory. *Journal of Economic Theory* 8: 149–160.

May, Kenneth O. 1952. A set of independent necessary and sufficient conditions for simple majority decision. *Econometrica* 20: 680–684.

Sen, Amartya K., and Prasanta K. Pattanaik. 1969. Necessary and sufficient conditions for rational choice under majority decision. *Journal of Economic Theory* 1: 178–202.

Woeginger, Gerhard J. 2003. A new characterization of the majority rule. *Economics Letters* 81: 89–94.

Under the strict majority rule, also called non-minority rule, an alternative $x$ is considered to be socially at least as good as some other alternative $y$ iff a majority of all individuals do not prefer $y$ to $x$. Thus, under the strict majority rule, between two alternatives social strict preference prevails iff one of the alternatives of the pair is strictly preferred by a majority of all individuals over the other alternative of the pair; otherwise the social indifference prevails. The strict majority rule is characterized by the following seven conditions: (i) independence of irrelevant alternatives, (ii) neutrality, (iii) monotonicity, (iv) weak Pareto-criterion, (v) anonymity, (vi) a set of individuals is a semidecisive set iff it is a strictly semidecisive set and (vii) every proper superset of a semidecisive set is a decisive set. A set of individuals constitutes a strictly semidecisive set iff whenever all the individuals in the set consider an alternative to be at least as good as another alternative, then the former is socially at least as good as the latter. In this chapter we will use the terms winning coalition, blocking coalition, strictly blocking coalition for decisive set, semidecisive set, strictly semidecisive set, respectively. The method of majority decision satisfies all the conditions listed above excepting condition (vi); and the strict majority rule satisfies all the conditions listed in Theorem 3.1 characterizing the method of majority decision excepting strict monotonicity.

Like the method of majority decision, under the strict majority rule also, it is possible to have social preferences that fail to satisfy transitivity, quasi-transitivity or acyclicity. Let $\mathscr{D}$ be a set of orderings of the set of alternatives $S$. It is shown in this chapter that the strict majority rule defined for an even number of individuals gives rise to transitive social weak preference relation for every profile of individual orderings belonging to $\mathscr{D}^n$ iff $\mathscr{D}$ satisfies the condition of Latin Square extremal value restriction; and that the strict majority rule defined for an odd number of individuals, number being greater than or equal to three, gives rise to transitive social weak preference relation for every profile of individual orderings belonging to $\mathscr{D}^n$ iff $\mathscr{D}$ satisfies the condition of weak Latin Square extremal value restriction. A set of orderings $\mathscr{D}$ satisfies Latin Square extremal value restriction over a triple of

© Springer Nature Singapore Pte Ltd. 2019
S. K. Jain, *Domain Conditions and Social Rationality*,
https://doi.org/10.1007/978-981-13-9672-4_4

alternatives $A$ iff there do not exist distinct $a, b, c \in A$ and $R^s, R^t \in \mathcal{D}|A$ such that (i) $R^s$ and $R^t$ belong to the same Latin Square; (ii) alternative $a$ is uniquely best in $R^s$, and medium in $R^t$ without being worst; and (iii) alternative $b$ is uniquely worst in $R^t$, and medium in $R^s$ without being best. Latin Square extremal value restriction is satisfied by a set of orderings iff it is satisfied over every triple of alternatives. A set of orderings $\mathcal{D}$ satisfies weak Latin Square extremal value restriction over a triple of alternatives $A$ iff there do not exist distinct $a, b, c \in A$ and $R^s, R^t, R^u \in \mathcal{D}|A$ such that (i) $R^s, R^t, R^u$ form a weak Latin Square; (ii) alternative $a$ is uniquely best in $R^s$, and medium in $R^t$ without being worst; and (iii) alternative $b$ is uniquely worst in $R^t$, and medium in $R^s$ without being best. Weak Latin Square extremal value restriction is satisfied by a set of orderings iff it is satisfied over every triple of alternatives. Weak Latin Square extremal value restriction requires what is required by Latin Square extremal value restriction only when a weak Latin Square is present. Thus the transitivity condition for the case of odd number of individuals is less stringent than the transitivity condition for the case of even number of individuals.

When the number of individuals is one or two then the strict majority rule coincides with the weak Pareto-rule. Under the weak Pareto-rule an alternative is socially preferred to another alternative iff everyone prefers the former to the latter. Between two alternatives social indifference prevails if neither of the two alternatives is unanimously preferred over the other. The weak Pareto-rule yields quasi-transitive social weak preference relation for every profile of individual orderings. Given a set of orderings $\mathcal{D}$, when the number of individuals is greater than or equal to three, the strict majority rule yields quasi-transitive social weak preference relation for every profile of individual orderings belonging to $\mathcal{D}^n$ iff $\mathcal{D}$ satisfies the condition of Latin Square unique value restriction. A set of orderings $\mathcal{D}$ satisfies Latin Square unique value restriction over a triple of alternatives $A$ iff there do not exist distinct $a, b, c \in A$ and $R^s, R^t, R^u \in \mathcal{D}|A$ such that (i) alternative $b$ is uniquely medium in $R^s$, uniquely best in $R^t$, uniquely worst in $R^u$, and (ii) $R^s, R^t, R^u$ form a Latin Square. Latin Square unique value restriction is satisfied by a set of orderings iff it is satisfied over every triple of alternatives.

This chapter is divided into seven sections, the sixth and the seventh sections constituting the Appendix. The first section contains some required definitions and notation. The next section provides the characterization of the strict majority rule mentioned above. Three restrictions of Latin Square extremal value restriction, weak Latin Square extremal value restriction and the Latin Square unique value restriction that turn out to be relevant for transitivity and quasi-transitivity under the strict majority rule are defined in Sect. 4.3. The two theorems providing characterizations of sets of orderings $\mathcal{D}$ such that every profile of individual orderings belonging to $\mathcal{D}^n$ gives rise to transitive social weak preference relation under the strict majority rule are stated and proved in Sect. 4.4. Section 4.5 states and proves the theorem providing characterization of sets of orderings $\mathcal{D}$ such that every profile of individual orderings belonging to $\mathcal{D}^n$ gives rise to quasi-transitive social weak preference relation under the strict majority rule. The appendix to the chapter contains decompositions of the three conditions figuring in the transitivity and quasi-transitivity theorems and brief notes on the literature. It is shown there that: (i) a set of orderings of a triple satisfies Latin Square extremal value restriction iff it satisfies extremal value restriction or conflictive preferences. (ii) A set of orderings of a triple satisfies weak Latin Square extremal value restriction iff it satisfies at least one of the three conditions of value restriction (1), extremal value restriction and conflictive preferences. (iii) A set of orderings of a triple satisfies Latin Square unique value restriction iff it satisfies at least one of the three conditions of value restriction (2), weakly conflictive preferences and unique value restriction.

## 4.1 Notation and Definitions

Let $f : \mathcal{T}^n \mapsto \mathcal{C}$ be a social decision rule. A set of individuals $V \subseteq N$ is called a winning coalition iff it is a decisive set; and a minimal winning coalition iff it is a minimal decisive set. The set of all winning coalitions will be denoted by $W$; and the set of all minimal winning coalitions by $W_m$. $V \subseteq N$ is called a blocking coalition iff it is a semidecisive set. The set of all blocking coalitions will be denoted by $B$. $V \subseteq N$ is called a strictly blocking coalition iff $(\forall(R_1, \ldots, R_n) \in \mathcal{T}^n)(\forall x, y \in S)[(\forall i \in V)(x R_i y) \rightarrow x R y]$. The set of all strictly blocking coalitions will be denoted by $B_s$.

*Remark 4.1* Let $f : \mathcal{T}^n \mapsto \mathcal{C}$ be a social decision rule. If $V_1, V_2 \in W$ then $V_1 \cap V_2$ must be nonempty, because $V_1 \cap V_2 = \emptyset$ would lead to a contradiction if we have for $x, y \in S : [(\forall i \in V_1)(x P_i y) \land (\forall i \in V_2)(y P_i x)]$, giving rise to $(x P y \land y P x)$. ◇

*Remark 4.2* Let $V \in W$. Then by the finiteness of $V$ and the fact that the empty set can never be winning, it follows that there exists a nonempty $V' \subseteq V$ such that $V' \in W_m$. ◇

*Remark 4.3* From the definitions of winning coalition, blocking coalition and strictly blocking coalition, it follows that if a coalition is winning or strictly blocking then it is blocking. ◇

Strict Majority Rule (Non-minority Rule): A social decision rule $f : \mathcal{T}^n \mapsto \mathcal{C}$ is the strict majority rule or non-minority rule iff $(\forall(R_1, \ldots, R_n) \in \mathcal{T}^n)(\forall x, y \in S)[x P y \leftrightarrow n(x P_i y) > \frac{n}{2}]$.

## 4.2 Characterization of the Strict Majority Rule

**Theorem 4.1** *A social decision rule $f : \mathcal{T}^n \mapsto \mathcal{C}$ is the strict majority rule iff it satisfies the conditions of (i) independence of irrelevant alternatives, (ii) neutrality, (iii) monotonicity, (iv) weak Pareto-criterion, (v) anonymity and its structure is such that (vi) a coalition is blocking iff it is strictly blocking and (vii) every proper superset of a blocking coalition is winning.*

*Proof* Let $f$ be the strict majority rule. Then from the definition of strict majority rule it is clear that $f$ satisfies conditions (i)–(v). Suppose $V$, $V \subseteq N$, is a blocking coalition. Then from the definitions of a blocking coalition and strict majority rule, it follows that $\#(N - V) \leq \frac{1}{2}n$. Therefore, for any $(R_1, \ldots, R_n) \in \mathscr{T}^n$ and any $x, y \in S$, $(\forall i \in V)(x R_i y)$ implies that $n(y P_i x) \leq \frac{1}{2}n$, which in turn implies $\sim y P x$, i.e., $x R y$. This establishes that $V$ is a strictly blocking coalition, thus proving that (vi) holds. As every coalition which has more than $\frac{1}{2}n$ individuals is winning, it follows that no coalition which has less than $\frac{1}{2}n$ individuals can be blocking. Thus every blocking coalition has at least $\frac{1}{2}n$ individuals. Consequently every proper superset of a blocking coalition has more than $\frac{1}{2}n$ individuals and is thus winning. Thus (vii) holds.

Next let social decision rule $f : \mathscr{T}^n \mapsto \mathscr{C}$ satisfy conditions (i)–(vii). As $f$ satisfies condition I we conclude that for any $x, y \in S$ the social $R$ over $\{x, y\}$ is completely determined by individual preferences over $\{x, y\}$. By neutrality the rule for determining social $R$ from individual preferences is the same for all ordered pairs of alternatives. Consider any profile $(R_1, \ldots, R_n) \in \mathscr{T}^n$ and any $x, y \in S$ such that $x P y$. Let $N_1, N_2, N_3$ designate the sets $\{i \in N \mid x P_i y\}$, $\{i \in N \mid x I_i y\}$, $\{i \in N \mid y P_i x\}$, respectively. Now consider any profile $(R_1', \ldots, R_n') \in \mathscr{T}^n$ such that $[(\forall i \in N_1)(x P_i' y) \wedge (\forall i \in N_2 \cup N_3)(y P_i' x)]$. Suppose $y R' x$. Then $N_2 \cup N_3$ is a blocking coalition as a consequence of conditions I, M and N. As every blocking coalition is strictly blocking we conclude that $N_2 \cup N_3$ is strictly blocking. But then in $(R_1, \ldots, R_n)$ situation we must have $y R x$, as we have $(\forall i \in N_2 \cup N_3)(y R_i x)$. As this contradicts $x P y$, we conclude that in $(R_1', \ldots, R_n')$ situation $y R' x$ is impossible, i.e., we must have $x P' y$. $x P' y$ in turn implies, in view of conditions I, M and N, that $N_1$ is a winning coalition. Thus we have shown that $(\forall (R_1, \ldots, R_n) \in \mathscr{T}^n)(\forall x, y \in S)[x P y \rightarrow (\exists V \in W)(\forall i \in V)(x P_i y)]$. If $V \in W$ then $(\forall (R_1, \ldots, R_n) \in \mathscr{T}^n)(\forall x, y \in S)[(\forall i \in V)(x P_i y) \rightarrow x P y]$, by the definition of a winning coalition. Thus, $(\forall (R_1, \ldots, R_n) \in \mathscr{T}^n)(\forall x, y \in S)[x P y \leftrightarrow (\exists V \in W)(\forall i \in V)(x P_i y)]$. Now, by the weak Pareto-criterion, $N$ is winning and thus $W$ is nonempty. If $V \in W$ and $\#V = k$, then by anonymity and the definition of a winning coalition we conclude that $(\forall V \subseteq N)(\#V \geq k \rightarrow V \in W)$. Let $\overline{k} = min \{k \mid (\exists V \in W)(\#V = k)\}$. Conditions I, M and N imply that if a coalition is not winning then its complement must be blocking. Therefore, it follows that every coalition which has at least $n - \overline{k} + 1$ individuals is blocking. As every proper superset of a blocking coalition is winning, it follows that every coalition which contains at least $n - \overline{k} + 2$ individuals must be winning. Therefore from the definition of $\overline{k}$ we conclude:

$$n - \overline{k} + 2 \geq \overline{k}$$
$$\rightarrow \overline{k} \leq \tfrac{1}{2}n + 1$$

Also, $(\forall V \subseteq N)(V \in W \rightarrow \#V > \frac{n}{2})$, otherwise as a consequence of anonymity, there will exist two nonempty disjoint winning coalitions leading to a contradiction (see Remark 4.1). Thus $\overline{k} > \frac{1}{2}n$.

$\frac{1}{2}n < \overline{k} \leq \frac{1}{2}n + 1$ establishes that $(\forall V \subseteq N)(V \in W \leftrightarrow \#V > \frac{1}{2}n)$. Consequently we have: $(\forall (R_1, \ldots, R_n) \in \mathscr{T}^n)(\forall x, y \in S)[xPy \leftrightarrow n(xP_iy) > \frac{1}{2}n]$, i.e., $f$ is the strict majority rule. $\square$

## 4.3   Restrictions on Preferences

Latin Square Extremal Value Restriction (LSEVR): Let $\mathscr{D} \subseteq \mathscr{T}$ be a set of orderings of $S$. Let $A = \{x, y, z\} \subseteq S$ be a triple of alternatives. $\mathscr{D}$ satisfies LSEVR over the triple $A$ iff there do not exist distinct $a, b, c \in A$ and $R^s, R^t \in D|A \cap T[LS(abca)]$ such that (i) alternative $a$ is uniquely best in $R^s$, and medium in $R^t$ without being worst; and (ii) alternative $b$ is uniquely worst in $R^t$, and medium in $R^s$ without being best. More formally, $\mathscr{D} \subseteq \mathscr{T}$ satisfies LSEVR over the triple $A$ iff $\sim$ $[(\exists$ distinct $a, b, c \in A)(\exists R^s, R^t \in D|A \cap T[LS(abca)])(aP^sbR^sc \wedge cR^taP^tb)]$. $\mathscr{D}$ satisfies LSEVR iff it satisfies LSEVR over every triple of alternatives contained in $S$.

Weak Latin Square Extremal Value Restriction (WLSEVR): Let $\mathscr{D} \subseteq \mathscr{T}$ be a set of orderings of $S$. Let $A = \{x, y, z\} \subseteq S$ be a triple of alternatives. $\mathscr{D}$ satisfies WLSEVR over the triple $A$ iff there do not exist distinct $a, b, c \in A$ and $R^s, R^t, R^u \in D|A \cap T[WLS(abca)]$ such that (i) $R^s, R^t, R^u$ form $WLS(abca)$, (ii) alternative $a$ is uniquely best in $R^s$, and medium in $R^t$ without being worst; and (iii) alternative $b$ is uniquely worst in $R^t$, and medium in $R^s$ without being best. More formally, $\mathscr{D} \subseteq \mathscr{T}$ satisfies WLSEVR over the triple $A$ iff $\sim [(\exists$ distinct $a, b, c \in A)(\exists R^s, R^t, R^u \in D|A \cap T[WLS(abca)])(aP^sbR^sc \wedge bR^ucR^ua \wedge cR^taP^tb)]$. $\mathscr{D}$ satisfies WLSEVR iff it satisfies WLSEVR over every triple of alternatives contained in $S$.

Latin Square Unique Value Restriction (LSUVR): Let $\mathscr{D} \subseteq \mathscr{T}$ be a set of orderings of $S$. Let $A = \{x, y, z\} \subseteq S$ be a triple of alternatives. $\mathscr{D}$ satisfies LSUVR over the triple $A$ iff there do not exist distinct $a, b, c \in A$ and $R^s, R^t, R^u \in D|A \cap T[LS(abca)]$ such that (i) alternative $b$ is uniquely medium in $R^s$, uniquely best in $R^t$, uniquely worst in $R^u$, and (ii) $R^s, R^t, R^u$ form $LS(abca)$. More formally, $\mathscr{D} \subseteq \mathscr{T}$ satisfies LSUVR over the triple $A$ iff $\sim [(\exists$ distinct $a, b, c \in A)(\exists R^s, R^t, R^u \in D|A \cap T[LS(abca)])(aP^sbP^sc \wedge bP^tcR^ta \wedge cR^uaP^ub)]$. $\mathscr{D}$ satisfies LSUVR iff it satisfies LSUVR over every triple of alternatives contained in $S$.

## 4.4   Transitivity Under the Strict Majority Rule

**Theorem 4.2** *Let $f : \mathscr{T}^n \mapsto \mathscr{C}$ be the strict majority rule defined for an even number of individuals; $n = 2k, k \geq 1$. Let $\mathscr{D} \subseteq \mathscr{T}$. Then $f$ yields transitive social $R$, $R = f(R_1, \ldots, R_n)$, for every $(R_1, \ldots, R_n) \in \mathscr{D}^n$ iff $\mathscr{D}$ satisfies the condition of Latin Square extremal value restriction.*

*Proof* Let $f : \mathcal{T}^n \mapsto \mathcal{C}$ be the strict majority rule defined for an even number of individuals, $n = 2k, k \geq 1$; and let $\mathcal{D} \subseteq \mathcal{T}$. Suppose $f$ does not yield transitive social $R$ for every $(R_1, \ldots, R_n) \in \mathcal{D}^n$. Then:

$$(\exists (R_1, \ldots, R_n) \in \mathcal{D}^n)(\exists x, y, z \in S)(xRy \wedge yRz \wedge zPx). \tag{T4.2-1}$$

$$zPx \rightarrow n(zP_ix) > \tfrac{n}{2}, \tag{T4.2-2}$$

$$xRy \rightarrow n(xR_iy) \geq \tfrac{n}{2}, \tag{T4.2-3}$$

as $n(xR_iy) < \tfrac{n}{2}$ would imply $n(yP_ix) > \tfrac{n}{2}$, which in turn would imply $yPx$, contradicting $xRy$

$$yRz \rightarrow n(yR_iz) \geq \tfrac{n}{2}, \tag{T4.2-4}$$

as $n(yR_iz) < \tfrac{n}{2}$ would imply $n(zP_iy) > \tfrac{n}{2}$, which in turn would imply $zPy$, contradicting $yRz$

$$(\text{T4.2-2}) \wedge (\text{T4.2-3}) \rightarrow (\exists j \in N)(zP_jxR_jy) \tag{T4.2-5}$$

$$(\text{T4.2-2}) \wedge (\text{T4.2-4}) \rightarrow (\exists k \in N)(yR_kzP_kx) \tag{T4.2-6}$$

(T4.2-1) implies that $x, y, z$ are distinct alternatives. $zP_jxR_jy$ and $yR_kzP_kx$ belong to $T[LS(xyzx)]$. In the triple $\{x, y, z\}$, $z$ is uniquely best according to $zP_jxR_jy$, and medium according to $yR_kzP_kx$ without being worst; furthermore $x$ is uniquely worst according to $yR_kzP_kx$ and medium according to $zP_jxR_jy$ without being best. Therefore LSEVR is violated over the triple $\{x, y, z\}$. Thus $\mathcal{D}$ violates LSEVR. It has been shown that violation of transitivity by $R = f(R_1, \ldots, R_n), (R_1, \ldots, R_n) \in \mathcal{D}^n$, implies violation of LSEVR by $\mathcal{D}$, i.e., if $\mathcal{D}$ satisfies LSEVR then every $(R_1, \ldots, R_n) \in \mathcal{D}^n$ yields transitive social $R$.

Suppose $\mathcal{D} \subseteq \mathcal{T}$ violates LSEVR. Then there exist distinct $x, y, z \in S$ such that $\mathcal{D}$ violates LSEVR over $\{x, y, z\}$. Violation of LSEVR by $\mathcal{D}$ over $\{x, y, z\}$ implies $(\exists \text{ distinct } a, b, c \in \{x, y, z\})(\exists R^s, R^t \in \mathcal{D})[aP^sbR^sc \wedge cR^taP^tb]$. As $n$ is even, there exists a partition of $N$, $(N_1, N_2)$ such that $\#N_1 = \#N_2 = \tfrac{n}{2} = k$. Now consider any $(R_1, \ldots, R_n) \in \mathcal{D}^n$ such that the restriction of $(R_1, \ldots, R_n)$ to $\{x, y, z\} = \{a, b, c\}, (R_1|\{x, y, z\}, \ldots, R_n|\{x, y, z\})$, is given by: $[(\forall i \in N_1)(aP_ibR_ic) \wedge (\forall i \in N_2)(cR_iaP_ib)]$. From $n(aP_ib) = n \wedge n(cP_ib) = \tfrac{n}{2} \wedge n(aP_ic) = \tfrac{n}{2}$ we conclude that $(aPb \wedge bIc \wedge aIc)$ holds, which violates transitivity. We have shown that if $\mathcal{D} \subseteq \mathcal{T}$ violates LSEVR then there exists a $(R_1, \ldots, R_n) \in \mathcal{D}^n$ such that $R = f(R_1, \ldots, R_n)$ is intransitive, i.e., if $f$ yields transitive $R$ for every $(R_1, \ldots, R_n) \in \mathcal{D}^n$ then $\mathcal{D}$ must satisfy LSEVR. This establishes the theorem. $\square$

**Theorem 4.3** *Let $f : \mathcal{T}^n \mapsto \mathcal{C}$ be the strict majority rule defined for an odd number of individuals; $n = 2k + 1, k \geq 1$. Let $\mathcal{D} \subseteq \mathcal{T}$. Then $f$ yields transitive social $R, R = f(R_1, \ldots, R_n)$, for every $(R_1, \ldots, R_n) \in \mathcal{D}^n$ iff $\mathcal{D}$ satisfies the condition of weak Latin Square extremal value restriction.*

*Proof* Let $f : \mathcal{T}^n \mapsto \mathcal{C}$ be the strict majority rule defined for an odd number of individuals, $n = 2k + 1, k \geq 1$; and let $\mathcal{D} \subseteq \mathcal{T}$. Suppose $f$ does not yield transitive social $R$ for every $(R_1, \ldots, R_n) \in \mathcal{D}^n$. Then:

$$(\exists (R_1, \ldots, R_n) \in \mathscr{D}^n)(\exists x, y, z \in S)(xRy \wedge yRz \wedge zPx). \tag{T4.3-1}$$

$$zPx \rightarrow n(zP_ix) > \frac{n}{2}, \tag{T4.3-2}$$

$$xRy \rightarrow n(xR_iy) \geq \frac{n}{2}, \tag{T4.3-3}$$

as $n(xR_iy) < \frac{n}{2}$ would imply $n(yP_ix) > \frac{n}{2}$, which in turn would imply $yPx$, contradicting $xRy$

$$yRz \rightarrow n(yR_iz) \geq \frac{n}{2}, \tag{T4.3-4}$$

as $n(yR_iz) < \frac{n}{2}$ would imply $n(zP_iy) > \frac{n}{2}$, which in turn would imply $zPy$, contradicting $yRz$

$$(\text{T4.3-3}) \rightarrow n(xR_iy) > \frac{n}{2}, \text{ as } n \text{ is odd} \tag{T4.3-5}$$

$$(\text{T4.3-4}) \rightarrow n(yR_iz) > \frac{n}{2}, \text{ as } n \text{ is odd} \tag{T4.3-6}$$

$$(\text{T4.3-4}) \wedge (\text{T4.3-5}) \rightarrow (\exists i \in N)(xR_iyR_iz) \tag{T4.3-7}$$

$$(\text{T4.3-2}) \wedge (\text{T4.3-6}) \rightarrow (\exists j \in N)(yR_jzP_jx) \tag{T4.3-8}$$

$$(\text{T4.3-2}) \wedge (\text{T4.3-5}) \rightarrow (\exists k \in N)(zP_kxR_ky) \tag{T4.3-9}$$

(T4.3-1) implies that $x, y, z$ are distinct alternatives. $xR_iyR_iz$, $yR_jzP_jx$, $zP_kxR_ky$ belong to $T[WLS(xyzx)]$ and form $WLS(xyzx)$. In the triple $\{x, y, z\}$, $z$ is uniquely best according to $zP_kxR_ky$, and medium according to $yR_jzP_jx$ without being worst; furthermore $x$ is uniquely worst according to $yR_jzP_jx$ and medium according to $zP_kxR_ky$ without being best. Therefore WLSEVR is violated over the triple $\{x, y, z\}$. Thus $\mathscr{D}$ violates WLSEVR. It has been shown that violation of transitivity by $R = f(R_1, \ldots, R_n)$, $(R_1, \ldots, R_n) \in \mathscr{D}^n$, implies violation of WLSEVR by $\mathscr{D}$, i.e., if $\mathscr{D}$ satisfies WLSEVR then every $(R_1, \ldots, R_n) \in \mathscr{D}^n$ yields transitive social $R$.

Suppose $\mathscr{D} \subseteq \mathscr{T}$ violates WLSEVR. Then there exist distinct $x, y, z \in S$ such that $\mathscr{D}$ violates WLSEVR over $\{x, y, z\}$. Violation of WLSEVR by $\mathscr{D}$ over $\{x, y, z\}$ implies ($\exists$ distinct $a, b, c \in \{x, y, z\}$)($\exists R^s, R^t, R^u \in \mathscr{D}$)$[aP^sbR^sc \wedge bR^tcR^ta \wedge cR^uaP^ub]$. As $n = 2k + 1, k \geq 1$, there exists a partition of $N$, $(N_1, N_2, N_3)$, such that $\#N_1 = \#N_3 = \frac{n-1}{2} = k \wedge \#N_2 = 1$. Now consider any $(R_1, \ldots, R_n) \in \mathscr{D}^n$ such that the restriction of $(R_1, \ldots, R_n)$ to $\{x, y, z\} = \{a, b, c\}$, $(R_1|\{x, y, z\}, \ldots, R_n|\{x, y, z\})$, is given by: $[(\forall i \in N_1)(aP_ibR_ic) \wedge (\forall i \in N_2)(bR_icR_ia) \wedge (\forall i \in N_3) (cR_iaP_ib)]$. From $n(aP_ib) = n - 1 \wedge n(cP_ib) = \frac{n-1}{2} \wedge n(aP_ic) = \frac{n-1}{2}$ we conclude that $(aPb \wedge bRc \wedge cRa)$ holds, which violates transitivity. We have shown that if $\mathscr{D} \subseteq \mathscr{T}$ violates WLSEVR then there exists a $(R_1, \ldots, R_n) \in \mathscr{D}^n$ such that $R = f(R_1, \ldots, R_n)$ is intransitive, i.e., if $f$ yields transitive $R$ for every $(R_1, \ldots, R_n) \in \mathscr{D}^n$ then $\mathscr{D}$ must satisfy WLSEVR. This establishes the theorem. □

## 4.5 Quasi-transitivity Under the Strict Majority Rule

Weak Pareto-Rule (WPR): A social decision rule $f : \mathscr{T}^n \mapsto \mathscr{C}$ is the weak Pareto-rule iff $(\forall (R_1, \ldots, R_n) \in \mathscr{T}^n)(\forall x, y \in S)[xPy \leftrightarrow (\forall i \in N)(xP_iy)]$.

Suppose $xPy \wedge yPz, x, y, z \in S$, under the weak Pareto-rule. $xPy$ and $yPz$ imply $(\forall i \in N)(xP_iy \wedge yP_iz)$. As individual preferences are transitive, from $(\forall i \in N)(xP_iy \wedge yP_iz)$ we obtain $(\forall i \in N)(xP_iz)$, which under the weak Pareto-rule implies $xPz$. Therefore it follows that the weak Pareto-rule yields quasi-transitive social $R$ for every $(R_1, \ldots, R_n) \in \mathcal{T}^n$.

*Remark 4.4* If $n \leq 2$ then the strict majority rule coincides with the weak Pareto-rule. Thus, if $n \leq 2$ then the strict majority rule yields quasi-transitive social $R$ for every $(R_1, \ldots, R_n) \in \mathcal{T}^n$.                                                    ◇

**Theorem 4.4** *Let $f : \mathcal{T}^n \mapsto \mathcal{C}$ be the strict majority rule. Let $\#N = n \geq 3$. Let $\mathcal{D} \subseteq \mathcal{T}$. Then $f$ yields quasi-transitive social $R$, $R = f(R_1, \ldots, R_n)$, for every $(R_1, \ldots, R_n) \in \mathcal{D}^n$ iff $\mathcal{D}$ satisfies the condition of Latin Square unique value restriction.*

*Proof* Let $f : \mathcal{T}^n \mapsto \mathcal{C}$ be the strict majority defined for $n \geq 3$; and let $\mathcal{D} \subseteq \mathcal{T}$. Suppose $f$ does not yield quasi-transitive social $R$ for every $(R_1, \ldots, R_n) \in \mathcal{D}^n$. Then:

$$(\exists (R_1, \ldots, R_n) \in \mathcal{D}^n)(\exists x, y, z \in S)(xPy \wedge yPz \wedge zRx). \qquad \text{(T4.4-1)}$$

$$xPy \rightarrow n(xP_iy) > \tfrac{n}{2}, \qquad \text{(T4.4-2)}$$

$$yPz \rightarrow n(yP_iz) > \tfrac{n}{2}, \qquad \text{(T4.4-3)}$$

$$zRx \rightarrow n(zR_ix) \geq \tfrac{n}{2}, \qquad \text{(T4.4-4)}$$

as $n(zR_ix) < \tfrac{n}{2}$ would imply $n(xP_iz) > \tfrac{n}{2}$, which in turn would imply $xPz$, contradicting $zRx$

$$(\text{T4.4-2}) \wedge (\text{T4.4-3}) \rightarrow (\exists i \in N)(xP_iyP_iz) \qquad \text{(T4.4-5)}$$

$$(\text{T4.4-3}) \wedge (\text{T4.4-4}) \rightarrow (\exists j \in N)(yP_jzR_jx) \qquad \text{(T4.4-6)}$$

$$(\text{T4.4-2}) \wedge (\text{T4.4-4}) \rightarrow (\exists k \in N)(zR_kxP_ky) \qquad \text{(T4.4-7)}$$

(T4.4-1) implies that $x, y, z$ are distinct alternatives. $xP_iyP_iz$, $yP_jzR_jx$, $zR_kxP_ky$ belong to $T[LS(xyzx)]$ and form $LS(xyzx)$. In the triple $\{x, y, z\}$, $y$ is uniquely medium according to $xP_iyP_iz$, uniquely best according to $yP_jzR_jx$ and uniquely worst according to $zR_kxP_ky$. Therefore LSUVR is violated over the triple $\{x, y, z\}$. Thus $\mathcal{D}$ violates LSUVR. It has been shown that violation of quasi-transitivity by $R = f(R_1, \ldots, R_n)$, $(R_1, \ldots, R_n) \in \mathcal{D}^n$, implies violation of LSUVR by $\mathcal{D}$, i.e., if $\mathcal{D}$ satisfies LSUVR then every $(R_1, \ldots, R_n) \in \mathcal{D}^n$ yields quasi-transitive social $R$.

Suppose $\mathcal{D} \subseteq \mathcal{T}$ violates LSUVR. Then there exist distinct $x, y, z \in S$ such that $\mathcal{D}$ violates LSUVR over $\{x, y, z\}$. Violation of LSUVR by $\mathcal{D}$ over $\{x, y, z\}$ implies $(\exists$ distinct $a, b, c \in \{x, y, z\})(\exists R^s, R^t, R^u \in \mathcal{D})[aP^sbP^sc \wedge bP^tcR^ta \wedge cR^uaP^ub]$. As $n \geq 3$, there exists a partition of $N$, $(N_1, N_2, N_3)$, such that $\#N_1 = \lfloor \tfrac{n}{2} \rfloor \wedge \#N_2 = n - \lfloor \tfrac{n}{2} \rfloor - 1 \wedge \#N_3 = 1$.[1] Now consider any $(R_1, \ldots, R_n) \in \mathcal{D}^n$ such

---

[1] Let $t$ be a real number. We use the standard notation $\lfloor t \rfloor$ to denote the largest integer less than or equal to $t$; and $\lceil t \rceil$ to denote the smallest integer greater than or equal to $t$.

that the restriction of $(R_1, \ldots, R_n)$ to $\{x, y, z\} = \{a, b, c\}$, $(R_1|\{x, y, z\}, \ldots, R_n| \{x, y, z\})$, is given by: $[(\forall i \in N_1)(a P_i b P_i c) \wedge (\forall i \in N_2)(b P_i c R_i a) \wedge (\forall i \in N_3) (c R_i a P_i b)]$. From $n(a P_i b) = \lfloor \frac{n}{2} \rfloor + 1 \wedge n(b P_i c) = n - 1 \wedge n(a P_i c) = \lfloor \frac{n}{2} \rfloor$ we conclude that $(a P b \wedge b P c \wedge c R a)$ holds, which violates quasi-transitivity. We have shown that if $\mathscr{D} \subseteq \mathscr{T}$ violates LSUVR then there exists a $(R_1, \ldots, R_n) \in \mathscr{D}^n$ such that $R = f(R_1, \ldots, R_n)$ is not quasi-transitive, i.e., if $f$ yields quasi-transitive $R$ for every $(R_1, \ldots, R_n) \in \mathscr{D}^n$ then $\mathscr{D}$ must satisfy LSUVR. This establishes the theorem. □

# Appendix

## 4.6 Interrelationships Among Restrictions on Preferences

In the context of the strict majority rule the following four domain conditions have been found to be relevant.

Extremal Value Restriction (EVR): Let $\mathscr{D} \subseteq \mathscr{T}$ be a set of orderings of $S$. Let $A = \{x, y, z\} \subseteq S$ be a triple of alternatives. $\mathscr{D}$ satisfies EVR over the triple $A$ iff (i) whenever an alternative is uniquely best in some $R \in \mathscr{D}|A$ then it is not medium in any $R \in \mathscr{D}|A$ unless it is worst also; or (ii) whenever an alternative is uniquely worst in some $R \in \mathscr{D}|A$ then it is not medium in any $R \in \mathscr{D}|A$ unless it is best also. More formally, $\mathscr{D}$ satisfies EVR over $A$ iff $(\forall$ distinct $a, b, c \in A)[(\exists R \in \mathscr{D}|A)(a P b \wedge a P c) \rightarrow (\forall R \in \mathscr{D}|A)[(b R a R c \rightarrow c R a) \wedge (c R a R b \rightarrow b R a)]] \vee (\forall$ distinct $a, b, c \in A)[(\exists R \in \mathscr{D}|A)(b P a \wedge c P a) \rightarrow (\forall R \in \mathscr{D}|A)[(b R a R c \rightarrow a R b) \wedge (c R a R b \rightarrow a R c)]]$. $\mathscr{D}$ satisfies EVR iff it satisfies EVR over every triple of alternatives contained in $S$.

Conflictive Preferences (CP): Let $\mathscr{D} \subseteq \mathscr{T}$ be a set of orderings of $S$. Let $A = \{x, y, z\} \subseteq S$ be a triple of alternatives. Let $\mathscr{D}_c$ be the set of concerned orderings in $\mathscr{D}|A$. $\mathscr{D}$ satisfies CP over the triple $A$ iff it is the case there exists a decomposition[2] $(\mathscr{D}_1, \mathscr{D}_2)$ of the set of concerned orderings in $\mathscr{D}|A$ such that for some distinct $a, b, c \in A$: every ordering $R \in \mathscr{D}_1$ is $a P b R c$ and every ordering $R \in \mathscr{D}_2$ is $c R b P a$. More formally, $\mathscr{D}$ satisfies CP over the triple $A$ iff $(\exists \mathscr{D}_1, \mathscr{D}_2 \subseteq \mathscr{D}_c)(\exists$ distinct $a, b, c \in A)[(\emptyset \subseteq \mathscr{D}_1 \wedge \emptyset \subseteq \mathscr{D}_2 \wedge \mathscr{D}_1 \cap \mathscr{D}_2 = \emptyset \wedge \mathscr{D}_1 \cup \mathscr{D}_2 = \mathscr{D}_c) \wedge (\forall R \in \mathscr{D}_1)(a P b R c) \wedge (\forall R \in \mathscr{D}_2)(c R b P a)]$. $\mathscr{D}$ satisfies CP iff it satisfies CP over every triple of alternatives contained in $S$.

Weakly Conflictive Preferences (WCP): Let $\mathscr{D} \subseteq \mathscr{T}$ be a set of orderings of $S$. Let $A = \{x, y, z\} \subseteq S$ be a triple of alternatives. $\mathscr{D}$ satisfies WCP over the triple $A$ iff it is the case that: (i) whenever in an $R \in \mathscr{D}|A$ an alternative best in some linear ordering belonging to $\mathscr{D}|A$ is worst, the alternative worst in the linear ordering is best in it; or (ii) whenever in an $R \in \mathscr{D}|A$ an alterna-

---

[2]Let $X$ be a set. $(X_1, \ldots, X_m)$ is a decomposition of $X$ iff $(\forall i, j \in \{1, \ldots, m\})[\emptyset \subseteq X_i \wedge (i \neq j \rightarrow X_i \cap X_j = \emptyset) \wedge \cup_{i=1}^m X_i = X]$.

tive worst in some linear ordering belonging to $\mathscr{D}|A$ is best, the alternative best in the linear ordering is worst in it. That is to say, $\mathscr{D}$ satisfies WCP over the triple $A$ iff ($\forall$ distinct $a, b, c \in A$)[($\exists R \in \mathscr{D}|A$)($aPbPc$) $\rightarrow$ ($\forall R \in \mathscr{D}|A$)($bRa \wedge cRa \rightarrow cRb$)] $\vee$ ($\forall$ distinct $a, b, c \in A$)[($\exists R \in \mathscr{D}|A$)($aPbPc$) $\rightarrow$ ($\forall R \in \mathscr{D}|A$) ($cRa \wedge cRb \rightarrow bRa$)]. $\mathscr{D}$ satisfies WCP iff it satisfies WCP over every triple of alternatives contained in $S$.

Unique Value Restriction (UVR): Let $\mathscr{D} \subseteq \mathscr{T}$ be a set of orderings of $S$. Let $A = \{x, y, z\} \subseteq S$ be a triple of alternatives. $\mathscr{D}$ satisfies UVR over the triple $A$ iff (i) there exist distinct $a, b \in A$ such that $a$ is not uniquely medium in any $R \in \mathscr{D}|A$, $b$ is not uniquely best in any $R \in \mathscr{D}|A$, and whenever $b$ is best in an $R \in \mathscr{D}|A$ $a$ is worst in it; or (ii) there exist distinct $a, b \in A$ such that $a$ is not uniquely medium in any $R \in \mathscr{D}|A$, $b$ is not uniquely worst in any $R \in \mathscr{D}|A$, and whenever $b$ is worst in an $R \in \mathscr{D}|A$ $a$ is best in it. More formally, $\mathscr{D} \subseteq \mathscr{T}$ satisfies UVR over the triple $A$ iff ($\exists$ distinct $a, b, c \in A$)($\forall R \in \mathscr{D}|A$)[[($aRb \wedge aRc$) $\vee$ ($bRa \wedge cRa$)] $\wedge$ [$aRb \vee cRb$] $\wedge$ [$bRa \wedge bRc \rightarrow cRa$]] $\vee$ ($\exists$ distinct $a, b, c \in A$)($\forall R \in \mathscr{D}|A$)[[($aRb \wedge aRc$) $\vee$ ($bRa \wedge cRa$)] $\wedge$ [$bRa \vee bRc$] $\wedge$ [$bRa \wedge cRb \rightarrow aRc$]]. $\mathscr{D}$ satisfies UVR iff it satisfies UVR over every triple of alternatives contained in $S$.

**Proposition 4.1**  *A set of orderings of a triple satisfies Latin Square extremal value restriction iff it satisfies extremal value restriction or conflictive preferences.*

*Proof* Let the set of orderings $\mathscr{D}|A$ over the triple $A = \{x, y, z\}$ violate LSEVR. Then we have:  [($\exists$ distinct $a, b, c \in A$)($\exists R^s, R^t \in \mathscr{D}|A \cap T[LS(abca)]$)($aP^sbR^sc \wedge cR^taP^tb$)].

$a$ is uniquely best in $aP^sbR^sc$ and medium in $cR^taP^tb$ without being worst; and $b$ is uniquely worst in $cR^taP^tb$ and medium in $aP^sbR^sc$ without being best. Therefore EVR is violated.

Given that $\mathscr{D}|A$ contains $aP^sbR^sc \wedge cR^taP^tb$, it is immediate that CP is violated. Thus we see that violation of LSEVR implies violation of both EVR and CP, implying that if EVR or CP holds then LSEVR also holds.

Next we show that if both CP and EVR are violated then LSEVR is also violated. It can be easily checked that the set of orderings $\mathscr{D}|A$ violates CP iff ($\exists$ distinct $a, b, c \in A$)($\exists R^s, R^t \in \mathscr{D}|A$)[($aP^sbP^sc \wedge bP^tcP^ta$) $\vee$ ($aP^sbP^sc \wedge bP^taP^tc$) $\vee$ ($aP^sbP^sc \wedge aP^tcP^tb$) $\vee$ ($aP^sbP^sc \wedge bP^tcI^ta$) $\vee$ ($aP^sbP^sc \wedge cI^taP^tb$) $\vee$ ($aP^sbI^sc \wedge bP^tcI^ta$) $\vee$ ($aP^sbI^sc \wedge aI^tbP^tc$) $\vee$ ($aI^sbP^sc \wedge bI^tcP^ta$)].

If ($\exists$ distinct $a, b, c \in A$)($\exists R^s, R^t \in \mathscr{D}|A$)[($aP^sbP^sc \wedge bP^tcP^ta$) $\vee$ ($aP^sbP^sc \wedge bP^tcI^ta$) $\vee$ ($aP^sbP^sc \wedge cI^taP^tb$) $\vee$ ($aP^sbI^sc \wedge aI^tbP^tc$)] then LSEVR is violated. Therefore it suffices to consider the remaining four cases.

Consider the case when $\mathscr{D}|A$ contains ($aP^sbP^sc \wedge bP^taP^tc$). $a$ is uniquely best in $aP^sbP^sc$ and medium in $bP^taP^tc$ without being worst. So one part of the EVR condition does not hold. The other part of EVR condition would be violated iff there is an ordering in $\mathscr{D}|A$ in which $a$ is uniquely worst or an ordering in $\mathscr{D}|A$ in which $b$ is uniquely worst or an ordering in $\mathscr{D}|A$ in which $c$ is medium without being best, i.e., it contains ($bP^ucP^ua \vee bI^ucP^ua \vee cP^ubP^ua \vee cP^uaP^ub \vee$

$cI^u a P^u b \vee a P^u c P^u b \vee a P^u c I^u b \vee b P^u c I^u a$). With the inclusion of the required ordering, in each case LSEVR is violated.

Next consider the case when $\mathscr{D}|A$ contains ($a P^s b P^s c \wedge a P^t c P^t b$). $c$ is uniquely worst in $a P^s b P^s c$ and medium in $a P^t c P^t b$ without being best. So one part of the EVR condition does not hold. The other part of EVR condition would be violated iff there is an ordering in $\mathscr{D}|A$ in which $b$ is uniquely best or an ordering in $\mathscr{D}|A$ in which $c$ is uniquely best or an ordering in $\mathscr{D}|A$ in which $a$ is medium without being worst, i.e., it contains ($b P^u c P^u a \vee b P^u c I^u a \vee b P^u a P^u c \vee c P^u a P^u b \vee c P^u a I^u b \vee c P^u b P^u a \vee b I^u a P^u c \vee c I^u a P^u b$). With the inclusion of the required ordering, in each case LSEVR is violated.

When $\mathscr{D}|A$ contains ($a P^s b I^s c \wedge b P^t c I^t a$) then EVR would be violated only if there is an ordering in $\mathscr{D}|A$ in which some alternative is uniquely worst, i.e., only if $\mathscr{D}|A$ contains ($b P^u c P^u a \vee b I^u c P^u a \vee c P^u b P^u a \vee a P^u c P^u b \vee a I^u c P^u b \vee c P^u a P^u b \vee a P^u b P^u c \vee a I^u b P^u c \vee b P^u a P^u c$). In all cases excepting when the ordering is ($b P^u c P^u a \vee a P^u c P^u b$), LSEVR is violated. When $\mathscr{D}|A$ contains [$a P^s b I^s c \wedge b P^t c I^t a \wedge (b P^u c P^u a \vee a P^u c P^u b)$)] then EVR would be violated iff an ordering in which $a$ is medium without being worst is included or an ordering in which $b$ is medium without being worst is included or an ordering in which $c$ is uniquely best is included, i.e., $\mathscr{D}|A$ contains ($b P^v a P^v c \vee b I^v a P^v c \vee c P^v a P^v b \vee c I^v a P^v b \vee a P^v b P^v c \vee c P^v b P^v a \vee c I^v b P^v a \vee c P^v a I^v b$). With the inclusion of the required ordering, in each case LSEVR is violated.

When $\mathscr{D}|A$ contains ($a I^s b P^s c \wedge b I^t c P^t a$) then EVR would be violated only if there is an ordering in $\mathscr{D}|A$ in which some alternative is uniquely best, i.e., only if $\mathscr{D}|A$ contains ($a P^u b P^u c \vee a P^u b I^u c \vee a P^u c P^u b \vee b P^u a P^u c \vee b P^u a I^u c \vee b P^u c P^u a \vee c P^u a P^u b \vee c P^u a I^u b \vee c P^u b P^u a$). In all cases excepting when the ordering is ($a P^u b P^u c \vee c P^u b P^u a$), LSEVR is violated. When $\mathscr{D}|A$ contains [$a I^s b P^s c \wedge b I^t c P^t a \wedge (a P^u b P^u c \vee c P^u b P^u a)$)] then EVR would be violated iff an ordering in which $a$ is medium without being best is included or an ordering in which $c$ is medium without being best is included or an ordering in which $b$ is uniquely worst is included, i.e., $\mathscr{D}|A$ contains ($b P^v a P^v c \vee b P^v a I^v c \vee c P^v a P^v b \vee c P^v a I^v b \vee a P^v c P^v b \vee a P^v c I^v b \vee b P^v c P^v a \vee c I^v a P^v b$). With the inclusion of the required ordering, in each case LSEVR is violated.

Thus, violation of both EVR and CP implies violation of LSEVR, i.e., if LSEVR holds then it must be the case that EVR or CP holds.

This, together with the earlier demonstration that if EVR or CP holds then LSEVR holds, establishes the proposition. □

**Proposition 4.2** *A set of orderings of a triple satisfies weak Latin Square extremal value restriction iff it satisfies at least one of the three conditions of value restriction (1), extremal value restriction and conflictive preferences.*

*Proof* Let the set of orderings $\mathscr{D}|A$ over the triple $A = \{x, y, z\}$ violate WLSEVR. Then we have: [($\exists$ distinct $a, b, c \in A)(\exists R^s, R^t, R^u \in \mathscr{D}|A \cap T[WLS(abca)])$ ($a P^s b R^s c \wedge b R^t c R^t a \wedge c R^u a P^u b$)].

$a P^s b R^s c \wedge b R^t c R^t a \wedge c R^u a P^u b$ form $WLS(abca)$, therefore VR(1) is violated.

$a P^s b R^s c \wedge c R^u a P^u b$ violate LSEVR. Then, In view of Proposition 4.1, it follows that both EVR and CP are violated.

Thus we see that violation of WLSEVR implies violation of all three conditions VR(1), EVR and CP, implying that if VR(1) or EVR or CP holds then WLSEVR also holds.

Next we show that if all three conditions VR(1), EVR and CP are violated then WLSEVR is also violated.

VR(1) is violated by $\mathscr{D}|A$ iff it contains $a R^s b R^s c \wedge b R^t c R^t a \wedge c R^u a R^u b$ for some distinct $a, b, c \in A$. Excepting the cases of $[(a P^s b I^s c \wedge b P^t c I^t a \wedge c P^u a I^u b) \vee (a I^s b P^s c \wedge b I^t c P^t a \wedge c I^u a P^u b) \vee (a I^s b I^s c)]$, in all other cases WLSEVR is violated.

$(a P^s b I^s c \wedge b P^t c I^t a \wedge c P^u a I^u b)$ violates CP. $\mathscr{D}|A$ containing $a P^s b I^s c \wedge b P^t c I^t a \wedge c P^u a I^u b$ would violate EVR only if an ordering of $A$ in which some alternative is uniquely worst is included, i.e., only if $(b P^v c P^v a \vee b I^v c P^v a \vee c P^v b P^v a \vee a P^v c P^v b \vee a I^v c P^v b \vee c P^v a P^v b \vee a P^v b P^v c \vee a I^v b P^v c \vee b P^v a P^v c)$ is included. In each case, with the inclusion of the required ordering, WLSEVR is violated.

$(a I^s b P^s c \wedge b I^t c P^t a \wedge c I^u a P^u b)$ violates CP. $\mathscr{D}|A$ containing $a I^s b P^s c \wedge b I^t c P^t a \wedge c I^u a P^u b$ would violate EVR only if an ordering of $A$ in which some alternative is uniquely best is included, i.e., only if $(a P^v b P^v c \vee a P^v b I^v c \vee a P^v c P^v b \vee b P^v a P^v c \vee b P^v a I^v c \vee b P^v c P^v a \vee c P^v a P^v b \vee c P^v a I^v b \vee c P^v b P^v a)$ is included. In each case, with the inclusion of the required ordering, WLSEVR is violated.

$a I^s b I^s c$ violates neither CP nor EVR. By Proposition 4.1, a set of orderings of a triple violates both CP and EVR iff it violates LSEVR. Thus, $\mathscr{D}|A$ containing $a I^s b I^s c$ would violate both CP and EVR iff for some distinct $a, b, c \in A$, it contains $a P^t b R^t c \wedge c R^u a P^u b$. $(a I^s b I^s c \wedge a P^t b R^t c \wedge c R^u a P^u b)$ violates WLSEVR.

Thus, violation of all three conditions of VR(1), EVR and CP implies violation of WLSEVR, i.e., if WLSEVR holds then it must be the case that VR(1) or EVR or CP holds.

This, together with the earlier demonstration that if VR(1) or EVR or CP holds then WLSEVR holds, establishes the proposition. $\qquad\square$

**Proposition 4.3** *A set of orderings of a triple satisfies Latin Square unique value restriction iff it satisfies at least one of the three conditions of value restriction (2), weakly conflictive preferences and unique value restriction.*

*Proof* Let the set of orderings $\mathscr{D}|A$ over the triple $A = \{x, y, z\}$ violate LSUVR. Then we have: $[(\exists$ distinct $a, b, c \in A)(\exists R^s, R^t, R^u \in \mathscr{D}|A \cap T[LS(abca)])(a P^s b P^s c \wedge b P^t c R^t a \wedge c R^u a P^u b)]$.

$a P^s b P^s c$, $b P^t c R^t a$, $c R^u a P^u b$ are concerned orderings and form $LS(abca)$. Therefore, VR(2) is violated.

$a P^s b P^s c$ is a linear ordering. In $b P^t c R^t a$, alternative best in the linear ordering $a P^s b P^s c$, namely $a$, is worst, without $c$, the worst alternative in the linear ordering

$a P^s b P^s c$, being best. Furthermore, in $c R^u a P^u b$ the worst alternative of the linear ordering $a P^s b P^s c$ is best without the best alternative of the linear ordering $a P^s b P^s c$ being worst in it. Thus WCP is violated.

$a$ is uniquely best in $a P^s b P^s c$, $b$ is uniquely best in $b P^t c R^t a$, $b$ is uniquely medium in $a P^s b P^s c$, $b$ is uniquely worst in $c R^u a P^u b$, and $c$ is uniquely worst in $a P^s b P^s c$. Thus if UVR is to hold then we must have: (i) $a$ is not uniquely medium in any $R \in \mathscr{D}|A$, $c$ is not uniquely best in any $R \in \mathscr{D}|A$, and $(\forall R \in \mathscr{D}|A)(cRa \wedge cRb \rightarrow bRa)$ or (ii) $c$ is not uniquely medium in any $R \in \mathscr{D}|A$, $a$ is not uniquely worst in any $R \in \mathscr{D}|A$, and $(\forall R \in \mathscr{D}|A)(bRa \wedge cRa \rightarrow cRb)$. In $c R^u a P^u b$, $c$ is best without $a$ being worst, so (i) cannot hold. In $b P^t c R^t a$, $a$ is worst without $c$ being best, so (ii) cannot hold. Thus UVR is violated.

We see that the violation of LSUVR implies violation of all three restrictions of VR(2), WCP and UVR. This establishes that if any of the three restrictions of VR(2), WCP and UVR is satisfied then LSUVR is also satisfied.

Next we show that if $\mathscr{D}|A$ violates all three restrictions of WCP, VR(2) and UVR, then it violates LSUVR.

It can be easily checked that the set of orderings $\mathscr{D}|A$ violates WCP iff ($\exists$ distinct $a, b, c \in A)(\exists R^s, R^t \in \mathscr{D}|A)(a P^s b P^s c \wedge b P^t c P^t a)$  $\vee$  ($\exists$ distinct $a, b, c \in A$) $(\exists R^s, R^t, R^u \in \mathscr{D}|A)(a P^s b P^s c \wedge b P^t c I^t a \wedge c I^u a P^u b) \vee$ ($\exists$ distinct $a, b, c \in A$) $(\exists R^s, R^t, R^u, R^v \in \mathscr{D}|A)(a P^s b P^s c \wedge a P^t c P^t b \wedge c I^u a P^u b \wedge c P^v b I^v a) \vee$ ($\exists$ distinct $a, b, c \in A)(\exists R^s, R^t, R^u, R^v \in \mathscr{D}|A)(a P^s b P^s c \wedge b P^t a P^t c \wedge c I^u a P^u b \wedge a P^v c I^v b)$.

First consider the case when $\mathscr{D}|A$ contains $a P^s b P^s c$ and $b P^t c P^t a$ for some distinct $a, b, c \in A$. $\mathscr{D}|A$ would violate UVR iff (i) there is an ordering in $\mathscr{D}|A$ in which $a$ is uniquely medium, or (ii) there is an ordering in $\mathscr{D}|A$ in which $c$ is uniquely best and there is an ordering in $\mathscr{D}|A$ in which $b$ is uniquely worst, or (iii) there is an ordering in $\mathscr{D}|A$ in which $c$ is uniquely best and there is an ordering in $\mathscr{D}|A$ in which $b$ is worst without $a$ being best, or (iv) there is an ordering in $\mathscr{D}|A$ in which $b$ is uniquely worst and there is an ordering in $\mathscr{D}|A$ in which $c$ is best without $a$ being worst, or (v) there is an ordering in $\mathscr{D}|A$ in which $b$ is worst without $a$ being best and there is an ordering in $\mathscr{D}|A$ in which $c$ is best without $a$ being worst. That is to say, $\mathscr{D}|A$ which includes $a P^s b P^s c$ and $b P^t c P^t a$ would violate UVR iff it includes $[(b P^u a P^u c \vee c P^u a P^u b) \vee [(c P^u a P^u b \vee c P^u a I^u b \vee c P^u b P^u a) \wedge (c P^v a P^v b \vee c I^v a P^v b \vee a P^v c P^v b)] \vee [(c P^u a P^u b \vee c P^u a I^u b \vee c P^u b P^u a) \wedge (c P^v a P^v b \vee c P^v a I^v b)] \vee [(c P^u a P^u b \vee c I^u a P^u b \vee a P^u c P^u b) \wedge (c P^v a P^v b \vee c I^v a P^v b)] \vee [(c P^u a P^u b \vee c P^u a I^u b) \wedge (c P^v a P^v b \vee c I^v a P^v b)]]$. Excepting the cases when we have $[(a P^s b P^s c \wedge b P^t c P^t a \wedge b P^u a P^u c) \vee (a P^s b P^s c \wedge b P^t c P^t a \wedge c P^u b P^u a \wedge a P^v c P^v b)]$, in all other cases LSUVR is violated. $(a P^s b P^s c \wedge b P^t c P^t a \wedge b P^u a P^u c)$ does not violate VR(2); VR(2) would be violated only if $\mathscr{D}|A$ includes [(concerned $c R^v a R^v b) \vee$ (concerned $a R^v c R^v b \wedge$ concerned $c R^w b R^w a$)]. With the required inclusion, in each case LSUVR is violated. $(a P^s b P^s c \wedge b P^t c P^t a \wedge c P^u b P^u a \wedge a P^v c P^v b)$ does not violate VR(2); VR(2) would be violated only if $\mathscr{D}|A$ includes (concerned $c R^v a R^v b \vee$ concerned $b R^v a R^v c)$. With the required inclusion, in each case LSUVR is violated. If $\mathscr{D}|A$ contains $(a P^s b P^s c \wedge b P^t c I^t a \wedge c I^u a P^u b)$ then LSUVR is violated.

If $\mathscr{D}|A$ contains $(aP^sbP^sc \wedge aP^tcP^tb \wedge cI^uaP^ub \wedge cP^vbI^va)$ then it would violate UVR iff an ordering in which $a$ is uniquely medium or an ordering in which $b$ is uniquely best or an ordering in which $b$ is best without $a$ being worst is included, i.e., $[bP^waP^wc \vee cP^waP^wb \vee bP^wcP^wa \vee bP^wcI^wa \vee bI^waP^wc]$ is included. In all cases other than that of $(aP^sbP^sc \wedge aP^tcP^tb \wedge cI^uaP^ub \wedge cP^vbI^va \wedge cP^waP^wb)$ LSUVR is violated. $(aP^sbP^sc \wedge aP^tcP^tb \wedge cI^uaP^ub \wedge cP^vbI^va \wedge cP^waP^wb)$ does not violate VR(2). VR(2) would be violated iff (concerned $bR^k cR^ka \vee$ concerned $bR^kaR^kc$) is included. With the inclusion of the required ordering LSUVR is violated.

If $\mathscr{D}|A$ contains $(aP^sbP^sc \wedge bP^taP^tc \wedge cI^uaP^ub \wedge aP^vcI^vb)$ then it would violate UVR iff an ordering in which $c$ is uniquely medium or an ordering in which $a$ is uniquely worst or an ordering in which $a$ is worst without $c$ being best is included, i.e., $[aP^wcP^wb \vee bP^wcP^wa \vee bI^wcP^wa \vee cP^wbP^wa \vee bP^wcI^wa]$ is included. In all cases other than that of $(aP^sbP^sc \wedge bP^taP^tc \wedge cI^uaP^ub \wedge aP^vcI^vb \wedge aP^wcP^wb)$ LSUVR is violated. $(aP^sbP^sc \wedge bP^taP^tc \wedge cI^uaP^ub \wedge aP^vcI^vb \wedge aP^wcP^wb)$ does not violate VR(2). VR(2) would be violated iff (concerned $bR^kcR^ka \vee$ concerned $cR^kbR^ka$) is included. With the inclusion of the required ordering LSUVR is violated.

Thus if all three restrictions of VR(2), WCP and UVR are violated then LSUVR must be violated.

This together with the earlier demonstration that violation of LSUVR implies violation of all three restrictions of VR(2), WCP and UVR, establishes the proposition.                                                                                      □

---

## 4.7   Notes on Literature

The characterization of non-minority rule given here is based on Jain (1994). The condition which ensures transitivity under the non-minority rule was formulated by Fine (1973). Fine's analysis is essentially equivalent to establishing Theorem 4.2 given here. The derivation of transitivity condition for the case of odd number of individuals is based on Jain (1989). The condition for quasi-transitivity was first derived in Jain (1984); in it, it was shown that the union of VR(2), WCP, and UVR is an Inada-type necessary and sufficient condition for quasi-transitivity under the non-minority rule.

# References

Fine, Kit. 1973. Conditions for the existence of cycles under majority and non-minority rules. *Econometrica* 41: 888–899.

Jain, Satish K. 1984. Non-minority rules: Characterization of configurations with rational social preferences. *Keio Economic Studies* 21: 45–54.

Jain, Satish K. 1989. Characterization theorems for social decision rules which are simple games. Paper presented at the IX world congress of the International Economic Association, economics research center, Athens School of Economics & Business, Athens, Greece, held on 28 August–September 1, 1989.

Jain, Satish K. 1994. Characterization of non-minority rules. DSA Working Paper 12/94. Centre for Economic Studies and Planning, Jawaharlal Nehru University.

# The Class of Semi-strict Majority Rules

**5**

For an alternative $x$ to be socially better than some other alternative $y$, under the method of majority decision only a majority of nonindifferent individuals are required to have $x$ preferred over $y$; but under the strict majority rule a majority of all individuals must have $x$ preferred over $y$ for $x$ to be socially better than $y$. The class of semi-strict majority rules falls in between these two extremes. Under $p$-semi-strict majority rule, where $0 < p < 1$, for $x$ to be socially better than $y$ the number of individuals preferring $x$ over $y$ must be greater than half of [$p$(number of individuals nonindifferent between $x$ and $y$) + $(1 - p)$ (number of all individuals)]. When $p$ is close to 1 then the $p$-semi-strict majority rule is close to the simple majority rule and when $p$ is close to 0 then the $p$-semi-strict majority rule is close to the strict majority rule.

In the context of transitivity under the class of semi-strict majority rules three conditions on preferences turn out to be relevant. These are: strict placement restriction, partial agreement and strongly antagonistic preferences (1). A set of orderings of a triple of alternatives satisfies the condition of strict placement restriction iff there is an alternative in the triple such that it is uniquely best in every concerned ordering; or there is an alternative such that it is uniquely worst in every concerned ordering; or there is an alternative such that it is uniquely medium in every concerned ordering; or there are two distinct alternatives such that in every ordering indifference holds between them. A set of orderings of a triple of alternatives satisfies the condition of partial agreement iff all orderings are concerned and there is an alternative such that it is best in every ordering; or all orderings are concerned and there is an alternative such that it is worst in every ordering. A set of orderings of a triple of alternatives satisfies the condition of strongly antagonistic preferences (1) iff it is a subset of $\{xPyPz, zPyPx, yPxIz\}$ or of $\{xPyPz, zPyPx, xIzPy\}$ for some distinct $x, y, z$ belonging to the triple. Let $\mathscr{D}$ be a set of orderings of the set of alternatives. It is shown in Theorem 5.1 of the chapter that if $\mathscr{D}$ is such that over every triple of alternatives at least one of three conditions of strict placement restriction, partial agreement and strongly antagonistic preferences (1) holds then

© Springer Nature Singapore Pte Ltd. 2019
S. K. Jain, *Domain Conditions and Social Rationality*,
https://doi.org/10.1007/978-981-13-9672-4_5

$p$-semi-strict majority rule yields transitive social weak preference relation for every profile of individual orderings belonging to $\mathscr{D}^n$. This holds regardless of the value of $p$ and regardless of the number of individuals. Furthermore, in Theorem 5.2 it is shown that for every $p$ there exists an $n$ such that $p$-semi-strict majority rule yields transitive social weak preference relation for every profile of individual orderings belonging to $\mathscr{D}^n$ iff $\mathscr{D}$ is such that over every triple of alternatives at least one of three conditions of strict placement restriction, partial agreement and strongly antagonistic preferences (1) holds. This implies that the satisfaction of at least one of the three conditions of strict placement restriction, partial agreement and strongly antagonistic preferences (1) over every triple of alternatives is maximally sufficient for transitivity under every semi-strict majority rule.

In the case of simple majority rule as well as in the case of strict majority rule we saw that the conditions that ensure transitivity when the number of individuals is odd are less stringent than the conditions that ensure transitivity when the number of individuals is even. A careful reading of the proof of Theorem 5.2 makes it clear that the same holds for the class of semi-strict majority rules as well. Theorem 5.2 does not merely show the existence of an $n$ for which satisfaction of at least one of the three conditions of strict placement restriction, partial agreement and strongly antagonistic preferences (1) over every triple of alternatives completely characterizes all $\mathscr{D}$ with the property that every profile of individual orderings belonging to $\mathscr{D}^n$ yields transitive social weak preference relation. What it shows is that for every $p$ there exists an even integer $n'$ such that for all even integers $n \geq n'$ satisfaction of at least one of the three conditions of strict placement restriction, partial agreement and strongly antagonistic preferences (1) over every triple of alternatives completely characterizes all $\mathscr{D}$ with the property that every profile of individual orderings belonging to $\mathscr{D}^n$ yields transitive social weak preference relation.

There are two conditions which turn out to be sufficient for quasi-transitivity under every semi-strict majority rule, namely, value restriction (2) and absence of unique extremal value. The satisfaction of the condition of absence of unique extremal value by a set of orderings of a triple requires that there be no alternative such that it is uniquely best in some ordering of the set or that there be no alternative such that it is uniquely worst in some ordering of the set. Also, for every $p \in (0, 1)$, there are infinitely many values of $n$ such that $p$-semi-strict majority rule yields quasi-transitive social weak preference relation for every profile of individual orderings belonging to $\mathscr{D}^n$ iff $\mathscr{D}$ is such that over every triple of alternatives at least one of the two conditions of value restriction (2) and absence of unique extremal value holds. Consequently it follows that for every $p$-semi-strict majority rule, satisfaction of at least one of the two conditions of value restriction (2) and absence of unique extremal value over every triple of alternatives is maximally sufficient for quasi-transitivity.

Conditions that ensure transitivity or quasi-transitivity under semi-strict majority rules display some interesting features. Dichotomous preferences are sufficient to ensure quasi-transitivity under both the simple majority rule and strict majority rule. However, there is no $p$-semi-strict majority rule for which satisfaction of the condition of dichotomous preferences is a sufficient condition for quasi-transitivity regardless of the value of $n$. Similar remarks apply to the case of transitivity as well.

The chapter is divided into three sections. Section 5.1 contains definitions of absence of unique extremal value, strict placement restriction, partial agreement and strongly antagonistic preferences (1). Sections 5.2 and 5.3 discuss for the class of semi-strict majority rules conditions for transitivity and quasi-transitivity, respectively.[1]

## 5.1  Restrictions on Preferences

The Class of Semi-Strict Majority Rules: Let $0 < p < 1$. $p$-semi-strict majority rule $f : \mathscr{T}^n \mapsto \mathscr{C}$ is defined by: $(\forall(R_1, \ldots, R_n) \in \mathscr{T}^n)(\forall x, y \in S)[x P y \leftrightarrow n(x P_i y) > \frac{1}{2}[p[n(x P_i y) + n(y P_i x)] + (1 - p)n]$. The class of semi-strict majority rules consists of all $p$-semi-strict majority rules defined for all possible values of $p \in (0, 1)$ and all positive integers $n$.

We define and discuss several restrictions on sets of orderings.

Absence of Unique Extremal Value (AUEV): $\mathscr{D} \subseteq \mathscr{T}$ satisfies AUEV over the triple $A \subseteq S$ iff $\sim (\exists$ distinct $x, y, z \in A)(\exists R \in \mathscr{D}|A)(x P y \wedge x P z) \vee \sim (\exists$ distinct $x, y, z \in A)(\exists R \in \mathscr{D}|A)(y P x \wedge z P x)$. $\mathscr{D}$ satisfies AUEV iff it satisfies AUEV over every triple contained in $S$.

Thus, the satisfaction of absence of unique extremal value by $\mathscr{D}$ over the triple $A$ requires that there be no alternative such that it is uniquely best in some $R \in \mathscr{D}|A$ or that there be no alternative such that it is uniquely worst in some $R \in \mathscr{D}|A$.

Strict Placement Restriction (SPR): $\mathscr{D} \subseteq \mathscr{T}$ satisfies SPR over the triple $A \subseteq S$ iff $(\exists$ distinct $x, y, z \in A)[(\forall$ concerned $R \in \mathscr{D}|A)(x P y \wedge x P z) \vee (\forall$ concerned $R \in \mathscr{D}|A)(y P x \wedge z P x) \vee (\forall$ concerned $R \in \mathscr{D}|A)(y P x P z \vee z P x P y) \vee (\forall R \in \mathscr{D}|A)(x I y)]$. $\mathscr{D}$ satisfies SPR iff it satisfies SPR over every triple contained in $S$.

Thus, the satisfaction of strict placement restriction by $\mathscr{D}$ over the triple $A$ requires that there be an alternative such that it is uniquely best in every concerned $R \in \mathscr{D}|A$; or there be an alternative such that it is uniquely worst in every concerned $R \in \mathscr{D}|A$; or there be an alternative such that it is uniquely medium in every concerned $R \in \mathscr{D}|A$; or there be two distinct alternatives such that in every $R \in \mathscr{D}|A$ indifference holds between them.

Partial Agreement (PA): $\mathscr{D} \subseteq \mathscr{T}$ satisfies PA over the triple $A \subseteq S$ iff $(\exists$ distinct $x, y, z \in A)[\sim (\exists R \in \mathscr{D}|A)(x I y I z) \wedge (\forall R \in \mathscr{D}|A)(x R y \wedge x R z)] \vee (\exists$ distinct $x, y, z \in A)[\sim (\exists R \in \mathscr{D}|A)(x I y I z) \wedge (\forall R \in \mathscr{D}|A)(y R x \wedge z R x)]$. $\mathscr{D}$ satisfies PA iff it satisfies PA over every triple contained in $S$.

Thus, $\mathscr{D}$ satisfies partial agreement over the triple $A$ iff all $R \in \mathscr{D}|A$ are concerned over $A$ and there is an alternative such that it is best in every $R \in \mathscr{D}|A$; or all $R \in \mathscr{D}|A$ are concerned over $A$ and there is an alternative such that it is worst in every $R \in \mathscr{D}|A$.

Strongly Antagonistic Preferences (1) (SAP(1)): $\mathscr{D} \subseteq \mathscr{T}$ satisfies SAP(1) over the triple $A \subseteq S$ iff $(\exists$ distinct $x, y, z \in A)[(\forall R \in \mathscr{D}|A)(x P y P z \vee z P y P x \vee y P x I z)] \vee$

---

[1]This chapter relies on Jain (1986).

($\exists$ distinct $x, y, z \in A)[(\forall R \in \mathscr{D}|A)(x P y P z \vee z P y P x \vee x I z P y)]$.   $\mathscr{D}$   satisfies SAP(1) iff it satisfies SAP(1) over every triple contained in $S$.

## 5.2   Conditions for Transitivity

**Theorem 5.1** *Let $f : \mathscr{T}^n \mapsto \mathscr{C}$ be a p-semi-strict majority rule. Let $\mathscr{D} \subseteq \mathscr{T}$. If for every triple $A \subseteq S$, $\mathscr{D} \mid A$ satisfies strict placement restriction or partial agreement or the condition of strongly antagonistic preferences (1) then $f$ yields transitive social $R$, $R = f(R_1, \ldots, R_n)$, for every $(R_1, \ldots, R_n) \in \mathscr{D}^n$.*

*Proof* Let $\mathscr{D} \subseteq \mathscr{T}$. Let $(R_1, \ldots, R_n) \in \mathscr{D}^n$. Suppose $R = f(R_1, \ldots, R_n)$ violates transitivity. Then we must have $x R y \wedge y R z \wedge \sim x R z$, i.e., $x R y \wedge y R z \wedge z P x$ for some $x, y, z \in S$. Let $\{x, y, z\}$ be designated by $A$.

$x R y \rightarrow n(y P_i x) \leq \frac{1}{2}[p(n(x P_i y) + n(y P_i x)) + (1 - p)n]$

$\rightarrow n(y P_i x) \leq \frac{1}{2} p[n(R_i$ concerned over $A) - n(R_i$ concerned over $A \wedge x I_i y)] + \frac{1}{2}$ $(1 - p)[n(R_i$ concerned over $A) + n(R_i$ unconcerned over $A)]$

$\rightarrow n(R_i$ concerned over $A \wedge x R_i y) \geq \frac{1}{2} n(R_i$ concerned over $A) + \frac{1}{2} pn$

$(R_i$ concerned over $A \wedge x I_i y)] - \frac{1}{2}(1 - p)n(R_i$ unconcerned over $A)$         (T5.1-1)

$\rightarrow n(x R_i y) \geq \frac{1}{2}n + \frac{1}{2}pn(R_i$ concerned over $A \wedge x I_i y) + \frac{1}{2}pn(R_i$ unconcerned over $A)$         (T5.1-2)

Similarly,

$y R z \rightarrow n(R_i$ concerned over $A \wedge y R_i z) \geq \frac{1}{2}n(R_i$ concerned over $A) + \frac{1}{2}pn(R_i$ concerned over $A \wedge y I_i z)] - \frac{1}{2}(1 - p)n(R_i$ unconcerned over $A)$

(T5.1-3)

$\rightarrow n(y R_i z) \geq \frac{1}{2}n + \frac{1}{2}pn(R_i$ concerned over $A \wedge y I_i z) + \frac{1}{2}pn(R_i$ unconcerned over $A)$         (T5.1-4)

$z P x \rightarrow n(z P_i x) > \frac{1}{2}[p(n(x P_i z) + n(z P_i x)) + (1 - p)n]$

$\rightarrow n(z P_i x) > \frac{1}{2}n(R_i$ concerned over $A) - \frac{1}{2}pn(R_i$ concerned over $A \wedge x I_i z) + \frac{1}{2}(1 - p)n(R_i$ unconcerned over $A)$

$\rightarrow n(R_i$ concerned over $A \wedge z R_i x) > \frac{1}{2}n(R_i$ concerned over $A) + (1 - \frac{1}{2}p)n(R_i$ concerned over $A \wedge x I_i z) + \frac{1}{2}(1 - p)n(R_i$ unconcerned over $A)$         (T5.1-5)

$\rightarrow n(z R_i x) > \frac{1}{2}n + (1 - \frac{1}{2}p)n(R_i$ concerned over $A \wedge x I_i z) + \frac{1}{2}(2 - p)n(R_i$ unconcerned over $A)$         (T5.1-6)

(T5.1-1)$\rightarrow n(R_i$ concerned over $A \wedge x R_i y) \geq \frac{1}{2}n(R_i$ concerned over $A) - \frac{1}{2}(1 - p)n(R_i$ unconcerned over $A)$         (T5.1-7)

(T5.1-3) $\to n(R_i$ concerned over $A \wedge yR_iz) \geq \frac{1}{2}n(R_i$ concerned over $A) - \frac{1}{2}(1 - p)n(R_i$ unconcerned over $A)$ (T5.1-8)

(T5.1-5) $\to n(R_i$ concerned over $A \wedge zR_ix) > \frac{1}{2}n(R_i$ concerned over $A) + \frac{1}{2}(1 - p)n(R_i$ unconcerned over $A)$ (T5.1-9)

(T5.1-8) $\wedge$ (T5.1-9) $\to (\exists i \in N)(i$ is concerned over $A \wedge yR_izR_ix)$ (T5.1-10)

(T5.1-7) $\wedge$ (T5.1-9) $\to (\exists i \in N)(i$ is concerned over $A \wedge zR_ixR_iy)$ (T5.1-11)

$zPx \to (\exists i \in N)(zP_ix)$ (T5.1-12)

(T5.1-10)–(T5.1-12) imply that SPR is violated. (T5.1-13)

(T5.1-2) $\wedge (\exists i \in N)(R_i$ is concerned over $A \wedge xI_iy) \to n(xR_iy) > \frac{1}{2}n$

(T5.1-14)

(T5.1-4) $\wedge (\exists i \in N)(R_i$ is concerned over $A \wedge yI_iz) \to n(yR_iz) > \frac{1}{2}n$

(T5.1-15)

From (T5.1-14) we obtain:

(T5.1-2) $\wedge$ (T5.1-4) $\wedge (\exists i \in N)(R_i$ is concerned over $A \wedge xI_iy) \to (\exists i \in N)$ $(xR_iyR_iz)$ (T5.1-16)

From (T5.1-15) we obtain:

(T5.1-2) $\wedge$ (T5.1-4) $\wedge (\exists i \in N)(R_i$ is concerned over $A \wedge yI_iz) \to (\exists i \in N)$ $(xR_iyR_iz)$ (T5.1-17)

(T5.1-16) $\to [\sim (\exists i \in N)(xP_iy) \to n(R_i$ is unconcerned over $A) \neq 0 \vee (\exists i \in N)$ $(xP_iz)]$ (T5.1-18)

(T5.1-17) $\to [\sim (\exists i \in N)(yP_iz) \to n(R_i$ is unconcerned over $A) \neq 0 \vee (\exists i \in N)$ $(xP_iz)]$ (T5.1-19)

From (T5.1-18) and (T5.1-19) we conclude:

$n(R_i$ is unconcerned over $A) = 0 \wedge \sim (\exists i \in N)(xP_iz) \to (\exists i \in N)(xP_iy) \wedge (\exists i \in N)(yP_iz)$ (T5.1-20)

(T5.1-20) is equivalent to:

$n(R_i$ is unconcerned over $A) \neq 0 \vee (\exists i \in N)(xP_iz) \vee [(\exists i \in N)(xP_iy) \wedge (\exists i \in N)(yP_iz)]$ (T5.1-21)

(T5.1-10), (T5.1-11), and (T5.1-21) imply that PA is violated. (T5.1-22)

Suppose SAP(1) holds. Then from (T5.1-10) and (T5.1-11) we conclude that $(\exists i \in N)(R_i$ is concerned over $A \wedge xI_iy) \vee (\exists i \in N)(R_i$ is concerned over $A \wedge yI_iz)$. By (T5.1-16) and (T5.1-17), this implies that $(\exists i \in N)(xR_iyR_iz)$. But this contradicts our supposition that SAP(1) holds, in view of (T5.1-10) and (T5.1-11).

This contradiction establishes that SAP(1) must be violated.                    (T5.1-23)

(T5.1-13), (T5.1-22), and (T5.1-23) establish the theorem.                               □

**Lemma 5.1** *Let $\mathscr{D} \subseteq \mathscr{T}$. Let $A \subseteq S$ be a triple of alternatives. Then $\mathscr{D}$ violates the condition of strict placement restriction over $A$ iff $\mathscr{D}|A$ contains one of the following eight sets of orderings of $A$, except for a formal interchange of alternatives.*

| | | | |
|---|---|---|---|
| $(i)$ | $x P^s y P^s z$ <br> $y P^t z P^t x$ | $(ii)$ | $x P^s y P^s z$ <br> $y I^t z P^t x$ |
| $(iii)$ | $x P^s y P^s z$ <br> $z P^t x I^t y$ | $(iv)$ | $x P^s y P^s z$ <br> $y P^t z I^t x$ |
| $(v)$ | $x P^s y P^s z$ <br> $z I^t x P^t y$ | $(vi)$ | $x P^s y I^s z$ <br> $y P^t z I^t x$ |
| $(vii)$ | $x I^s y P^s z$ <br> $y I^t z P^t x$ | $(viii)$ | $x P^s y I^s z$ <br> $x I^t y P^t z$ |

*Proof* It can be easily checked that there does not exist an alternative such that it is uniquely best in every $R^i \in \mathscr{D}|A$ iff $\mathscr{D}|A$ contains one of the following seven sets of orderings of $A$, except for a formal interchange of alternatives:

| | | | |
|---|---|---|---|
| $(A)$ | $x P^s y P^s z$ <br> $y P^t z P^t x$ | $(B)$ | $x P^s y P^s z$ <br> $y P^t x P^t z$ |
| $(C)$ | $x P^s y P^s z$ <br> $z P^t y P^t x$ | $(D)$ | $x P^s y P^s z$ <br> $y P^t z I^t x$ |
| $(E)$ | $x P^s y P^s z$ <br> $z P^t x I^t y$ | $(F)$ | $x P^s y I^s z$ <br> $y P^t z I^t x$ |
| $(G)$ | $x I^s y P^s z$ | | |

A, D, E and F are the same as (i), (iv), (iii) and (vi), respectively. Therefore it suffices to consider only the remaining three sets.

(B) does not violate SPR as $z$ is uniquely worst in both the orderings; it would be violated iff a concerned ordering in which $z$ is not uniquely worst is included. With the inclusion of such an ordering, except for a formal interchange of alternatives, one of the sets (i)–(viii) is contained in the set of orderings.

(C) does not violate SPR as $y$ is uniquely medium in both the orderings; it would be violated iff a concerned ordering in which $y$ is not uniquely medium is included. With the inclusion of such an ordering, except for a formal interchange of alternatives, one of the sets (i)–(viii) is contained in the set of orderings.

In the single ordering of (G) $z$ is uniquely worst. So, SPR would be violated only if a concerned ordering in which $z$ is not uniquely worst is included. With the inclusion of such an ordering, excepting the case when the included ordering is $zP^txI^ty$, we obtain one of the sets (i)–(viii), possibly with a formal interchange of alternatives. The set $\{xI^syP^sz, zP^txI^ty\}$ does not violate SPR as in both the orderings we have $xIy$. SPR would be violated iff an ordering in which $x$ is not indifferent to $y$ is included. With the inclusion of such an ordering one of the sets (i)–(viii) is contained in the set of orderings, possibly with a formal interchange of alternatives.

The proof of the lemma is completed by noting that all the eight sets (i)–(viii) violate SPR.                                                                                    □

**Lemma 5.2** *Let $\mathscr{D} \subseteq \mathscr{T}$. Let $A \subseteq S$ be a triple of alternatives. Then $\mathscr{D}$ violates all three conditions of strict placement restriction, partial agreement and strongly antagonistic preferences (1) over $A$ iff $\mathscr{D}|A$ contains one of the following 13 sets of orderings of $A$, except for a formal interchange of alternatives.*

$$
\begin{array}{llll}
(i) & \begin{array}{l} xP^syP^sz \\ yP^tzP^tx \end{array} &
(ii) & \begin{array}{l} xP^syP^sz \\ yI^tzP^tx \end{array} \\[4ex]

(iii) & \begin{array}{l} xP^syP^sz \\ zP^txI^ty \end{array} &
(iv) & \begin{array}{l} xP^syP^sz \\ yP^tzI^tx \\ xI^uyI^uz \end{array} \\[6ex]

(v) & \begin{array}{l} xP^syP^sz \\ zI^txP^ty \\ xI^uyI^uz \end{array} &
(vi) & \begin{array}{l} xP^syP^sz \\ yP^tzI^tx \\ zI^uxP^uy \end{array} \\[6ex]

(vii) & \begin{array}{l} xP^syI^sz \\ yP^tzI^tx \\ xI^uyI^uz \end{array} &
(viii) & \begin{array}{l} xP^syI^sz \\ yP^tzI^tx \\ zP^uxI^uy \end{array} \\[6ex]

(ix) & \begin{array}{l} xI^syP^sz \\ yI^tzP^tx \\ xI^uyI^uz \end{array} &
(x) & \begin{array}{l} xI^syP^sz \\ yI^tzP^tx \\ zI^uxP^uy \end{array} \\[6ex]

(xi) & \begin{array}{l} xP^syI^sz \\ xI^tyP^tz \\ xI^uyI^uz \end{array} &
(xii) & \begin{array}{l} xP^syI^sz \\ yP^tzI^tx \\ zI^uxP^uy \end{array} \\[6ex]

(xiii) & \begin{array}{l} xP^syI^sz \\ yI^tzP^tx \\ zI^uxP^uy \end{array}
\end{array}
$$

*Proof* From Lemma 5.1, $\mathscr{D} \mid A$ would violate SPR iff it contains one of the following eight sets of orderings of $A$, except for a formal interchange of alternatives.

$(A)$ $\;x P^s y P^s z$
$\qquad\; y P^t z P^t x$

$(B)$ $\;x P^s y P^s z$
$\qquad\; y I^t z P^t x$

$(C)$ $\;x P^s y P^s z$
$\qquad\; z P^t x I^t y$

$(D)$ $\;x P^s y P^s z$
$\qquad\; y P^t z I^t x$

$(E)$ $\;x P^s y P^s z$
$\qquad\; z I^t x P^t y$

$(F)$ $\;x P^s y I^s z$
$\qquad\; y P^t z I^t x$

$(G)$ $\;x I^s y P^s z$
$\qquad\; y I^t z P^t x$

$(H)$ $\;x P^s y I^s z$
$\qquad\; x I^t y P^t z$

The sets A, B and C are the same as the first three sets, (i)–(iii), in the statement of the lemma, so it suffices to consider the remaining five sets.

The set D does not violate PA as $z$ is worst in both the orderings. PA would be violated iff an ordering in which $z$ is not worst is included or the unconcerned ordering $x I^s y I^s z$ is included, i.e., iff $(y P^u z P^u x \vee z P^u x P^u y \vee x P^u z P^u y \vee z P^u y P^u x \vee z P^u x I^u y \vee y I^u z P^u x \vee z I^u x P^u y \vee x I^u y I^u z)$ is included. Excepting the case of inclusion of $z P^u y P^u x$, in all other cases with the inclusion of the required ordering the enlarged set contains one of the sets, possibly with a formal interchange of alternatives, mentioned in the statement of the lemma. The set $(x P^s y P^s z \wedge y P^t z I^t x \wedge z P^u y P^u x)$ does not violate SAP(1); it would do so iff any other ordering is included in the set. With the inclusion of the required ordering the enlarged set contains one of the sets mentioned in the statement of the lemma, possibly with a formal interchange of alternatives.

Set $E = \{x P^s y P^s z, z I^t x P^t y\}$ does not violate PA as $x$ is best in both the orderings. PA would be violated iff an ordering in which $x$ is not best is included or the unconcerned ordering $x I^s y I^s z$ is included, i.e., iff $(y P^u z P^u x \vee z P^u x P^u y \vee z P^u y P^u x \vee y P^u x P^u z \vee y P^u z I^u x \vee z P^u x I^u y \vee y I^u z P^u x \vee x I^u y I^u z)$ is included. Excepting the case of inclusion of $z P^u y P^u x$, in all other cases with the inclusion of the required ordering the enlarged set contains one of the sets, possibly with a formal interchange of alternatives, mentioned in the statement of the lemma. The set $\{x P^s y P^s z, z I^t x P^t y, z P^u y P^u x\})$ does not violate SAP(1); it would do so iff any other ordering is included in the set. With the inclusion of the required ordering the enlarged set contains one of the sets mentioned in the statement of the lemma, possibly with a formal interchange of alternatives.

Set $F = \{x P^s y I^s z, y P^t z I^t x\}$ does not violate PA as $z$ is worst in both the orderings. PA would be violated iff an ordering in which $z$ is not worst is included or the unconcerned ordering $x I^s y I^s z$ is included, i.e., iff $(y P^u z P^u x \vee z P^u x P^u y \vee x P^u z P^u y \vee z P^u y P^u x \vee z P^u x I^u y \vee y I^u z P^u x \vee z I^u x P^u y \vee x I^u y I^u z)$ is included. With the inclusion of the required ordering the enlarged set contains one of the sets, possibly with a formal interchange of alternatives, mentioned in the statement of the lemma. $G = \{xI^s yP^s z, yI^t zP^t x\}$ does not violate PA as $y$ is best in both the orderings. PA would be violated iff an ordering in which $y$ is not best is included or the unconcerned ordering $xI^s yI^s z$ is included, i.e., iff $(xP^u yP^u z \vee zP^u xP^u y \vee xP^u zP^u y \vee zP^u yP^u x \vee xP^u yI^u z \vee zP^u xI^u y \vee zI^u xP^u y \vee xI^u yI^u z)$ is included. With the inclusion of the required ordering the enlarged set contains one of the sets, possibly with a formal interchange of alternatives, mentioned in the statement of the lemma. Set $H = \{xP^s yI^s z, xI^t yP^t z\}$ does not violate PA as $x$ is best and $z$ is worst in both the orderings. PA would be violated only if an ordering in which $x$ is not best is included or the unconcerned ordering $xI^s yI^s z$ is included, i.e., only if $(yP^u zP^u x \vee zP^u xP^u y \vee zP^u yP^u x \vee yP^u xP^u z \vee yP^u zI^u x \vee zP^u xI^u y \vee yI^u z P^u x \vee xI^u yI^u z)$ is included. Excepting the cases of inclusion of $yP^u xP^u z$ or of $yP^u zI^u x$, in all other cases with the inclusion of the required ordering the enlarged set contains one of the sets, possibly with a formal interchange of alternatives, mentioned in the statement of the lemma. The sets $\{xP^s yI^s z, xI^t yP^t z, yP^u xP^u z\}$ and $\{xP^s yI^s z, xI^t yP^t z, yP^u zI^u x\}$ do not violate PA as $z$ is worst in every ordering of each set. PA would be violated iff an ordering in which $z$ is not worst is included or the unconcerned ordering $xI^v yI^v z$ is included, i.e., $(yP^v zP^v x \vee zP^v xP^v y \vee xP^v zP^v y \vee zP^v yP^v x \vee zP^v xI^v y \vee yI^v zP^v x \vee zI^v xP^v y \vee xI^v yI^v z)$ is included. With the inclusion of the required ordering, in each of the two cases, the enlarged set contains one of the sets mentioned in the statement of the lemma, possibly with a formal interchange of alternatives.

The proof of the lemma is completed by noting that all the 13 sets mentioned in the statement of the lemma violate all the three conditions of SPR, PA and SAP(1).  □

**Theorem 5.2** *Let $f : \mathscr{T}^n \mapsto \mathscr{C}$ be a p-semi-strict majority rule. Let $n$ be an even positive integer greater than $\frac{(4+p)(2+p)}{p(1-p)^2}$. Let $\mathscr{D} \subseteq \mathscr{T}$. Then $f$ yields transitive social $R$, $R = f(R_1, \ldots, R_n)$, for every $(R_1, \ldots, R_n) \in \mathscr{D}^n$ iff for every triple $A \subseteq S$, $\mathscr{D} \mid A$ satisfies at least one of the three conditions of strict placement restriction, partial agreement and strongly antagonistic preferences (1).*

*Proof* The assertion that if for every triple $A \subseteq S$, $\mathscr{D} \mid A$ satisfies SPR or PA or SAP(1) then every $(R_1, \ldots, R_n) \in \mathscr{D}^n$ yields transitive social $R$ follows from Theorem 5.1.

Let $\mathscr{D}$ violate all three restrictions SPR, PA, and SAP(1) over the triple $A$. Then $\mathscr{D} \mid A$ contains, for some distinct $x$, $y$, $z \in A$, one of the 13 sets mentioned in Lemma 5.2. We show that in each of the 13 cases there exists a profile $(R_1 \mid A, \ldots, R_n \mid A) \in (\mathscr{D} \mid A)^n$ which yields an $R$ violating transitivity.

(i) Let $\mathscr{D} \mid A \supseteq \{x P^s y P^s z, y P^t z P^t x\}$. Consider any $(R_1, \ldots, R_n) \in \mathscr{D}^n$ such that: $n(x P_i y P_i z) = n(y P_i z P_i x) = \frac{n}{2}$.

We obtain: $n(x P_i y) = n(y P_i x) = \frac{n}{2}$; $n(y P_i z) = n, n(z P_i y) = 0$; $n(x P_i z) = n(z P_i x)$ $= \frac{n}{2}$.

$n(x P_i y) = n(y P_i x) = \frac{n}{2} \to x I y$; $n(y P_i z) = n \to y P z$; $n(x P_i z) = n(z P_i x)$ $= \frac{n}{2} \to x I z$.

Thus $R \mid A = (x I y \wedge y P z \wedge x I z)$.

(ii) Let $\mathscr{D} \mid A \supseteq \{x P^s y P^s z, y I^t z P^t x\}$. Consider any $(R_1, \ldots, R_n) \in \mathscr{D}^n$ such that: $n(x P_i y P_i z) = n(y I^t z P^t x) = \frac{n}{2}$.

We obtain: $n(x P_i y) = n(y P_i x) = \frac{n}{2}$; $n(y P_i z) = \frac{n}{2}, n(z P_i y) = 0$; $n(x P_i z) = n(z P_i x)$ $= \frac{n}{2}$.

We          have:          $n(y P_i z) = \frac{n}{2} > \frac{1}{2}[p[n(y P_i z) + n(z P_i y)] + (1 - p)n] = \frac{1}{2}[p\frac{n}{2} +$ $(1 - p)n] = \frac{n}{2} - p\frac{n}{4}$. Therefore $y P z$.

$n(x P_i y) = n(y P_i x) = \frac{n}{2} \to x I y$; and $n(x P_i z) = n(z P_i x) = \frac{n}{2} \to x I z$.

Thus $R \mid A = (x I y \wedge y P z \wedge x I z)$.

(iii) Let $\mathscr{D} \mid A \supseteq \{x P^s y P^s z, z P^t x I^t y\}$. Consider any $(R_1, \ldots, R_n) \in \mathscr{D}^n$ such that: $n(x P_i y P_i z) = n(z P^t x I^t y) = \frac{n}{2}$.

We obtain: $n(x P_i y) = \frac{n}{2}, n(y P_i x) = 0$; $n(y P_i z) = n(z P_i y) = \frac{n}{2}$; $n(x P_i z) = n(z P_i x)$ $= \frac{n}{2}$.

We have: $n(x P_i y) = \frac{n}{2} > \frac{1}{2}[p[n(x P_i y) + n(y P_i x)] + (1 - p)n] = \frac{1}{2}[p\frac{n}{2} + (1 - p)n]$ $= \frac{n}{2} - p\frac{n}{4}$. Therefore $x P y$.

$n(y P_i z) = n(z P_i y) = \frac{n}{2} \to y I z$; and $n(x P_i z) = n(z P_i x) = \frac{n}{2} \to x I z$.

Thus $R \mid A = (x P y \wedge y I z \wedge x I z)$.

(iv) Let $\mathscr{D} \mid A \supseteq \{x P^s y P^s z, y P^t z I^t x, x I^u y I^u z\}$. Consider any $(R_1, \ldots, R_n) \in \mathscr{D}^n$ such that: $n(x P^s y P^s z) = \lfloor \frac{1-p}{2(2-p)}n \rfloor = \frac{1-p}{2(2-p)}n - \varepsilon, 0 \le \varepsilon < 1$; $n(x I^u y I^u z) = \lceil \frac{1}{(2-p)}n \rceil -$ $1 = \frac{1}{(2-p)}n + \delta - 1, 0 \le \delta < 1$; $n(y P^t z I^t x) = n - \lfloor \frac{1-p}{2(2-p)}n \rfloor - \lceil \frac{1}{(2-p)}n \rceil$ $+ 1 = \frac{1-p}{2(2-p)}n + \varepsilon - \delta + 1$.

We   obtain:   $n(x P_i y) = \frac{1-p}{2(2-p)}n - \varepsilon, n(y P_i x) = \frac{1-p}{2(2-p)}n + \varepsilon - \delta + 1, n(y P_i z) = \frac{1-p}{(2-p)}n - \delta + 1, n(z P_i y) = 0, n(x P_i z) = \frac{1-p}{2(2-p)}n - \varepsilon, n(z P_i x) = 0$.

We have:

$\frac{1}{2}[p[n(xP_iy) + n(yP_ix)] + (1-p)n] = \frac{1}{2}[p[\frac{1-p}{(2-p)}n + 1 - \delta] + (1-p)n] = \frac{1-p}{(2-p)}n + \frac{1}{2}p(1-\delta)$.

$n(xP_iy) - \frac{1}{2}[p[n(xP_iy) + n(yP_ix)] + (1-p)n] = \frac{1-p}{2(2-p)}n - \varepsilon - [\frac{1-p}{(2-p)}n + \frac{1}{2}p(1-\delta)] = -\frac{1-p}{2(2-p)}n - \varepsilon - \frac{1}{2}p(1-\delta) < 0$. Therefore $\sim xPy$.

$n(yP_ix) - \frac{1}{2}[p[n(xP_iy) + n(yP_ix)] + (1-p)n] = \frac{1-p}{2(2-p)}n + \varepsilon - \delta + 1 - [\frac{1-p}{(2-p)}n + \frac{1}{2}p(1-\delta)] = -\frac{1-p}{2(2-p)}n + \varepsilon + (1 - \frac{1}{2}p)(1-\delta) < 0$, as $n > \frac{(4-p)(2-p)}{1-p}$.
Therefore $\sim yPx$.

$\frac{1}{2}[p[n(yP_iz) + n(zP_iy)] + (1-p)n] = \frac{1}{2}[p[\frac{1-p}{(2-p)}n + 1 - \delta] + (1-p)n] = \frac{1-p}{(2-p)}n + \frac{1}{2}p(1-\delta)$.

$n(yP_iz) - \frac{1}{2}[p[n(yP_iz) + n(zP_iy)] + (1-p)n] = \frac{1-p}{(2-p)}n + 1 - \delta - \frac{1-p}{(2-p)}n - \frac{1}{2}p(1-\delta) = (1 - \frac{1}{2}p)(1-\delta) > 0$. Therefore $yPz$.

$\frac{1}{2}[p[n(xP_iz) + n(zP_ix)] + (1-p)n] = \frac{1}{2}[p[\frac{1-p}{2(2-p)}n - \varepsilon] + (1-p)n] = \frac{(1-p)(4-p)}{4(2-p)}n - \frac{1}{2}p\varepsilon$.

$n(xP_iz) - \frac{1}{2}[p[n(xP_iz) + n(zP_ix)] + (1-p)n] = \frac{1-p}{2(2-p)}n - \varepsilon - \frac{(1-p)(4-p)}{4(2-p)}n + \frac{1}{2}p\varepsilon = -\frac{(1-p)}{4}n - (1 - \frac{1}{2}p)\varepsilon < 0$. Therefore $\sim xPz$.

$n(zP_ix) = 0 \rightarrow \sim zPx$.

Thus $R \mid A = (xIy \wedge yPz \wedge xIz)$.

(v) Let $\mathscr{D} \mid A \supseteq \{xP^syP^sz, zI^txP^ty, xI^uyI^uz\}$. Consider any $(R_1, \ldots, R_n) \in \mathscr{D}^n$ such that: $n(xP^syP^sz) = \lfloor\frac{1-p}{2(2-p)}n\rfloor = \frac{1-p}{2(2-p)}n - \varepsilon, 0 \le \varepsilon < 1; n(xI^uyI^uz) = \lceil\frac{1}{(2-p)}n\rceil - 1 = \frac{1}{(2-p)}n + \delta - 1, 0 \le \delta < 1; n(zI^txP^ty) = n - \lfloor\frac{1-p}{2(2-p)}n\rfloor - \lceil\frac{1}{(2-p)}n\rceil + 1 = \frac{1-p}{2(2-p)}n + \varepsilon - \delta + 1$.

We obtain: $n(xP_iy) = \frac{1-p}{(2-p)}n - \delta + 1, n(yP_ix) = 0, n(yP_iz) = \frac{1-p}{2(2-p)}n - \varepsilon$, $n(zP_iy) = \frac{1-p}{2(2-p)}n + \varepsilon - \delta + 1, n(xP_iz) = \frac{1-p}{2(2-p)}n - \varepsilon, n(zP_ix) = 0$.

We have:

$\frac{1}{2}[p[n(xP_iy) + n(yP_ix)] + (1-p)n] = \frac{1}{2}[p[\frac{1-p}{(2-p)}n + 1 - \delta] + (1-p)n] = \frac{1-p}{(2-p)}n + \frac{1}{2}p(1-\delta)$.

$n(xP_iy) - \frac{1}{2}[p[n(xP_iy) + n(yP_ix)] + (1-p)n] = \frac{1-p}{(2-p)}n + 1 - \delta - \frac{1-p}{(2-p)}n - \frac{1}{2}p(1-\delta) = (1 - \frac{1}{2}p)(1-\delta) > 0$. Therefore $xPy$.

$\frac{1}{2}[p[n(yP_iz) + n(zP_iy)] + (1-p)n] = \frac{1}{2}[p[\frac{1-p}{(2-p)}n + 1 - \delta] + (1-p)n] = \frac{1-p}{(2-p)}n + \frac{1}{2}p(1-\delta)$.

$n(y P_i z) - \frac{1}{2}[p[n(y P_i z) + n(z P_i y)] + (1 - p)n] = \frac{1-p}{2(2-p)}n - \varepsilon - [\frac{1-p}{(2-p)}n + \frac{1}{2}$
$p(1 - \delta)] = -\frac{1-p}{2(2-p)}n - \varepsilon - \frac{1}{2}p(1 - \delta) < 0.$ Therefore $\sim y P z.$

$n(z P_i y) - \frac{1}{2}[p[n(y P_i z) + n(z P_i y)] + (1 - p)n] = \frac{1-p}{2(2-p)}n + \varepsilon - \delta + 1 - [\frac{1-p}{(2-p)}n$
$+ \frac{1}{2}p(1 - \delta)] = -\frac{1-p}{2(2-p)}n + \varepsilon + (1 - \frac{1}{2}p)(1 - \delta) < 0,$ as $n > \frac{(4-p)(2-p)}{1-p}.$ Therefore $\sim z P y.$

$\frac{1}{2}[p[n(x P_i z) + n(z P_i x)] + (1 - p)n] = \frac{1}{2}[p[\frac{1-p}{2(2-p)}n - \varepsilon] + (1 - p)n] = \frac{(1-p)(4-p)}{4(2-p)}$
$n - \frac{1}{2}p\varepsilon.$

$n(x P_i z) - \frac{1}{2}[p[n(x P_i z) + n(z P_i x)] + (1 - p)n] = \frac{1-p}{2(2-p)}n - \varepsilon - \frac{(1-p)(4-p)}{4(2-p)}n + \frac{1}{2}p\varepsilon = -\frac{(1-p)}{4}n - (1 - \frac{1}{2}p)\varepsilon < 0.$ Therefore $\sim x P z.$

$n(z P_i x) = 0 \rightarrow\, \sim z P x.$

Thus $R \mid A = (x P y \wedge y I z \wedge x I z).$

(vi) Let $\mathscr{D} \mid A \supseteq \{x P^s y P^s z, y P^t z I^t x, z I^u x P^u y\}.$ Consider any $(R_1, \ldots, R_n) \in \mathscr{D}^n$
such that: $n(x P^s y P^s z) = \lfloor \frac{1-p}{(2-p)}n \rfloor = \frac{1-p}{(2-p)}n - \varepsilon, 0 \le \varepsilon < 1; n(y P^t z I^t x) = \frac{n}{2} - \lfloor \frac{1-p}{(2-p)}n \rfloor = \frac{p}{2(2-p)}n + \varepsilon; n(z I^u x P^u y) = \frac{n}{2}.$

We obtain: $n(x P_i y) = \frac{n}{2} + \frac{1-p}{(2-p)}n - \varepsilon, n(y P_i x) = \frac{p}{2(2-p)}n + \varepsilon, n(y P_i z) = \frac{n}{2},$
$n(z P_i y) = \frac{n}{2}, n(x P_i z) = \frac{1-p}{(2-p)}n - \varepsilon, n(z P_i x) = 0.$

We have:

$n(x P_i y) = \frac{n}{2} + \frac{1-p}{(2-p)}n - \varepsilon > \frac{n}{2},$ as $n > \frac{2-p}{(1-p)}.$ Therefore $x P y.$

$n(y P_i z) = n(z P_i y) = \frac{n}{2} \rightarrow y I z.$

$\frac{1}{2}[p[n(x P_i z) + n(z P_i x)] + (1 - p)n] = \frac{1}{2}[p[\frac{1-p}{(2-p)}n - \varepsilon] + (1 - p)n] = \frac{(1-p)}{(2-p)}n - \frac{1}{2}p\varepsilon.$

$n(x P_i z) - \frac{1}{2}[p[n(x P_i z) + n(z P_i x)] + (1 - p)n] = \frac{1-p}{(2-p)}n - \varepsilon - \frac{(1-p)}{(2-p)}n + \frac{1}{2}p\varepsilon = -(1 - \frac{1}{2}p)\varepsilon < 0.$ Therefore $\sim x P z.$

$n(z P_i x) = 0 \rightarrow\, \sim z P x.$

Thus $R \mid A = (x P y \wedge y I z \wedge x I z).$

(vii) Let $\mathscr{D} \mid A \supseteq \{x P^s y I^s z, y P^t z I^t x, x I^u y I^u z\}.$ Consider any $(R_1, \ldots, R_n) \in \mathscr{D}^n$
such that: $n(x P^s y I^s z) = \lfloor \frac{1-p}{(2-p)}n \rfloor + 1 = \frac{1-p}{(2-p)}n - \varepsilon + 1, 0 \le \varepsilon < 1; n(y P^t z I^t x) = \lfloor \frac{1-p}{(2-p)}n \rfloor = \frac{1-p}{(2-p)}n - \varepsilon; n(x I^u y I^u z) = n - 2\lfloor \frac{1-p}{(2-p)}n \rfloor - 1 = \frac{p}{(2-p)}n + 2\varepsilon - 1.$

We obtain: $n(x P_i y) = \frac{1-p}{(2-p)}n - \varepsilon + 1, n(y P_i x) = \frac{1-p}{(2-p)}n - \varepsilon, n(y P_i z) = \frac{1-p}{(2-p)}n - \varepsilon, n(z P_i y) = 0, n(x P_i z) = \frac{1-p}{(2-p)}n - \varepsilon + 1, n(z P_i x) = 0.$

We have:

$\frac{1}{2}[p[n(xP_iy) + n(yP_ix)] + (1-p)n] = \frac{1}{2}[p[\frac{2(1-p)}{(2-p)}n + 1 - 2\varepsilon] + (1-p)n] = \frac{(1-p)(2+p)}{2(2-p)}n + \frac{1}{2}p(1-2\varepsilon)$.

$n(xP_iy) - \frac{1}{2}[p[n(xP_iy) + n(yP_ix)] + (1-p)n] = \frac{1-p}{(2-p)}n - \varepsilon + 1 - [\frac{(1-p)(2+p)}{2(2-p)}n + \frac{1}{2}p(1-2\varepsilon)] = -\frac{p(1-p)}{2(2-p)}n - \varepsilon(1-p) + (1-\frac{1}{2}p) < 0$, as $n > \frac{(2-p)^2}{p(1-p)}$. Therefore $\sim xPy$.

$n(yP_ix) < n(xP_iy) \to \sim yPx$.

$\frac{1}{2}[p[n(yP_iz) + n(zP_iy)] + (1-p)n] = \frac{1}{2}[p[\frac{(1-p)}{(2-p)}n - \varepsilon] + (1-p)n] = \frac{(1-p)}{(2-p)}n - \frac{1}{2}p\varepsilon$.

$n(yP_iz) - \frac{1}{2}[p[n(yP_iz) + n(zP_iy)] + (1-p)n] = \frac{1-p}{(2-p)}n - \varepsilon - \frac{1-p}{(2-p)}n + \frac{1}{2}p\varepsilon = -(1-\frac{1}{2}p)\varepsilon \leq 0$. Therefore $\sim yPz$.

$n(zP_iy) = 0 \to \sim zPy$.

$\frac{1}{2}[p[n(xP_iz) + n(zP_ix)] + (1-p)n] = \frac{1}{2}[p[\frac{(1-p)}{(2-p)}n - \varepsilon + 1] + (1-p)n] = \frac{(1-p)}{(2-p)}n + \frac{1}{2}p(1-\varepsilon)$.

$n(xP_iz) - \frac{1}{2}[p[n(xP_iz) + n(zP_ix)] + (1-p)n] = \frac{1-p}{(2-p)}n - \varepsilon + 1 - [\frac{(1-p)}{(2-p)}n + \frac{1}{2}p(1-\varepsilon)] = (1-\frac{1}{2}p)(1-\varepsilon) > 0$. Therefore $xPz$.

Thus $R \mid A = (xIy \wedge yIz \wedge xPz)$.

(viii) Let $\mathscr{D} \mid A \supseteq \{xP^syI^sz, yP^tzI^tx, zP^uxI^uy\}$. Consider any $(R_1, \ldots, R_n) \in \mathscr{D}^n$ such that: $n(xP^syI^sz) = \lceil\frac{1}{(2+p)}n\rceil + 1 = \frac{1}{(2+p)}n + \varepsilon + 1, 0 \leq \varepsilon < 1; n(yP^tzI^tx) = \lceil\frac{p}{(2+p)}n\rceil = \frac{p}{(2+p)}n + \delta, 0 \leq \delta < 1; n(zP^uxI^uy) = n - \lceil\frac{1}{(2+p)}n\rceil - \lceil\frac{p}{(2+p)}n\rceil - 1 = \frac{1}{(2+p)}n - 1 - \varepsilon - \delta$.

We obtain: $n(xP_iy) = \frac{1}{(2+p)}n + \varepsilon + 1, n(yP_ix) = \frac{p}{(2+p)}n + \delta, n(yP_iz) = \frac{p}{(2+p)}n + \delta, n(zP_iy) = \frac{1}{(2+p)}n - 1 - \varepsilon - \delta, n(xP_iz) = \frac{1}{(2+p)}n + \varepsilon + 1, n(zP_ix) = \frac{1}{(2+p)}n - 1 - \varepsilon - \delta$.

We have:

$\frac{1}{2}[p[n(xP_iy) + n(yP_ix)] + (1-p)n] = \frac{1}{2}[p[\frac{1+p}{(2+p)}n + 1 + \varepsilon + \delta] + (1-p)n] = \frac{1}{(2+p)}n + \frac{1}{2}p(1+\varepsilon+\delta)$.

$n(xP_iy) - \frac{1}{2}[p[n(xP_iy) + n(yP_ix)] + (1-p)n] = \frac{1}{(2+p)}n + \varepsilon + 1 - [\frac{1}{(2+p)}n + \frac{1}{2}p(1+\varepsilon+\delta)] = (1-\frac{1}{2}p)(1+\varepsilon) - \frac{1}{2}p\delta > 0$. Therefore $xPy$ holds.

$\frac{1}{2}[p[n(yP_iz) + n(zP_iy)] + (1-p)n] = \frac{1}{2}[p[\frac{1+p}{(2+p)}n - 1 - \varepsilon] + (1-p)n] = \frac{1}{(2+p)}n - \frac{1}{2}p(\varepsilon+1)$.

$n(y P_i z) - \frac{1}{2}[p[n(y P_i z) + n(z P_i y)] + (1-p)n] = \frac{p}{(2+p)}n + \delta - [\frac{1}{(2+p)}n -$

$\frac{1}{2}p(\varepsilon + 1)] = -\frac{1-p}{(2+p)}n + \delta + \frac{1}{2}p(\varepsilon + 1) < 0$, as $n > \frac{(1+p)(2+p)}{(1-p)}$. Therefore $\sim y P z$.

$n(z P_i y) - \frac{1}{2}[p[n(y P_i z) + n(z P_i y)] + (1-p)n] = \frac{1}{(2+p)}n - 1 - \varepsilon - \delta - [\frac{1}{(2+p)}n$

$- \frac{1}{2}p(\varepsilon + 1)] = -(1 - \frac{1}{2}p)(\varepsilon + 1) - \delta < 0$. Therefore $\sim z P y$.

$\frac{1}{2}[p[n(x P_i z) + n(z P_i x)] + (1-p)n] = \frac{1}{2}[p[\frac{2}{(2+p)}n - \delta] + (1-p)n] = \frac{p+2-p^2}{2(2+p)}n$

$- \frac{1}{2}p\delta$

$n(x P_i z) - \frac{1}{2}[p[n(x P_i z) + n(z P_i x)] + (1-p)n] = \frac{1}{(2+p)}n + \varepsilon + 1 - [\frac{p+2-p^2}{2(2+p)}n -$

$\frac{1}{2}p\delta] = -\frac{p(1-p)}{2(2+p)}n + 1 + \varepsilon + \frac{1}{2}p\delta < 0$, as $n > \frac{(4+p)(2+p)}{p(1-p)}$. Therefore $\sim x P z$.

$n(z P_i x) < n(x P_i z) \rightarrow \sim z P x$.

Thus $R \mid A = (x P y \wedge y I z \wedge x I z)$.

(ix) Let $\mathscr{D} \mid A \supseteq \{x I^s y P^s z, y I^t z P^t x, x I^u y I^u z\}$. Consider any $(R_1, \ldots, R_n) \in \mathscr{D}^n$ such that: $n(x I^s y P^s z) = \lfloor \frac{1-p}{(2-p)}n \rfloor + 1 = \frac{1-p}{(2-p)}n - \varepsilon + 1, 0 \le \varepsilon < 1; n(y I^t z P^t x) = \lfloor \frac{1-p}{(2-p)}n \rfloor = \frac{1-p}{(2-p)}n - \varepsilon; n(x I^u y I^u z) = n - 2\lfloor \frac{1-p}{(2-p)}n \rfloor - 1 = \frac{p}{(2-p)}n + 2\varepsilon - 1$.

We obtain: $n(x P_i y) = 0, n(y P_i x) = \frac{1-p}{(2-p)}n - \varepsilon, n(y P_i z) = \frac{1-p}{(2-p)}n - \varepsilon + 1$,

$n(z P_i y) = 0, n(x P_i z) = \frac{1-p}{(2-p)}n - \varepsilon + 1, n(z P_i x) = \frac{1-p}{(2-p)}n - \varepsilon$.

We have:

$\frac{1}{2}[p[n(x P_i y) + n(y P_i x)] + (1-p)n] = \frac{1}{2}[p[\frac{1-p}{(2-p)}n - \varepsilon] + (1-p)n] = \frac{(1-p)}{(2-p)}n - \frac{1}{2}p\varepsilon$.

$n(y P_i x) - \frac{1}{2}[p[n(x P_i y) + n(y P_i x)] + (1-p)n] = \frac{1-p}{(2-p)}n - \varepsilon - [\frac{(1-p)}{(2-p)}n - \frac{1}{2}p\varepsilon]$

$= -(1 - \frac{1}{2}p)\varepsilon \le 0$. Therefore $\sim y P x$.

$n(x P_i y) = 0 \rightarrow \sim x P y$.

$\frac{1}{2}[p[n(y P_i z) + n(z P_i y)] + (1-p)n] = \frac{1}{2}[p[\frac{(1-p)}{(2-p)}n - \varepsilon + 1] + (1-p)n] = \frac{(1-p)}{(2-p)}n + \frac{1}{2}p(1-\varepsilon)$.

$n(y P_i z) - \frac{1}{2}[p[n(y P_i z) + n(z P_i y)] + (1-p)n] = \frac{1-p}{(2-p)}n - \varepsilon + 1 - [\frac{(1-p)}{(2-p)}n + \frac{1}{2}p(1-\varepsilon)] = (1 - \frac{1}{2}p)(1-\varepsilon) > 0$. Therefore $y P z$.

$\frac{1}{2}[p[n(x P_i z) + n(z P_i x)] + (1-p)n] = \frac{1}{2}[p[\frac{2(1-p)}{(2-p)}n - 2\varepsilon + 1] + (1-p)n] = \frac{(1-p)(2+p)}{2(2-p)}n + \frac{1}{2}p(1-2\varepsilon)$.

$n(x P_i z) - \frac{1}{2}[p[n(x P_i z) + n(z P_i x)] + (1-p)n] = \frac{1-p}{(2-p)}n - \varepsilon + 1 - [\frac{(1-p)(2+p)}{2(2-p)}n + \frac{1}{2}p(1-2\varepsilon)] = -\frac{p(1-p)}{2(2-p)}n + (1 - \frac{1}{2}p)(1-\varepsilon) + \frac{1}{2}p\varepsilon < 0$, as $n > \frac{2(2-p)}{p(1-p)}$. Therefore $\sim x P z$.

$n(z P_i x) < n(x P_i z) \rightarrow \sim z P x$.

Thus $R \mid A = (x I y \wedge y P z \wedge x I z)$.

(x) Let $\mathscr{D} \mid A \supseteq \{xI^syP^sz, yI^tzP^tx, zI^uxP^uy\}$. Consider any $(R_1, \ldots, R_n) \in \mathscr{D}^n$ such that: $n(xI^syP^sz) = \lceil \frac{1}{(2+p)}n \rceil + 1 = \frac{1}{(2+p)}n + \varepsilon + 1, 0 \le \varepsilon < 1; n(yI^tzP^tx)$ $= \lceil \frac{p}{(2+p)}n \rceil = \frac{p}{(2+p)}n + \delta, 0 \le \delta < 1; n(zI^uxP^uy) = n - \lceil \frac{1}{(2+p)}n \rceil - \lceil \frac{p}{(2+p)}n \rceil - 1$ $= \frac{1}{(2+p)}n - 1 - \varepsilon - \delta$.

We obtain: $n(xP_iy) = \frac{1}{(2+p)}n - 1 - \varepsilon - \delta, n(yP_ix) = \frac{p}{(2+p)}n + \delta, n(yP_iz) = \frac{1}{(2+p)}n$ $+ \varepsilon + 1, n(zP_iy) = \frac{1}{(2+p)}n - 1 - \varepsilon - \delta, n(xP_iz) = \frac{1}{(2+p)}n + \varepsilon + 1, n(zP_ix) =$ $\frac{p}{(2+p)}n + \delta$.

We have:

$\frac{1}{2}[p[n(xP_iy) + n(yP_ix)] + (1 - p)n] = \frac{1}{2}[p[\frac{1+p}{(2+p)}n - 1 - \varepsilon] + (1 - p)n] = \frac{1}{(2+p)}n$ $- \frac{1}{2}p(\varepsilon + 1)$.

$n(xP_iy) - \frac{1}{2}[p[n(xP_iy) + n(yP_ix)] + (1 - p)n] = \frac{1}{(2+p)}n - 1 - \varepsilon - \delta - [\frac{1}{(2+p)}n$ $- \frac{1}{2}p(\varepsilon + 1)] = -(1 - \frac{1}{2}p)(\varepsilon + 1) - \delta < 0$. Therefore $\sim xPy$.

$n(yP_ix) - \frac{1}{2}[p[n(xP_iy) + n(yP_ix)] + (1 - p)n] = \frac{p}{(2+p)}n + \delta - [\frac{1}{(2+p)}n - \frac{1}{2}$ $p(\varepsilon + 1)] = -\frac{1-p}{(2+p)}n + \delta + \frac{1}{2}p(\varepsilon + 1) < 0$, as $n > \frac{(1+p)(2+p)}{(1-p)}$. Therefore $\sim yPx$.

$\frac{1}{2}[p[n(yP_iz) + n(zP_iy)] + (1 - p)n] = \frac{1}{2}[p[\frac{2}{(2+p)}n - \delta] + (1 - p)n] = \frac{p+2-p^2}{2(2+p)}n$ $- \frac{1}{2}p\delta$

$n(yP_iz) - \frac{1}{2}[p[n(yP_iz) + n(zP_iy)] + (1 - p)n] = \frac{1}{(2+p)}n + \varepsilon + 1 - [\frac{p+2-p^2}{2(2+p)}n -$ $\frac{1}{2}p\delta] = -\frac{p(1-p)}{2(2+p)}n + 1 + \varepsilon + \frac{1}{2}p\delta < 0$, as $n > \frac{(4+p)(2+p)}{p(1-p)}$. Therefore $\sim yPz$.

$n(zP_iy) < n(yP_iz) \rightarrow \sim zPy$.

$\frac{1}{2}[p[n(xP_iz) + n(zP_ix)] + (1 - p)n] = \frac{1}{2}[p[\frac{1+p}{(2+p)}n + 1 + \varepsilon + \delta] + (1 - p)n] =$ $\frac{1}{(2+p)}n + \frac{1}{2}p(1 + \varepsilon + \delta)$.

$n(xP_iz) - \frac{1}{2}[p[n(xP_iz) + n(zP_ix)] + (1 - p)n] = \frac{1}{(2+p)}n + \varepsilon + 1 - [\frac{1}{(2+p)}n +$ $\frac{1}{2}p(1 + \varepsilon + \delta)] = (1 - \frac{1}{2}p)(1 + \varepsilon) - \frac{1}{2}p\delta > 0$. Therefore $xPz$ holds.

Thus $R \mid A = (xIy \wedge yIz \wedge xPz)$.

(xi) Let $\mathscr{D} \mid A \supseteq \{xP^syI^sz, xI^tyP^tz, xI^uyI^uz\}$. Consider any $(R_1, \ldots, R_n) \in \mathscr{D}^n$ such that: $n(xP^syI^sz) = \lfloor \frac{1-p}{2(2-p)}n \rfloor + 1 = \frac{1-p}{2(2-p)}n - \varepsilon + 1, 0 \le \varepsilon < 1; n(xI^tyP^tz)$ $= \lceil \frac{1-p}{2(2-p)}n \rceil = \frac{1-p}{2(2-p)}n + \delta, 0 \le \delta < 1; n(xI^uyI^uz) = n - \lfloor \frac{1-p}{2(2-p)}n \rfloor - 1 -$ $\lceil \frac{1-p}{2(2-p)}n \rceil = \frac{1}{(2-p)}n + \varepsilon - \delta - 1$.

It should be noted that: $(\varepsilon = 0 \rightarrow \delta = 0) \wedge (\varepsilon > 0 \rightarrow \delta = 1 - \varepsilon)$.

We obtain: $n(xP_iy) = \frac{1-p}{2(2-p)}n - \varepsilon + 1, n(yP_ix) = 0, n(yP_iz) = \frac{1-p}{2(2-p)}n + \delta,$ $n(zP_iy) = 0, n(xP_iz) = \frac{1-p}{(2-p)}n + 1 + \delta - \varepsilon, n(zP_ix) = 0$.

We have:

$\frac{1}{2}[p[n(xP_iy)+n(yP_ix)]+(1-p)n]=\frac{1}{2}[p[\frac{1-p}{2(2-p)}n-\varepsilon+1]+(1-p)n]=$
$\frac{(1-p)(4-p)}{4(2-p)}n+\frac{1}{2}p(1-\varepsilon).$

$n(xP_iy)-\frac{1}{2}[p[n(xP_iy)+n(yP_ix)]+(1-p)n]=\frac{1-p}{2(2-p)}n-\varepsilon+1-[\frac{(1-p)(4-p)}{4(2-p)}n+$
$\frac{1}{2}p(1-\varepsilon)]=-\frac{1-p}{4}n+(1-\frac{1}{2}p)(1-\varepsilon)<0,$ as $n>\frac{2(2-p)}{1-p}.$ Therefore $\sim xPy.$

$n(yP_ix)=0\rightarrow\sim yPx.$

$\frac{1}{2}[p[n(yP_iz)+n(zP_iy)]+(1-p)n]=\frac{1}{2}[p[\frac{1-p}{2(2-p)}n+\delta]+(1-p)n]=\frac{(1-p)(4-p)}{4(2-p)}n$
$+\frac{1}{2}p\delta.$

$n(yP_iz)-\frac{1}{2}[p[n(yP_iz)+n(zP_iy)]+(1-p)n]=\frac{1-p}{2(2-p)}n+\delta-[\frac{(1-p)(4-p)}{4(2-p)}n+$
$\frac{1}{2}p\delta]=-\frac{1-p}{4}n+(1-\frac{1}{2}p)\delta<0,$ as $n>\frac{2(2-p)}{1-p}.$ Therefore $\sim yPz.$

$n(zP_iy)=0\rightarrow\sim zPy.$

$\frac{1}{2}[p[n(xP_iz)+n(zP_ix)]+(1-p)n]=\frac{1}{2}[p[\frac{1-p}{(2-p)}n+1+\delta-\varepsilon]+(1-p)n]=$
$\frac{1-p}{(2-p)}n+\frac{1}{2}p(1+\delta-\varepsilon).$

$n(xP_iz)-\frac{1}{2}[p[n(xP_iz)+n(zP_ix)]+(1-p)n]=\frac{1-p}{(2-p)}n+1+\delta-\varepsilon-[\frac{1-p}{(2-p)}n+$
$\frac{1}{2}p(1+\delta-\varepsilon)]=(1-\frac{1}{2}p)(1+\delta-\varepsilon).$

If $\varepsilon=0$ then $(1-\frac{1}{2}p)(1+\delta-\varepsilon)=(1-\frac{1}{2}p)>0;$ and if $\varepsilon>0$ then $(1-\frac{1}{2}p)(1+$
$\delta-\varepsilon)=2(1-\frac{1}{2}p)(1-\varepsilon)>0.$

Thus, $n(xP_iz)-\frac{1}{2}[p[n(xP_iz)+n(zP_ix)]+(1-p)n]>0$ regardless of whether
$\varepsilon=0$ or $\varepsilon>0.$ Thus $xPz$ holds.

Thus $R\mid A=(xIy\wedge yIz\wedge xPz).$

(xii) Let $\mathscr{D}\mid A\supseteq\{xP^syI^sz,yP^tzI^tx,zI^uxP^uy\}.$ Consider any $(R_1,\ldots,R_n)\in\mathscr{D}^n$
such     that:     $n(xP^syI^sz)=\lfloor\frac{1-p}{(2-p)}n\rfloor=\frac{1-p}{(2-p)}n-\delta,0\le\delta<1;n(yP^tzI^tx)=$
$\lfloor\frac{1}{2(2-p)}n\rfloor=\frac{1}{2(2-p)}n-\varepsilon,0\le\varepsilon<1;n(zI^uxP^uy)=n-\lfloor\frac{1-p}{(2-p)}n\rfloor-\lfloor\frac{1}{2(2-p)}n\rfloor=$
$\frac{1}{2(2-p)}n+\varepsilon+\delta.$

We obtain: $n(xP_iy)=\frac{3-2p}{2(2-p)}n+\varepsilon,n(yP_ix)=\frac{1}{2(2-p)}n-\varepsilon,n(yP_iz)=\frac{1}{2(2-p)}n-$
$\varepsilon,n(zP_iy)=\frac{1}{2(2-p)}n+\varepsilon+\delta,n(xP_iz)=\frac{1-p}{(2-p)}n-\delta,n(zP_ix)=0.$

We have:

$n(xP_iy)=\frac{3-2p}{2(2-p)}n+\varepsilon>\frac{1}{2}n\rightarrow xPy.$

$\frac{1}{2}[p[n(yP_iz)+n(zP_iy)]+(1-p)n]=\frac{1}{2}[p[\frac{1}{(2-p)}n+\delta]+(1-p)n]=\frac{2-2p+p^2}{2(2-p)}n$
$+\frac{1}{2}p\delta.$

$n(zP_iy)-\frac{1}{2}[p[n(yP_iz)+n(zP_iy)]+(1-p)n]=\frac{1}{2(2-p)}n+\varepsilon+\delta-[\frac{2-2p+p^2}{2(2-p)}n+$
$\frac{1}{2}p\delta]=-\frac{(1-p)^2}{2(2-p)}n+\varepsilon+(1-\frac{1}{2}p)\delta<0,$ as $n>\frac{(4-p)(2-p)}{(1-p)^2}.$ Therefore $\sim zPy.$

$n(yP_iz)\le n(zP_iy)\rightarrow\sim yPz.$

$\frac{1}{2}[p[n(xP_iz)+n(zP_ix)]+(1-p)n]=\frac{1}{2}[p[\frac{1-p}{(2-p)}n-\delta]+(1-p)n]=\frac{1-p}{(2-p)}n-\frac{1}{2}p\delta.$

$n(x P_i z) - \frac{1}{2}[p[n(x P_i z) + n(z P_i x)] + (1-p)n] = \frac{1-p}{(2-p)}n - \delta - [\frac{1-p}{(2-p)}n - \frac{1}{2}p\delta] = -(1 - \frac{1}{2}p)\delta \leq 0$. Therefore $\sim x P z$.

$n(z P_i x) = 0 \to \sim z P x$.

Thus $R \mid A = (x P y \wedge y I z \wedge x I z)$.

(xiii) Let $\mathcal{D} \mid A \supseteq \{x P^s y I^s z, y I^t z P^t x, z I^u x P^u y\}$. Consider any $(R_1, \ldots, R_n) \in \mathcal{D}^n$ such that: $n(y I^t z P^t x) = \lfloor \frac{1}{2(2-p)}n \rfloor = \frac{1}{2(2-p)}n - \varepsilon, 0 \leq \varepsilon < 1; n(z I^u x P^u y) = \lfloor \frac{1-p}{(2-p)}n \rfloor = \frac{1-p}{(2-p)}n - \delta, 0 \leq \delta < 1; n(x P^s y I^s z) = n - \lfloor \frac{1}{2(2-p)}n \rfloor - \lfloor \frac{1-p}{(2-p)}n \rfloor = \frac{1}{2(2-p)}n + \varepsilon + \delta$.

We obtain: $n(x P_i y) = \frac{3-2p}{2(2-p)}n + \varepsilon, n(y P_i x) = \frac{1}{2(2-p)}n - \varepsilon, n(y P_i z) = 0, n(z P_i y) = \frac{1-p}{(2-p)}n - \delta, n(x P_i z) = \frac{1}{2(2-p)}n + \varepsilon + \delta, n(z P_i x) = \frac{1}{2(2-p)}n - \varepsilon$.

We have:

$n(x P_i y) = \frac{3-2p}{2(2-p)}n + \varepsilon > \frac{1}{2}n \to x P y$.

$\frac{1}{2}[p[n(y P_i z) + n(z P_i y)] + (1-p)n] = \frac{1}{2}[p[\frac{1-p}{(2-p)}n - \delta] + (1-p)n] = \frac{(1-p)}{(2-p)}n - \frac{1}{2}p\delta$.

$n(z P_i y) - \frac{1}{2}[p[n(y P_i z) + n(z P_i y)] + (1-p)n] = \frac{1-p}{(2-p)}n - \delta - \frac{(1-p)}{(2-p)}n + \frac{1}{2}p\delta = -(1 - \frac{1}{2}p)\delta \leq 0$. Therefore $\sim z P y$.

$n(y P_i z) = 0 \to \sim y P z$.

$\frac{1}{2}[p[n(x P_i z) + n(z P_i x)] + (1-p)n] = \frac{1}{2}[p[\frac{1}{(2-p)}n + \delta] + (1-p)n] = \frac{2-2p+p^2}{2(2-p)}n + \frac{1}{2}p\delta$.

$n(x P_i z) - \frac{1}{2}[p[n(x P_i z) + n(z P_i x)] + (1-p)n] = \frac{1}{2(2-p)}n + \varepsilon + \delta - [\frac{2-2p+p^2}{2(2-p)}n + \frac{1}{2}p\delta] = -\frac{(1-p)^2}{2(2-p)}n + \varepsilon + (1 - \frac{1}{2}p)\delta < 0$, as $n > \frac{(4-p)(2-p)}{(1-p)^2}$. Therefore $\sim x P z$.

$n(z P_i x) \leq n(x P_i z) \to \sim z P x$.

Thus $R \mid A = (x P y \wedge y I z \wedge x I z)$.  □

*Remark 5.1* Thus Theorem 5.2 establishes that for every $p$-semi-strict majority rule there exists a positive even integer $n'$ such that for every even integer $n \geq n'$, every $(R_1, \ldots, R_n) \in \mathcal{D}^n$ yields transitive social $R$ iff for every triple $A \subseteq S, \mathcal{D} \mid A$ satisfies at least one of the three conditions of strict placement restriction, partial agreement and strongly antagonistic preferences (1).  ◇

## 5.3  Conditions for Quasi-transitivity

**Lemma 5.3** *Let $f$ be a $p$-semi-strict majority rule, $0 < p < 1$. Let $\mathcal{D} \subseteq \mathcal{T}$ satisfy value restriction (2). Then $f$ yields quasi-transitive social $R, R = f(R_1, \ldots, R_n)$, for every $(R_1, \ldots, R_n) \in \mathcal{D}^n$.*

*Proof* Let $\mathscr{D} \subseteq \mathscr{T}$. Let $(R_1, \ldots, R_n) \in \mathscr{D}^n$. Suppose $R = f(R_1, \ldots, R_n)$ violates quasi-transitivity. Then we must have $xPy \wedge yPz \wedge \sim xPz$ for some $x, y, z \in S$. Let $\{x, y, z\}$ be designated by $A$.

$xPy \rightarrow n(xP_iy) > \frac{1}{2}[p(n(xP_iy) + n(yP_ix)) + (1-p)n]$

$\rightarrow n(xP_iy) > \frac{1}{2}n - \frac{1}{2}pn(xI_iy)$

$\rightarrow n(xR_iy \wedge R_i \text{ concerned over } A) > \frac{1}{2}n(R_i \text{ concerned over}A) + (1 - \frac{1}{2}p)$

$n(xI_iy \wedge R_i \text{ concerned over } A) + \frac{1}{2}(1-p)n(xI_iyI_iz)$

$\rightarrow n(xR_iy \wedge R_i \text{ concerned over } A) > \frac{1}{2}n(R_i \text{ concerned over}A) + \frac{1}{2}(1-p)n$

$(xI_iyI_iz)$. $\hfill$ (L5.3-1)

Analogously,

$yPz \rightarrow n(yR_iz \wedge R_i \text{ concerned over } A) > \frac{1}{2}n(R_i \text{ concerned over}A) + (1 - \frac{1}{2}p)$

$n(yI_iz \wedge R_i \text{ concerned over } A) + \frac{1}{2}(1-p)n(xI_iyI_iz)$

$\rightarrow n(yR_iz \wedge R_i \text{ concerned over } A) > \frac{1}{2}n(R_i \text{ concerned over}A) + \frac{1}{2}(1-p)n$

$(xI_iyI_iz)$. $\hfill$ (L5.3-2)

$zRx \rightarrow n(xP_iz) \leq \frac{1}{2}n - \frac{1}{2}pn(xI_iz)$

$\rightarrow n(zR_ix) \geq \frac{1}{2}n + \frac{1}{2}pn(xI_iz)$

$\rightarrow n(zR_ix \wedge R_i \text{ concerned over } A) \geq \frac{1}{2}n + \frac{1}{2}pn(xI_iz) - n(xI_iyI_iz)$

$\rightarrow n(zR_ix \wedge R_i \text{ concerned over } A) \geq \frac{1}{2}n(R_i \text{ concerned over}A) + \frac{1}{2}pn(xI_iz \wedge R_i$

concerned over $A) - \frac{1}{2}(1-p)n(xI_iyI_iz)$

$\rightarrow n(zR_ix \wedge R_i \text{ concerned over } A) \geq \frac{1}{2}n(R_i \text{ concerned over}A) - \frac{1}{2}(1-p)$

$n(xI_iyI_iz)$. $\hfill$ (L5.3-3)

(L5.3-1) $\wedge$ (L5.3-2) $\rightarrow (\exists R_i \text{ concerned over } A)(xR_iyR_iz)$. $\hfill$ (L5.3-4)

(L5.3-2) $\wedge$ (L5.3-3) $\rightarrow (\exists R_i \text{ concerned over } A)(yR_izR_ix)$. $\hfill$ (L5.3-5)

(L5.3-3) $\wedge$ (L5.3-1) $\rightarrow (\exists R_i \text{ concerned over } A)(zR_ixR_iy)$. $\hfill$ (L5.3-6)

(L5.3-4)–(L5.3-6) imply that VR(2) is violated.

This establishes the lemma. $\hfill \square$

**Lemma 5.4** *Let $f$ be a p-semi-strict majority rule, $0 < p < 1$. Let $\mathscr{D} \subseteq \mathscr{T}$ satisfy absence of unique extremal value. Then $f$ yields quasi-transitive social $R$, $R = f(R_1, \ldots, R_n)$, for every $(R_1, \ldots, R_n) \in \mathscr{D}^n$.*

*Proof* Let $\mathscr{D} \subseteq \mathscr{T}$ satisfy AUEV. Let $A = \{x, y, z\} \subseteq S$ be a triple. Let $\{xPyIz, yPzIx, zPxIy, xIyIz\}$ and $\{xIyPz, yIzPx, zIxPy, xIyIz\}$ be denoted by $\mathscr{D}_1$ and $\mathscr{D}_2$, respectively. Then we must have: $(\mathscr{D} \mid A \subseteq \mathscr{D}_1 \vee \mathscr{D} \mid A \subseteq \mathscr{D}_2)$.
Let $(R_1, \ldots, R_n) \in \mathscr{D}^n$ and $R = f(R_1, \ldots, R_n)$. Let $\mathscr{D} \mid A \subseteq \mathscr{D}_1$. Suppose $xPy \wedge yPz$.

$xPy \rightarrow n(xP_iy) > \frac{1}{2}[p(n(xP_iy) + n(yP_ix)) + (1-p)n]$
$\rightarrow n(xP_iyI_iz) > \frac{1}{2}[p(n(xP_iyI_iz) + n(yP_izI_ix)) + (1-p)n]$          (L5.4-1)
Analogously, $yPz \rightarrow$
$n(yP_izI_ix) > \frac{1}{2}[p(n(yP_izI_ix) + n(zP_ixI_iy)) + (1-p)n]$          (L5.4-2)

Adding (L5.4-1) and (L5.4-2) we obtain:
$n(xP_iyI_iz) + n(yP_izI_ix) > \frac{1}{2}p(n(xP_iyI_iz) + 2n(yP_izI_ix) + n(zP_ixI_iy)) +$
$\frac{1}{2}(1-p)n + \frac{1}{2}(1-p)n$
$\rightarrow n(xP_iyI_iz) > \frac{1}{2}p(n(xP_iyI_iz) + n(zP_ixI_iy)) + \frac{1}{2}(1-p)n + \frac{1}{2}(1-p)(n(xP_i$
$yI_iz) + n(zP_ixI_iy) + n(xI_iyI_iz) - n(yP_izI_ix))$          (L5.4-3)
(L5.4-1) $\rightarrow n(xP_iyI_iz) + n(zP_ixI_iy) + n(xI_iyI_iz) > \frac{1}{2}p(n(xP_iyI_iz) + n(yP_izI_ix)$
$+ n(zP_ixI_iy) + n(xI_iyI_iz)) + \frac{1}{2}(1-p)n + (1-\frac{1}{2}p)(n(zP_ixI_iy) + n(xI_iyI_iz))$
$\rightarrow n(xP_iyI_iz) + n(zP_ixI_iy) + n(xI_iyI_iz) > \frac{1}{2}n$          (L5.4-4)

From (L5.4-3) and (L5.4-4) we conclude:
$n(xP_iyI_iz) > \frac{1}{2}p(n(xP_iyI_iz) + n(zP_ixI_iy)) + \frac{1}{2}(1-p)n$
which implies that $xPz$ holds.
As permuting $x, y, z$ leaves $\mathscr{D}_1$ unchanged, the demonstration that $xPy \wedge yPz$
implies $xPz$ suffices to conclude that if $\mathscr{D} \mid A \subseteq \mathscr{D}_1$ then $R \mid A$ would be quasi-transitive.
The proof that if $\mathscr{D} \mid A \subseteq \mathscr{D}_2$ then $R \mid A$ would be quasi-transitive is analogous.
Therefore it follows that if $\mathscr{D}$ satisfies AUEV then $R$ would be quasi-transitive.   $\square$

**Lemma 5.5** *Let $\mathscr{D}$ be a set of orderings of the triple A. $\mathscr{D}$ violates both value restriction (2) and absence of unique extremal value iff $\mathscr{D}$ contains one of the following nine sets of orderings of A, except for a formal interchange of alternatives.*

(i)   $xP^syP^sz$                    (ii)   $xP^syP^sz$
      $yP^tzP^tx$                           $yP^tzP^tx$
      $zP^uxP^uy$                           $zP^uxI^uy$

(iii) $xP^syP^sz$                    (iv)   $xP^syP^sz$
      $yP^tzP^tx$                           $yP^tzI^tx$
      $zI^uxP^uy$                           $zP^uxI^uy$

(v)   $xP^syP^sz$                    (vi)   $xP^syP^sz$
      $yI^tzP^tx$                           $yI^tzP^tx$
      $zI^uxP^uy$                           $zP^uxI^uy$

(vii) $xP^syP^sz$                    (viii) $xP^syI^sz$
      $yP^tzI^tx$                           $yP^tzI^tx$
      $zI^uxP^uy$                           $zI^uxP^uy$

(ix)  $xP^syI^sz$
      $yI^tzP^tx$
      $zI^uxP^uy$

*Proof* $\mathscr{D}$ would violate VR(2) iff ($\exists$ distinct $x, y, z \in A$)($\exists$ concerned $R^s, R^t, R^u \in \mathscr{D}$)$[xR^syR^sz \wedge yR^tzR^tx \wedge zR^uxR^uy]$, i.e., it contains one of the following 11 sets of orderings of $A$, except for a formal interchange of alternatives:

$$
\begin{array}{ll}
(A) \quad \begin{array}{l} xP^syP^sz \\ yP^tzP^tx \\ zP^uxP^uy \end{array} &
(B) \quad \begin{array}{l} xP^syP^sz \\ yP^tzP^tx \\ zP^uxI^uy \end{array} \\[3em]
(C) \quad \begin{array}{l} xP^syP^sz \\ yP^tzP^tx \\ zI^uxP^uy \end{array} &
(D) \quad \begin{array}{l} xP^syP^sz \\ yP^tzI^tx \\ zP^uxI^uy \end{array} \\[3em]
(E) \quad \begin{array}{l} xP^syP^sz \\ yI^tzP^tx \\ zI^uxP^uy \end{array} &
(F) \quad \begin{array}{l} xP^syP^sz \\ yI^tzP^tx \\ zP^uxI^uy \end{array} \\[3em]
(G) \quad \begin{array}{l} xP^syP^sz \\ yP^tzI^tx \\ zI^uxP^uy \end{array} &
(H) \quad \begin{array}{l} xP^syI^sz \\ yP^tzI^tx \\ zI^uxP^uy \end{array} \\[3em]
(I) \quad \begin{array}{l} xP^syI^sz \\ yI^tzP^tx \\ zI^uxP^uy \end{array} &
(J) \quad \begin{array}{l} xP^syI^sz \\ yP^tzI^tx \\ zP^uxI^uy \end{array} \\[3em]
(K) \quad \begin{array}{l} xI^syP^sz \\ yI^tzP^tx \\ zI^uxP^uy \end{array} &
\end{array}
$$

(A)–(I) are the same as (i)–(ix), respectively. It therefore suffices to consider only sets (J) and (K). (J) does not violate AUEV; it would be violated iff an ordering in which some alternative is uniquely worst is included. With the required inclusion $\mathscr{D}$ would contain one of the sets mentioned in the statement of the lemma, except for a formal interchange of alternatives. (K) does not violate AUEV; it would be violated iff an ordering in which some alternative is uniquely best is included. With the required inclusion $\mathscr{D}$ would contain one of the sets mentioned in the statement of the lemma, except for a formal interchange of alternatives. The proof of the lemma is completed by noting that all the nine sets mentioned in the statement of the lemma violate both VR(2) and AUEV. $\qquad\square$

**Theorem 5.3** *Let* $f : \mathscr{T}^n \mapsto \mathscr{C}$ *be a p-semi-strict majority rule. Let* $n = 4m + 1 - \lceil pm \rceil$, *where* $m$ *is a positive integer greater than* $\frac{2(2-p)}{p^2(1-p)}$. *Let* $\mathscr{D} \subseteq \mathscr{T}$. *Then* $f$ *yields quasi-transitive social* $R$, $R = f(R_1, \ldots, R_n)$, *for every* $(R_1, \ldots, R_n) \in \mathscr{D}^n$ *iff for every triple* $A \subseteq S$, $\mathscr{D} \mid A$ *satisfies value restriction (2) or absence of unique extremal value.*

*Proof* The assertion that if for every triple $A \subseteq S$ $\mathscr{D} \mid A$ satisfies VR(2) or AUEV then every $(R_1, \ldots, R_n) \in \mathscr{D}^n$ yields quasi-transitive social $R$ follows from Lemmas 5.3 and 5.4.

Let $\mathscr{D}$ violate both VR(2) and AUEV over the triple $A$. Then $\mathscr{D}$ contains, for some distinct $x, y, z \in A$, one of the nine sets mentioned in Lemma 5.5. We show that in each of the nine cases there exists a profile $(R_1, \ldots, R_n) \in \mathscr{D}^n$ which yields an $R$ violating quasi-transitivity.

(i) Let $\mathscr{D} \mid A \supseteq \{x P^s y P^s z \wedge y P^t z P^t x \wedge z P^u x P^u y\}$. Consider any $(R_1, \ldots, R_n) \in \mathscr{D}^n$ such that: $n(x P_i y P_i z) = n(y P_i z P_i x) = n(z P_i x P_i y) = k$, if $n = 3k$, $k$ a positive integer; $n(x P_i y P_i z) = k + 1, n(y P_i z P_i x) = n(z P_i x P_i y) = k$, if $n = 3k + 1$; $n(x P_i y P_i z) = n(y P_i z P_i x) = k + 1, n(z P_i x P_i y) = k$, if $n = 3k + 2$. In each case we obtain: $n(x P_i y) > \frac{n}{2} \wedge n(y P_i z) > \frac{n}{2} \wedge n(x P_i z) \leq n(z P_i x)$. This results in $R \mid A = (x P y \wedge y P z \wedge \sim x P z)$.

(ii) $n = 4m + 1 - \lceil pm \rceil \wedge m > \frac{2(2-p)}{p^2(1-p)} \to n > 9$. Let $\mathscr{D} \mid A \supseteq \{x P^s y P^s z \wedge y P^t z P^t x \wedge z P^u x I^u y\}$. Consider any $(R_1, \ldots, R_n) \in \mathscr{D}^n$ such that: $n(x P_i y P_i z) = n(y P_i z P_i x) = n(z P_i x I_i y) = k$, if $n = 3k$; $n(x P_i y P_i z) = n(y P_i z P_i x) = k, n(z P_i x I_i y) = k + 1$, if $n = 3k + 1$; $n(x P_i y P_i z) = n(y P_i z P_i x) = k + 1, n(z P_i x I_i y) = k$, if $n = 3k + 2$. In each case we obtain: $n(y P_i z) > \frac{n}{2} \wedge n(z P_i x) > \frac{n}{2} \wedge n(y P_i x) = n(x P_i y)$, as $k > 2$. This results in $R \mid A = (y P z \wedge z P x \wedge \sim y P x)$.

(iii) Let $\mathscr{D} \mid A \supseteq \{x P^s y P^s z \wedge y P^t z P^t x \wedge z I^u x P^u y\}$. Consider any $(R_1, \ldots, R_n) \in \mathscr{D}^n$ such that: $n(x P_i y P_i z) = n(y P_i z P_i x) = n(z I_i x P_i y) = k$, if $n = 3k$; $n(x P_i y P_i z) = n(y P_i z P_i x) = k, n(z I_i x P_i y) = k + 1$, if $n = 3k + 1$; $n(x P_i y P_i z) = n(y P_i z P_i x) = k + 1, n(z I_i x P_i y) = k$, if $n = 3k + 2$. In each case we obtain: $n(x P_i y) > \frac{n}{2} \wedge n(y P_i z) > \frac{n}{2} \wedge n(x P_i z) = n(x P_i z)$, as $k > 2$. This results in $R \mid A = (x P y \wedge y P z \wedge \sim x P z)$.

(iv) Let $\mathscr{D} \mid A \supseteq \{x P^s y P^s z \wedge y P^t z I^t x \wedge z P^u x I^u y\}$. Consider any $(R_1, \ldots, R_n) \in \mathscr{D}^n$ such that: $n(x P_i y P_i z) = n(y P_i z I_i x) = m, n(z P_i x I_i y) = n - 2m = 2m + 1 - \lceil pm \rceil$, as $n = 4m + 1 - \lceil pm \rceil$.

We have: $n(x P_i y) = m, n(y P_i x) = m, n(y P_i z) = 2m, n(z P_i y) = n - 2m, n(x P_i z) = m, n(z P_i x) = n - 2m$.

$n(x P_i y) = n(y P_i x) = m \to x I y$.

$m > \frac{2(2-p)}{p^2(1-p)} \to m > \frac{1}{p}$

$\to \lceil pm \rceil > 1$

$\to n < 4m$.

As $n(y P_i z) = 2m > \frac{n}{2}$, we conclude that $y P z$ holds.

$\frac{1}{2}[p(n(x P_i z) + n(z P_i x)) + (1 - p)n] = \frac{1}{2}[p(n - m) + (1 - p)n] = \frac{1}{2}(n - pm)$

We have: $n(z P_i x) - \frac{1}{2}[p(n(x P_i z) + n(z P_i x)) + (1 - p)n] = n - 2m - \frac{1}{2}n + \frac{1}{2}pm$

$= \frac{1}{2}n - (2 - \frac{1}{2}p)m = 2m + \frac{1}{2} - \frac{1}{2}\lceil pm \rceil - (2 - \frac{1}{2}p)m = \frac{1}{2}[1 + pm - \lceil pm \rceil] > 0$.

Therefore $z P x$ holds.

$y P z \wedge z P x \wedge x I y$ is violative of quasi-transitivity.

(v) Let $\mathscr{D} \mid A \supseteq \{x P^s y P^s z \wedge y I^t z P^t x \wedge z I^u x P^u y\}$. Consider any $(R_1, \ldots, R_n) \in \mathscr{D}^n$ such that: $n(x P_i y P_i z) = m, n(y I_i z P_i x) = n - 2m = 2m + 1 - \lceil pm \rceil$, $n(z I_i x P_i y) = m$.

We have: $n(x P_i y) = 2m, n(y P_i x) = n - 2m, n(y P_i z) = m, n(z P_i y) = m$, $n(x P_i z) = m, n(z P_i x) = n - 2m$.

$n(y P_i z) = n(z P_i y) = m \rightarrow y I z$.

As $n(x P_i y) = 2m > \frac{n}{2}$, we conclude that $x P y$ holds.

$\frac{1}{2}[p(n(x P_i z) + n(z P_i x)) + (1 - p)n] = \frac{1}{2}[p(n - m) + (1 - p)n] = \frac{1}{2}(n - pm)$

We have: $n(z P_i x) - \frac{1}{2}[p(n(x P_i z) + n(z P_i x)) + (1 - p)n] = n - 2m - \frac{1}{2}n + \frac{1}{2}pm$

$= \frac{1}{2}n - (2 - \frac{1}{2}p)m = 2m + \frac{1}{2} - \frac{1}{2}\lceil pm \rceil - (2 - \frac{1}{2}p)m = \frac{1}{2}[1 + pm - \lceil pm \rceil] > 0$.

Therefore $z P x$ holds.

$z P x \wedge x P y \wedge z I y$ is violative of quasi-transitivity.

(vi) Let $\mathscr{D} \mid A \supseteq \{x P^s y P^s z \wedge y I^t z P^t x \wedge z P^u x I^u y\}$. Consider any $(R_1, \ldots, R_n) \in \mathscr{D}^n$ such that: $n(x P_i y P_i z) = \lceil \frac{n}{2} \rceil - 1, n(y I_i z P_i x) = n - \lceil \frac{n}{2} \rceil + 1 - \lceil \frac{1}{2} pn \rceil$, $n(z P_i x I_i y) = \lceil \frac{1}{2} pn \rceil$.

We have: $n(x P_i y) = \lceil \frac{n}{2} \rceil - 1, n(y P_i x) = n - \lceil \frac{n}{2} \rceil + 1 - \lceil \frac{1}{2} pn \rceil, n(y P_i z) = \lceil \frac{n}{2} \rceil - 1, n(z P_i y) = \lceil \frac{1}{2} pn \rceil, n(x P_i z) = \lceil \frac{n}{2} \rceil - 1, n(z P_i x) = n - \lceil \frac{n}{2} \rceil + 1$.

$n(z P_i x) = n - \lceil \frac{n}{2} \rceil + 1 \rightarrow z P x$, as $n - \lceil \frac{n}{2} \rceil + 1 > \frac{n}{2}$.

We have: $n(x P_i y) - \frac{1}{2}[p[n(x P_i y) + n(y P_i x)] + (1 - p)n] = \lceil \frac{n}{2} \rceil - 1 - \frac{1}{2}[p[n - \lceil \frac{1}{2} pn \rceil] + (1 - p)n] = [\lceil \frac{n}{2} \rceil - \frac{n}{2}] + [\frac{1}{2} p \lceil \frac{1}{2} pn \rceil - 1] > 0$, as $n > \frac{4}{p^2}$ and therefore $\frac{1}{2} p \lceil \frac{1}{2} pn \rceil - 1 > 0$. Thus $x P y$ holds.

Let $\lceil \frac{n}{2} \rceil = \frac{n}{2} + \varepsilon_1, 0 \leq \varepsilon_1 < 1$; and $\lceil \frac{1}{2} pn \rceil = \frac{1}{2} pn + \varepsilon_2, 0 \leq \varepsilon_2 < 1$.

We have: $\frac{1}{2}[p[n(y P_i z) + n(z P_i y)] + (1 - p)n] = \frac{1}{2}[p[\frac{n}{2} + \varepsilon_1 - 1 + \frac{1}{2} pn + \varepsilon_2] + (1 - p)n] = [\frac{1}{2} - \frac{1}{4} p(1 - p)]n + \frac{1}{2} p[\varepsilon_1 + \varepsilon_2 - 1]$.

Now, $n(z P_i y) - \frac{1}{2}[p[n(y P_i z) + n(z P_i y)] + (1 - p)n] = \frac{1}{2} pn + \varepsilon_2 - [\frac{1}{2} - \frac{1}{4} p(1 - p)]n - \frac{1}{2} p[\varepsilon_1 + \varepsilon_2 - 1] = -\frac{1}{2}(1 - \frac{1}{2} p)(1 - p)n - \frac{1}{2} p\varepsilon_1 + (1 - \frac{1}{2} p)\varepsilon_2 + \frac{1}{2} p < 0$, as $n > \frac{4}{(2-p)(1-p)}$. Thus $\sim z P y$.

$z P x \wedge x P y \wedge \sim z P y$ is violative of quasi-transitivity.

(vii) Let $\mathscr{D} \mid A \supseteq \{x P^s y P^s z \wedge y P^t z I^t x \wedge z I^u x P^u y\}$. Consider any $(R_1, \ldots, R_n) \in \mathscr{D}^n$ such that: $n(x P_i y P_i z) = n - 2\lceil \frac{n}{2(2-p)} \rceil, n(y P_i z I_i x) = \lceil \frac{n}{2(2-p)} \rceil, n(z P_i x I_i y) = \lceil \frac{n}{2(2-p)} \rceil$.

We have: $n(x P_i y) = n - \lceil \frac{n}{2(2-p)} \rceil, n(y P_i x) = \lceil \frac{n}{2(2-p)} \rceil, n(y P_i z) = n - \lceil \frac{n}{2(2-p)} \rceil$, $n(z P_i y) = \lceil \frac{n}{2(2-p)} \rceil, n(x P_i z) = n - 2\lceil \frac{n}{2(2-p)} \rceil, n(z P_i x) = 0$.

Let $\lceil \frac{n}{2(2-p)} \rceil = \frac{n}{2(2-p)} + \varepsilon$.

We have: $n(xP_iy) - \frac{1}{2}[p[n(xP_iy) + n(yP_ix)] + (1-p)n] = n - \lceil \frac{n}{2(2-p)} \rceil - \frac{1}{2}n =$
$\frac{1}{2}n - \frac{1}{2(2-p)}n - \varepsilon = \frac{1-p}{2(2-p)}n - \varepsilon > 0$, as $n > \frac{2(2-p)}{(1-p)}$. Therefore $xPy$ holds.
Similarly, $yPz$ holds.

Now, $n(xP_iz) - \frac{1}{2}[p[n(xP_iz) + n(zP_ix)] + (1-p)n] = (1 - \frac{1}{2}p)[n - \frac{1}{2-p}n - 2\varepsilon]$
$- \frac{1}{2}(1-p)n = -(2-p)\varepsilon < 0$. Therefore $\sim xPz$ holds.

$xPy \wedge yPz \wedge \sim xPz$ violates quasi-transitivity.

(viii) Let $\mathscr{D} \mid A \supseteq \{xP^syI^sz \wedge yP^tzI^tx \wedge zI^uxP^uy\}$. Consider any $(R_1, \ldots, R_n)$
$\in \mathscr{D}^n$ such that: $n(xP_iyI_iz) = \lfloor \frac{1-p}{(2-p)}n \rfloor, n(yP_izI_ix) = \lceil \frac{n}{2} \rceil - 1, n(zI_ixP_iy) = n - \lfloor \frac{1-p}{(2-p)}n \rfloor - \lceil \frac{n}{2} \rceil + 1$.

We have: $n(xP_iy) = n - \lceil \frac{n}{2} \rceil + 1, n(yP_ix) = \lceil \frac{n}{2} \rceil - 1, n(yP_iz) = \lceil \frac{n}{2} \rceil - 1, n(zP_iy)$
$= n - \lfloor \frac{1-p}{(2-p)}n \rfloor - \lceil \frac{n}{2} \rceil + 1, n(xP_iz) = \lfloor \frac{1-p}{(2-p)}n \rfloor, n(zP_ix) = 0$.
Let $\lfloor \frac{1-p}{(2-p)}n \rfloor = \frac{1-p}{(2-p)}n - \varepsilon_1, 0 \leq \varepsilon_1 < 1$; and $\lceil \frac{n}{2} \rceil = \frac{n}{2} + \varepsilon_2, 0 \leq \varepsilon_2 < 1$.
We have: $n(xP_iy) = n - \lceil \frac{n}{2} \rceil + 1 > \frac{n}{2}$. Therefore $xPy$.
$n(yP_iz) - \frac{1}{2}[p[n(yP_iz) + n(zP_iy)] + (1-p)n] = \lceil \frac{n}{2} \rceil - 1 - \frac{1}{2}[p[n - \lfloor \frac{1-p}{2-p}n \rfloor]] + (1-p)n] = \frac{n}{2} + \varepsilon_2 - 1 - \frac{1}{2}[p[n - \frac{1-p}{2-p}n + \varepsilon_1] + (1-p)n] = \frac{p(1-p)}{2(2-p)}n - \frac{1}{2}p\varepsilon_1 + \varepsilon_2 - 1 > 0$, as $n > \frac{4(2-p)}{p(1-p)}$. Therefore $yPz$.
Now, $n(xP_iz) - \frac{1}{2}[p[n(xP_iz) + n(zP_ix)] + (1-p)n] = (1 - \frac{1}{2}p)[\frac{1-p}{2-p}n - \varepsilon_1] - \frac{1}{2}(1-p)n = -(1 - \frac{1}{2}p)\varepsilon_1 < 0$. Therefore $\sim xPz$ holds.

$xPy \wedge yPz \wedge \sim xPz$ violates quasi-transitivity.

(ix) Let $\mathscr{D} \mid A \supseteq \{xP^syI^sz \wedge yI^tzP^tx \wedge zI^uxP^uy\}$. Consider any $(R_1, \ldots, R_n) \in$
$\mathscr{D}^n$ such that: $n(xP_iyI_iz) = n - \lfloor \frac{1-p}{(2-p)}n \rfloor - \lceil \frac{n}{2} \rceil + 1, n(yI_izP_ix) = \lceil \frac{n}{2} \rceil - 1, n(zI_ixP_iy) = \lfloor \frac{1-p}{(2-p)}n \rfloor$.
We  have:  $n(xP_iy) = n - \lceil \frac{n}{2} \rceil + 1, n(yP_ix) = \lceil \frac{n}{2} \rceil - 1, n(yP_iz) = 0, n(zP_iy)$
$= \lfloor \frac{1-p}{(2-p)}n \rfloor, n(xP_iz) = n - \lfloor \frac{1-p}{(2-p)}n \rfloor - \lceil \frac{n}{2} \rceil + 1, n(zP_ix) = \lceil \frac{n}{2} \rceil - 1$.
Let $\lfloor \frac{1-p}{(2-p)}n \rfloor = \frac{1-p}{(2-p)}n - \varepsilon_1, 0 \leq \varepsilon_1 < 1$; and $\lceil \frac{n}{2} \rceil = \frac{n}{2} + \varepsilon_2, 0 \leq \varepsilon_2 < 1$.
We have: $n(xP_iy) = n - \lceil \frac{n}{2} \rceil + 1 > \frac{n}{2}$. Therefore $xPy$.
$n(zP_ix) - \frac{1}{2}[p[n(zP_ix) + n(xP_iz)] + (1-p)n] = \lceil \frac{n}{2} \rceil - 1 - \frac{1}{2}[p[n - \lfloor \frac{1-p}{(2-p)}n \rfloor]] + (1-p)n] = \frac{n}{2} + \varepsilon_2 - 1 - \frac{1}{2}[p[n - \frac{1-p}{2-p}n + \varepsilon_1] + (1-p)n] = \frac{p(1-p)}{2(2-p)}n - \frac{1}{2}p\varepsilon_1 + \varepsilon_2 - 1 > 0$, as $n > \frac{4(2-p)}{p(1-p)}$. Therefore $zPx$.
Now, $n(zP_iy) - \frac{1}{2}[p[n(zP_iy) + n(yP_iz)] + (1-p)n] = (1 - \frac{1}{2}p)[\frac{1-p}{2-p}n - \varepsilon_1] - \frac{1}{2}(1-p)n = -(1 - \frac{1}{2}p)\varepsilon_1 < 0$. Therefore $\sim zPy$ holds.

$zPx \wedge xPy \wedge \sim zPy$ violates quasi-transitivity.                        $\square$

*Remark 5.2*  Let positive integer $m$ be greater than $\frac{2(2-p)}{p^2(1-p)}$. Then the above theorem holds for $n = 4m + 1 - \lceil pm \rceil$. Now let $m'$ be any positive integer greater than $m$. Then the above theorem holds for $n' = 4m' + 1 - \lceil pm' \rceil$, as $m' > \frac{2(2-p)}{p^2(1-p)}$. Thus the above theorem holds for infinitely many values of $n$.                                    $\Diamond$

## Reference

Jain, Satish K. 1986. Semi-strict majority rules: Necessary and sufficient conditions for quasi-transitivity and transitivity. Unpublished manuscript. Centre for Economic Studies and Planning, Jawaharlal Nehru University.

# Special Majority Rules

The simple majority rule is a member of the class of majority rules. Let $\frac{1}{2} \leq p < 1$. Under $p$-majority rule an alternative $x$ is socially at least as good as another alternative $y$ iff the number of individuals preferring $y$ over $x$ is less than or equal to $p$ multiplied by the number of individuals nonindifferent between $x$ and $y$. Thus, under $p$-majority rule, for an alternative $x$ to be socially better than some other alternative $y$, the number of individuals preferring $x$ over $y$ must be greater than $p$ fraction of the number of individuals nonindifferent between $x$ and $y$. If $p = \frac{1}{2}$, then the rule is the simple majority rule; and if $p > \frac{1}{2}$, then the rule is a special majority rule. The most important special majority rule is the two-thirds majority rule, often used for deciding on special matters like the constitutional amendments.

In the context of transitivity under the class of special majority rules a condition called the placement restriction turns out to be relevant. A set of orderings of a triple of alternatives satisfies the condition of placement restriction iff there is an alternative in the triple such that it is best in every ordering; or there is an alternative such that it is worst in every ordering; or there is an alternative such that it is uniquely medium in every concerned ordering; or there are two distinct alternatives such that in every ordering indifference holds between them. Let $\mathscr{D}$ be a set of orderings of the set of alternatives. It is shown in Theorem 6.1 of the chapter that if $\mathscr{D}$ is such that over every triple of alternatives the condition of placement restriction holds then $p$-majority rule, $\frac{1}{2} < p < 1$, yields transitive social weak preference relation for every profile of individual orderings belonging to $\mathscr{D}^n$. This holds regardless of the value of $p$ and regardless of the number of individuals. Furthermore, in Theorem 6.2 it is shown that for every $p \in (\frac{1}{2}, 1)$ there exists an $n$ such that $p$-majority rule yields transitive social weak preference relation for every profile of individual orderings belonging to $\mathscr{D}^n$ iff $\mathscr{D}$ is such that it satisfies the condition of placement restriction over every triple of alternatives. This implies that the satisfaction of placement restriction over every triple of alternatives is maximally sufficient for transitivity under every $p$-majority rule, $\frac{1}{2} < p < 1$. In fact what is shown is a much stronger result, namely, that there are infinitely many values of $n$ such that $p$-majority rule, $\frac{1}{2} < p < 1$, yields transitive social weak preference relation for every profile of individual orderings belonging

© Springer Nature Singapore Pte Ltd. 2019
S. K. Jain, *Domain Conditions and Social Rationality*,
https://doi.org/10.1007/978-981-13-9672-4_6

to $\mathscr{D}^n$ iff $\mathscr{D}$ is such that it satisfies the condition of placement restriction over every triple of alternatives.

For the simple majority rule we have the result that if the number of individuals is greater than or equal to five then every profile of individual orderings belonging to $\mathscr{D}^n$ yields quasi-transitive social weak preference relation iff $\mathscr{D}$ is such that it satisfies the condition of Latin Square partial agreement over every triple of alternatives. A similar result holds for every special majority rule. It is shown in this chapter that for every $p \in (\frac{1}{2}, 1)$ there exists an $n'$ such that if the number of individuals $n$ is greater than or equal to $n'$ then $p$-majority rule yields quasi-transitive social weak preference relation for every profile of individual orderings belonging to $\mathscr{D}^n$ iff $\mathscr{D}$ is such that over every triple of alternatives the condition of Latin Square partial agreement holds.

For the two-thirds majority rule more complete results are derived. It is shown that if the number of individuals is greater than or equal to 10 then: (i) The two-thirds majority rule yields transitive social weak preference relation for every profile of individual orderings belonging to $\mathscr{D}^n$ iff $\mathscr{D}$ is such that over every triple of alternatives the condition of placement restriction holds; and (ii) the two-thirds majority rule yields quasi-transitive social weak preference relation for every profile of individual orderings belonging to $\mathscr{D}^n$ iff $\mathscr{D}$ is such that over every triple of alternatives the condition of Latin Square partial agreement holds.

The chapter is divided into four sections. The first section defines and discusses the placement restriction. Sections 6.2 and 6.3 are concerned, respectively, with conditions for transitivity and quasi-transitivity under the class of special majority rules. The fourth section analyses the domain conditions for the two-thirds majority rule.[1]

## 6.1 Placement Restriction

The Class of Majority Rules:

Let $p \in [\frac{1}{2}, 1)$. $p$-majority rule $f : \mathscr{T}^n \mapsto \mathscr{C}$ is defined by: $(\forall (R_1, \ldots, R_n) \in \mathscr{T}^n)(\forall x, y \in S)[xPy \leftrightarrow n(xP_iy) > p[n(xP_iy) + n(yP_ix)]]$; or equivalently by $(\forall (R_1, \ldots, R_n) \in \mathscr{T}^n)(\forall x, y \in S)[xRy \leftrightarrow n(xP_iy) \geq \frac{1-p}{p}n(yP_ix)]$. The class of majority rules consists of all $p$-majority rules, $p \in [\frac{1}{2}, 1)$. If $p = \frac{1}{2}$, the definition of $p$-majority rule reduces to that of simple majority rule. Excluding the simple majority rule from the class of majority rules we obtain the class of special majority rules. Thus the class of special majority rules consists of all $p$-majority rules where $p \in (\frac{1}{2}, 1)$.

Placement Restriction (PR): $\mathscr{D} \subseteq \mathscr{T}$ satisfies PR over the triple $A \subseteq S$ iff ($\exists$ distinct $x, y, z \in A)[(\forall R \in \mathscr{D}|A)(xRy \wedge xRz) \vee (\forall$ concerned $R \in \mathscr{D}|A)(zPxPy \vee yPx$ $Pz) \vee (\forall R \in \mathscr{D}|A)(yRx \wedge zRx) \vee (\forall R \in \mathscr{D}|A)(xIy)]$. $\mathscr{D}$ satisfies PR iff it satisfies PR over every triple of alternatives contained in $S$.

---

[1]This chapter relies on Jain (1983).

Less formally, a set of orderings $\mathcal{D}$ satisfies PR over a triple of alternatives $A$ iff (i) there exists an alternative in the triple such that it is best in every ordering belonging to $\mathcal{D}|A$; or (ii) there exists an alternative in the triple such that it is uniquely medium in every concerned ordering belonging to $\mathcal{D}|A$; or (iii) there exists an alternative in the triple such that it is worst in every ordering belonging to $\mathcal{D}|A$; or (iv) there exists a pair of distinct alternatives belonging to the triple such that indifference holds between them in every ordering belonging to $\mathcal{D}|A$.

**Lemma 6.1** *Let $\mathcal{D} \subseteq \mathcal{T}$. Let $A \subseteq S$ be a triple of alternatives. Then $\mathcal{D}$ violates the condition of placement restriction over $A$ iff $\mathcal{D}|A$ contains one of the following 10 sets of orderings of $A$, except for a formal interchange of alternatives.*

| | | | |
|---|---|---|---|
| (i) | $x P^s y P^s z$<br>$y P^t z P^t x$ | (ii) | $x P^s y P^s z$<br>$z P^t x I^t y$ |
| (iii) | $x P^s y P^s z$<br>$y I^t z P^t x$ | (iv) | $x P^s y P^s z$<br>$z P^t y P^t x$<br>$y P^u z I^u x$ |
| (v) | $x P^s y P^s z$<br>$z P^t y P^t x$<br>$z I^u x P^u y$ | (vi) | $x P^s y P^s z$<br>$y P^t z I^t x$<br>$z I^u x P^u y$ |
| (vii) | $x P^s y I^s z$<br>$y P^t z I^t x$<br>$z I^u x P^u y$ | (viii) | $x I^s y P^s z$<br>$y P^t z I^t x$<br>$z I^u x P^u y$ |
| (ix) | $x P^s y I^s z$<br>$y P^t z I^t x$<br>$z P^u x I^u y$ | (x) | $x I^s y P^s z$<br>$y I^t z P^t x$<br>$z I^u x P^u y$ |

*Proof* It can be easily checked that there does not exist an alternative such that it is best in every $R^i \in \mathcal{D}|A$ iff $\mathcal{D}|A$ contains one of the following nine sets of orderings of $A$, except for a formal interchange of alternatives:

| | | | |
|---|---|---|---|
| (A) | $x P^s y P^s z$<br>$y P^t z P^t x$ | (B) | $x P^s y P^s z$<br>$y P^t x P^t z$ |
| (C) | $x P^s y P^s z$<br>$z P^t y P^t x$ | (D) | $x P^s y P^s z$<br>$y P^t z I^t x$ |
| (E) | $x P^s y P^s z$<br>$z P^t x I^t y$ | (F) | $x P^s y P^s z$<br>$y I^t z P^t x$ |
| (G) | $x P^s y I^s z$<br>$y P^t z I^t x$ | (H) | $x P^s y I^s z$<br>$y I^t z P^t x$ |
| (I) | $x I^s y P^s z$<br>$y I^t z P^t x$<br>$z I^u x P^u y$ | | |

A, E, F and I are the same as (i), (ii), (iii) and (x), respectively. Therefore it suffices to consider only the remaining five sets.

(B) would violate PR iff an ordering in which $z$ is not worst is included. With the inclusion of such an ordering, except for a formal interchange of alternatives, one of the sets (i)–(x) is contained in the set of orderings. (C) would violate PR iff a concerned ordering in which $y$ is not uniquely medium is included. With the inclusion of such an ordering, except for a formal interchange of alternatives, one of the sets (i)–(x) is contained in the set of orderings. (D) would violate PR iff an ordering in which $z$ is not worst is included. With the inclusion of such an ordering, except for a formal interchange of alternatives, one of the sets (i)–(x) is contained in the set of orderings. (G) would violate PR iff an ordering in which $z$ is not worst is included. With the inclusion of such an ordering, except for a formal interchange of alternatives, one of the sets (i)–(x) is contained in the set of orderings. (H) would violate PR iff an ordering in which $y$ is not indifferent to $z$ is included. With the inclusion of such an ordering, except for a formal interchange of alternatives, one of the sets (i)–(x) is contained in the set of orderings. Proof of the lemma is completed by noting that all 10 sets (i)–(x) violate PR.                                              □

## 6.2  Transitivity Under Special Majority Rules

**Theorem 6.1** *Let $f : \mathscr{T}^n \mapsto \mathscr{C}$ be a p-majority rule, $p \in (\frac{1}{2}, 1)$. Let $\mathscr{D} \subseteq \mathscr{T}$. Then $f$ yields transitive social $R$, $R = f(R_1, \ldots, R_n)$, for every $(R_1, \ldots, R_n) \in \mathscr{D}^n$ if $\mathscr{D}$ satisfies the condition of placement restriction.*

*Proof* Suppose $f$ does not yield transitive social $R$ for every $(R_1, \ldots, R_n) \in \mathscr{D}^n$. Then $(\exists (R_1, \ldots, R_n) \in \mathscr{D}^n)(\exists x, y, z \in S)(xRy \wedge yRz \wedge zPx)$.

Let $A = \{x, y, z\}$ and let $n_A$ denote the number of individuals concerned over $A$.

$xRy \rightarrow n(yP_ix) \leq p[n(xP_iy) + n(yP_ix)]$

$\rightarrow n(xP_iy) \geq (1 - p)[n(xP_iy) + n(yP_ix)]$

$\rightarrow n(xP_iy) + n(R_i$ concerned over $A \wedge xI_iy) \geq (1 - p)n_A + p\, n(R_i$ concerned over $A \wedge xI_iy)$

$\rightarrow n(R_i$ concerned over $A \wedge xR_iy) \geq (1 - p)n_A$           (T6.1-1)

Similarly, $yRz \rightarrow n(R_i$ concerned over $A \wedge yR_iz) \geq (1 - p)n_A$           (T6.1-2)

$zPx \rightarrow n(zP_ix) > p[n(zP_ix) + n(xP_iz)]$

$\rightarrow n(R_i$ concerned over $A \wedge zR_ix) > pn_A$           (T6.1-3)

(T6.1-1) $\wedge$ (T6.1-3) $\rightarrow$ $(\exists i \in N)[R_i$ is concerned over $A \wedge zR_ixR_iy]$           (T6.1-4)

(T6.1-2) $\wedge$ (T6.1-3) $\rightarrow$ $(\exists j \in N)[R_j$ is concerned over $A \wedge yR_jzR_jx]$           (T6.1-5)

(T6.1-4) $\rightarrow$ $(\exists i \in N)[zP_iy]$           (T6.1-6)

(T6.1-6) $\wedge\ yRz \rightarrow (\exists k \in N)[yP_kz]$ (T6.1-7)

(T6.1-5) $\rightarrow (\exists j \in N)[yP_jx]$ (T6.1-8)

(T6.1-8) $\wedge\ xRy \rightarrow (\exists l \in N)[xP_ly]$ (T6.1-9)

$zPx \rightarrow (\exists m \in N)[zP_mx]$ (T6.1-10)

(T6.1-6), (T6.1-8) and (T6.1-10) imply that there does not exist distinct $a, b \in A$
such that $(\forall R \in \mathscr{D}|A)(aIb)$ (T6.1-11)

(T6.1-8), (T6.1-6) and (T6.1-7) imply that there does not exist an alternative in $A$
such that it is best in every $R \in \mathscr{D}|A$ (T6.1-12)

(T6.1-9), (T6.1-7) and (T6.1-10) imply that there does not exist an alternative in $A$
such that it is worst in every $R \in \mathscr{D}|A$ (T6.1-13)

(T6.1-4) and (T6.1-5) imply that there does not exist an alternative in $A$ such that it
is uniquely medium in every concerned $R \in \mathscr{D}|A$ (T6.1-14)

(T6.1-11)–(T6.1-14) imply that PR is violated; establishing the theorem. $\square$

**Theorem 6.2** *Let $f : \mathscr{T}^n \mapsto \mathscr{C}$ be a p-majority rule, $p \in (\frac{1}{2}, 1)$. Let $\mathscr{D} \subseteq \mathscr{T}$.
Then there exists a positive integer n such that $f$ yields transitive social $R$, $R = f(R_1, \ldots, R_n)$, for every $(R_1, \ldots, R_n) \in \mathscr{D}^n$ only if $\mathscr{D}$ satisfies the condition of
placement restriction.*

*Proof* Let $f : \mathscr{T}^n \mapsto \mathscr{C}$ be a p-majority rule, $p \in (\frac{1}{2}, 1)$; and let $\mathscr{D}$ violate PR. Then
$\mathscr{D}$ violates PR over some triple $A \subseteq S$. By Lemma 6.1, $\mathscr{D}|A$ must contain one of the
10 sets mentioned in the statement of the lemma, except for a formal interchange of
alternatives. In what follows we consider each of these 10 cases.

Choose a positive integer $m > \frac{p}{1-p}$. Choose a positive integer $k > \frac{1}{2p-1}(m+1)$.
Let $n = 2k + 2 + \lfloor\frac{p}{1-p}k\rfloor$.
Let $\lfloor\frac{p}{1-p}k\rfloor = \frac{p}{1-p}k - \varepsilon$, where $0 \le \varepsilon < 1$.

We now show that for such a choice of $n$, for every $\mathscr{D} \mid A$ containing one of the
10 sets (i)–(x) there exists a profile $(R_1, \ldots, R_n) \in \mathscr{D}^n$ which yields non-transitive
$R = f(R_1, \ldots, R_n)$ under the $p$-majority rule.

If $\mathscr{D} \mid A$ contains set (i) then consider any $(R_1, \ldots, R_n) \in \mathscr{D}^n$ such that $n(xP^sy P^sz)$
$= \lfloor pn \rfloor$ and $n(yP^tzP^tx) = n - \lfloor pn \rfloor$. We have: $n(xP_iy) = \lfloor pn \rfloor$, $n(yP_ix) = n - \lfloor pn \rfloor$, $n(yP_iz) = n$, $n(zP_iy) = 0$, $n(xP_iz) = \lfloor pn \rfloor$, $n(zP_ix) = n - \lfloor pn \rfloor$. As $\lfloor pn \rfloor$
$\le pn$ and $n - \lfloor pn \rfloor < pn$ in view of $n > \frac{1}{2p-1}$, we obtain: $xIy \wedge yPz \wedge xIz$.

If $\mathscr{D} \mid A$ contains set (ii) then consider any $(R_1, \ldots, R_n) \in \mathscr{D}^n$ such that $n(xP^sy P^sz)$
$= \lfloor pn \rfloor$ and $n(zP^txI^ty) = n - \lfloor pn \rfloor$. We have: $n(xP_iy) = \lfloor pn \rfloor$, $n(yP_ix) = 0$,
$n(yP_iz) = \lfloor pn \rfloor$, $n(zP_iy) = n - \lfloor pn \rfloor$, $n(xP_iz) = \lfloor pn \rfloor$, $n(zP_ix) = n - \lfloor pn \rfloor$.
Consequently we obtain: $xPy \wedge yIz \wedge xIz$.

If $\mathscr{D} \mid A$ contains set (iii) then consider any $(R_1, \ldots, R_n) \in \mathscr{D}^n$ such that $n(x P^s y P^s z)$ $= \lfloor pn \rfloor$ and $n(y I^t z P^t x) = n - \lfloor pn \rfloor$. We have: $n(x P_i y) = \lfloor pn \rfloor, n(y P_i x) = n - \lfloor pn \rfloor, n(y P_i z) = \lfloor pn \rfloor, n(z P_i y) = 0, n(x P_i z) = \lfloor pn \rfloor, n(z P_i x) = n - \lfloor pn \rfloor$. Consequently we obtain: $x I y \wedge y P z \wedge x I z$.

If $\mathscr{D} \mid A$ contains set (iv) then consider any $(R_1, \ldots, R_n) \in \mathscr{D}^n$ such that $n(x P^s y P^s z)$ $= n - m - \lceil(1 - p)n\rceil$, $n(z P^t y P^t x) = \lceil(1 - p)n\rceil - 1$ and $n(y P^u z I^u x) = m + 1$. We have: $n(x P_i y) = n - m - \lceil(1 - p)n\rceil, n(y P_i x) = m + \lceil(1 - p)n\rceil, n(y P_i z) = n + 1 - \lceil(1 - p)n\rceil, n(z P_i y) = \lceil(1 - p)n\rceil - 1, n(x P_i z) = n - m - \lceil(1 - p)n\rceil, n(z P_i x) = \lceil(1 - p)n\rceil - 1$.

As $m + \lceil(1 - p)n\rceil > (1 - p)n$ it follows that $n(x P_i y) < pn$, therefore $\sim x P y$. We have: $m + \lceil(1 - p)n\rceil < m + 1 + (1 - p)n < (2p - 1)k + (1 - p)n < (2p - 1)n + (1 - p)n = pn$, therefore $\sim y P x$.
As $\lceil(1 - p)n\rceil < (1 - p)n + 1, n + 1 - \lceil(1 - p)n\rceil > pn$. Consequently $y P z$.
$n(x P_i z) = n - m - \lceil(1 - p)n\rceil \le n - m - (1 - p)n = pn - m = p(n - m - 1) - (1 - p)m + p < p(n - m - 1) - (1 - p)\frac{p}{1-p} + p = p(n - m - 1) = p[n(x P_i z) + n(z P_i x)]$. Therefore $\sim x P z$.

$p(n - m - 1) = pn - p(m + 1) > pn - (2p - 1)pk > pn - (2p - 1)pn = 2p(1 - p)n > (1 - p)n > \lceil(1 - p)n\rceil - 1 = n(z P_i x)$. Consequently, $\sim z P x$. Thus we have: $x I y \wedge y P z \wedge x I z$.

If $\mathscr{D} \mid A$ contains set (v) then consider any $(R_1, \ldots, R_n) \in \mathscr{D}^n$ such that $n(x P^s y P^s z)$ $= n - m - \lceil(1 - p)n\rceil$, $n(z P^t y P^t x) = \lceil(1 - p)n\rceil - 1$ and $n(z I^u x P^u y) = m + 1$. We have: $n(x P_i y) = n + 1 - \lceil(1 - p)n\rceil, n(y P_i x) = \lceil(1 - p)n\rceil - 1, n(y P_i z) = n - m - \lceil(1 - p)n\rceil, n(z P_i y) = m + \lceil(1 - p)n\rceil, n(x P_i z) = n - m - \lceil(1 - p)n\rceil, n(z P_i x) = \lceil(1 - p)n\rceil - 1$.
$n(x P_i y) = n + 1 - \lceil(1 - p)n\rceil \to x P y$
$n(y P_i z) = n - m - \lceil(1 - p)n\rceil \wedge n(z P_i y) = m + \lceil(1 - p)n\rceil \to y I z$
$n(x P_i z) = n - m - \lceil(1 - p)n\rceil \wedge n(z P_i x) = \lceil(1 - p)n\rceil - 1 \to x I z$.

If $\mathscr{D} \mid A$ contains set (vi) then consider any $(R_1, \ldots, R_n) \in \mathscr{D}^n$ such that $n(x P^s y P^s z) = n - 2 - 2\lfloor(1 - p)n\rfloor, n(y P^t z I^t x) = \lfloor(1 - p)n\rfloor + 1$ and $n(z I^u x P^u y) = \lfloor(1 - p)n\rfloor + 1$. We have: $n(x P_i y) = n - \lfloor(1 - p)n\rfloor - 1, n(y P_i x) = \lfloor(1 - p)n\rfloor + 1, n(y P_i z) = n - \lfloor(1 - p)n\rfloor - 1, n(z P_i y) = \lfloor(1 - p)n\rfloor + 1, n(x P_i z) = n - 2 - 2\lfloor(1 - p)n\rfloor, n(z P_i x) = 0$.
$n(x P_i y) = n - \lfloor(1 - p)n\rfloor - 1 \wedge n(y P_i x) = \lfloor(1 - p)n\rfloor + 1 \to x I y$
$n(y P_i z) = n - \lfloor(1 - p)n\rfloor - 1 \wedge n(z P_i y) = \lfloor(1 - p)n\rfloor + 1 \to y I z$
$n(x P_i z) = n - 2 - 2\lfloor(1 - p)n\rfloor \wedge n(z P_i x) = 0 \to x P z$.

If $\mathscr{D} \mid A$ contains set (vii) then consider any $(R_1, \ldots, R_n) \in \mathscr{D}^n$ such that $n(x P^s y I^s z) = n - 2 - 2\lfloor(1 - p)n\rfloor, n(y P^t z I^t x) = \lfloor(1 - p)n\rfloor + 1$ and $n(z I^u x P^u y) = \lfloor(1 - p)n\rfloor + 1$. We have: $n(x P_i y) = n - \lfloor(1 - p)n\rfloor - 1, n(y P_i x) = \lfloor(1 - p)n\rfloor + 1, n(y P_i z) = \lfloor(1 - p)n\rfloor + 1, n(z P_i y) = \lfloor(1 - p)n\rfloor + 1, n(x P_i z) = n - 2 - 2\lfloor(1 - p)n\rfloor, n(z P_i x) = 0$.
$n(x P_i y) = n - \lfloor(1 - p)n\rfloor - 1 \wedge n(y P_i x) = \lfloor(1 - p)n\rfloor + 1 \to x I y$
$n(y P_i z) = \lfloor(1 - p)n\rfloor + 1, n(z P_i y) = \lfloor(1 - p)n\rfloor + 1 \to y I z$
$n(x P_i z) = n - 2 - 2\lfloor(1 - p)n\rfloor, n(z P_i x) = 0 \to x P z$.

If $\mathscr{D} \mid A$ contains set (viii) then consider any $(R_1, \ldots, R_n) \in \mathscr{D}^n$ such that $n(xI^s y P^s z) = n - 2 - 2\lfloor(1-p)n\rfloor, n(yP^t zI^t x) = \lfloor(1-p)n\rfloor + 1$ and $n(zI^u xP^u y) = \lfloor(1-p)n\rfloor + 1$. We have: $n(xP_i y) = \lfloor(1-p)n\rfloor + 1, n(yP_i x) = \lfloor(1-p)n\rfloor + 1, n(yP_i z) = n - \lfloor(1-p)n\rfloor - 1, n(zP_i y) = \lfloor(1-p)n\rfloor + 1, n(xP_i z) = n - 2 - 2\lfloor(1-p)n\rfloor, n(zP_i x) = 0.$

$n(xP_i y) = \lfloor(1-p)n\rfloor + 1 \wedge n(yP_i x) = \lfloor(1-p)n\rfloor + 1 \rightarrow xIy$

$n(yP_i z) = n - \lfloor(1-p)n\rfloor - 1 \wedge n(zP_i y) = \lfloor(1-p)n\rfloor + 1 \rightarrow yIz$

$n(xP_i z) = n - 2 - 2\lfloor(1-p)n\rfloor \wedge n(zP_i x) = 0 \rightarrow xPz.$

If $\mathscr{D} \mid A$ contains set (ix) then consider any $(R_1, \ldots, R_n) \in \mathscr{D}^n$ such that $n(xP^s yI^s z) = \lfloor\frac{p}{1-p}k\rfloor + 1 = \frac{p}{1-p}k - \varepsilon + 1, 0 \leq \varepsilon < 1, n(yP^t zI^t x) = k$ and $n(zP^u xI^u y) = k + 1$. We have: $n(xP_i y) = \lfloor\frac{p}{1-p}k\rfloor + 1, n(yP_i x) = k, n(yP_i z) = k, n(zP_i y) = k + 1, n(xP_i z) = \lfloor\frac{p}{1-p}k\rfloor + 1, n(zP_i x) = k + 1.$

$p[n(xP_i y) + n(yP_i x)] = p[\lfloor\frac{p}{1-p}k\rfloor + 1 + k] = p[\frac{p}{1-p}k - \varepsilon + 1 + k] = \frac{p}{1-p}k - p\varepsilon + p = \lfloor\frac{p}{1-p}k\rfloor + \varepsilon - p\varepsilon + p = \lfloor\frac{p}{1-p}k\rfloor + \varepsilon(1 - p) + p < \lfloor\frac{p}{1-p}k\rfloor + (1 - p) + p = \lfloor\frac{p}{1-p}k\rfloor + 1 = n(xP_i y)$. Therefore $xPy$.

$p[n(yP_i z) + n(zP_i y)] = p(2k + 1) = 2pk + p = k + (2p - 1)k + p > k + (m + 1) + p > k + 1$. Therefore $yIz$.

$p[n(xP_i z) + n(zP_i x)] = p[\lfloor\frac{p}{1-p}k\rfloor + 2 + k] = p[\frac{p}{1-p}k - \varepsilon + 2 + k] = \frac{p}{1-p}k - p\varepsilon + 2p = \lfloor\frac{p}{1-p}k\rfloor + \varepsilon - p\varepsilon + 2p = \lfloor\frac{p}{1-p}k\rfloor + \varepsilon(1 - p) + 2p > \lfloor\frac{p}{1-p}k\rfloor + 1 = n(xP_i z)$. Therefore $xIz$.

If $\mathscr{D} \mid A$ contains set (x) then consider any $(R_1, \ldots, R_n) \in \mathscr{D}^n$ such that $n(xI^s yP^s z) = \lfloor\frac{p}{1-p}k\rfloor + 1, n(yI^t zP^t x) = k$ and $n(zI^u xP^u y) = k + 1$. We have: $n(xP_i y) = k + 1, n(yP_i x) = k, n(yP_i z) = \lfloor\frac{p}{1-p}k\rfloor + 1, n(zP_i y) = k + 1, n(xP_i z) = \lfloor\frac{p}{1-p}k\rfloor + 1, n(zP_i x) = k.$

$n(xP_i y) = k + 1 \wedge n(yP_i x) = k \rightarrow xIy.$

$n(yP_i z) = \lfloor\frac{p}{1-p}k\rfloor + 1 \wedge n(zP_i y) = k + 1 \rightarrow yIz.$

$n(xP_i z) = \lfloor\frac{p}{1-p}k\rfloor + 1 \wedge n(zP_i x) = k \rightarrow xPz.$ $\qquad\square$

*Remark 6.1* Let $f$ be a $p$-majority rule, $p \in (\frac{1}{2}, 1)$. Theorem 6.2 asserts the existence of an $n \in \mathbb{N}$ such that $f$ yields transitive social $R$ for every $(R_1, \ldots, R_n) \in \mathscr{D}^n$ only if $\mathscr{D}$ satisfies the condition of placement restriction. If $n$ is such that $n = 2k + 2 + \lfloor\frac{p}{1-p}k\rfloor, k > \frac{1}{2p-1}(m + 1), m > \frac{p}{1-p}$, then by Theorem 6.2 $f$ yields transitive social $R$ for every $(R_1, \ldots, R_n) \in \mathscr{D}^n$ only if $\mathscr{D}$ satisfies the condition of placement restriction. Now consider any $m' > m, k' > \frac{1}{2p-1}(m' + 1), n' = 2k' + 2 + \lfloor\frac{p}{1-p}k'\rfloor$. Then it follows that $f$ yields transitive social $R$ for every $(R_1, \ldots, R_{n'}) \in \mathscr{D}^{n'}$ only if $\mathscr{D}$ satisfies the condition of placement restriction. Thus there are an infinite number of values of $n$ for which the assertion '$f$ yields transitive social $R$ for every $(R_1, \ldots, R_n) \in \mathscr{D}^n$ only if $\mathscr{D}$ satisfies the condition of placement restriction' holds. $\qquad\Diamond$

Combining Theorems 6.1 and 6.2, we obtain:

**Theorem 6.3** *Let $f : \mathcal{T}^n \mapsto \mathcal{C}$ be a p-majority rule, $p \in (\frac{1}{2}, 1)$. Let $n = 2k + 2 + \lfloor \frac{p}{1-p} k \rfloor$, where $k$ is a positive integer greater than $\frac{1}{2p-1}(m+1)$, $m$ a positive integer greater than $\frac{p}{1-p}$. Let $\mathcal{D} \subseteq \mathcal{T}$. Then $f$ yields transitive social $R$, $R = f(R_1, \ldots, R_n)$, for every $(R_1, \ldots, R_n) \in \mathcal{D}^n$ iff $\mathcal{D}$ satisfies the condition of placement restriction.*

## 6.3  Quasi-transitivity Under Special Majority Rules

**Theorem 6.4** *Let $f : \mathcal{T}^n \mapsto \mathcal{C}$ be a p-majority rule, $p \in [\frac{1}{2}, 1)$. Let $\mathcal{D} \subseteq \mathcal{T}$. Then $f$ yields quasi-transitive social $R$, $R = f(R_1, \ldots, R_n)$, for every $(R_1, \ldots, R_n) \in \mathcal{D}^n$ if $\mathcal{D}$ satisfies the condition of Latin Square partial agreement.*

*Proof* Suppose $f$ does not yield quasi-transitive social $R$ for every $(R_1, \ldots, R_n) \in \mathcal{D}^n$. Then $(\exists (R_1, \ldots, R_n) \in \mathcal{D}^n)(\exists x, y, z \in S)(xPy \wedge yPz \wedge zRx)$.

$$xPy \rightarrow n(xP_iy) > \tfrac{p}{1-p}n(yP_ix) \tag{T6.4-1}$$

$$yPz \rightarrow n(yP_iz) > \tfrac{p}{1-p}n(zP_iy) \tag{T6.4-2}$$

$$zRx \rightarrow n(zP_ix) \geq \tfrac{1-p}{p}n(xP_iz) \tag{T6.4-3}$$

(T6.4-1), (T6.4-2) and (T6.4-3) imply, respectively:

$$n(zP_ixP_iy) + n(zI_ixP_iy) + n(xP_izP_iy) + n(xP_izI_iy) + n(xP_iyP_iz)$$
$$> \tfrac{p}{1-p}[n(zP_iyP_ix) + n(zI_iyP_ix) + n(yP_izP_ix) + n(yP_izI_ix) +$$
$$n(yP_ixP_iz)] \tag{T6.4-4}$$

$$n(xP_iyP_iz) + n(xI_iyP_iz) + n(yP_ixP_iz) + n(yP_ixI_iz) + n(yP_izP_ix)$$
$$> \tfrac{p}{1-p}[n(xP_izP_iy) + n(xI_izP_iy) + n(zP_ixP_iy) + n(zP_ixI_iy) +$$
$$n(zP_iyP_ix)] \tag{T6.4-5}$$

$$n(yP_izP_ix) + n(yI_izP_ix) + n(zP_iyP_ix) + n(zP_iyI_ix) + n(zP_ixP_iy)$$
$$\geq \tfrac{1-p}{p}[n(yP_ixP_iz) + n(yI_ixP_iz) + n(xP_iyP_iz) + n(xP_iyI_iz) +$$
$$n(xP_izP_iy)] \tag{T6.4-6}$$

Multiplying (T6.4-6) by $\frac{p}{1-p}$ we obtain:

$$\tfrac{p}{1-p}[n(yP_izP_ix) + n(yI_izP_ix) + n(zP_iyP_ix) + n(zP_iyI_ix) + n(zP_i$$
$$xP_iy)] \geq n(yP_ixP_iz) + n(yI_ixP_iz) + n(xP_iyP_iz) + n(xP_iyI_iz) +$$
$$n(xP_izP_iy) \tag{T6.4-7}$$

Adding (T6.4-4) and (T6.4-5), adding (T6.4-5) and (T6.4-7), and adding (T6.4-7) and (T6.4-4), we obtain, respectively:

$2n(x P_i y P_i z) + n(x P_i y I_i z) + n(x I_i y P_i z) > \frac{2p-1}{1-p}[n(y P_i z P_i x) +$
$n(z P_i x P_i y) + n(y P_i z I_i x) + n(z I_i x P_i y) + n(x P_i z P_i y) + n(y P_i x P_i z)]$
$+ \frac{2p}{1-p} n(z P_i y P_i x) + \frac{p}{1-p}[n(z P_i y I_i x) + n(z I_i y P_i x)]$           (T6.4-8)

$\frac{1}{1-p} n(y P_i z P_i x) + n(y P_i z I_i x) + \frac{p}{1-p} n(y I_i z P_i x) > \frac{1}{1-p} n(x P_i z P_i y) + n(x P_i z I_i y) +$
$\frac{p}{1-p} n(x I_i z P_i y)$           (T6.4-9)

$\frac{1}{1-p} n(z P_i x P_i y) + \frac{p}{1-p} n(z P_i x I_i y) + n(z I_i x P_i y) > \frac{1}{1-p} n(y P_i x P_i z) + \frac{p}{1-p}$
$n(y P_i x I_i z) + n(y I_i x P_i z)$           (T6.4-10)

Adding (T6.4-4), (T6.4-5) and (T6.4-7) we obtain:

$n(x P_i y P_i z) + n(y P_i z P_i x) + n(z P_i x P_i y) > \frac{p}{1-p}[n(x P_i z P_i y) + n(z P_i$
$y P_i x) + n(y P_i x P_i z)] + \frac{2p-1}{1-p}[n(x I_i z P_i y) + n(y P_i x I_i z)]$           (T6.4-11)

Adding (T6.4-11) to (T6.4-4), (T6.4-5), (T6.4-7) we obtain, respectively:

$2n(x P_i y P_i z) + 2n(z P_i x P_i y) + n(x P_i y I_i z) + \frac{2-3p}{1-p} n(z I_i x P_i y) > \frac{2p}{1-p}$
$[n(z P_i y P_i x) + n(y P_i x P_i z)] + \frac{p}{1-p} n(z I_i y P_i x) + \frac{3p-1}{1-p} n(y P_i x I_i z) +$
$\frac{2p-1}{1-p}[n(x P_i z P_i y) + n(y P_i z P_i x)]$           (T6.4-12)

$2n(x P_i y P_i z) + 2n(y P_i z P_i x) + n(x I_i y P_i z) + \frac{2-3p}{1-p} n(y P_i z I_i x) > \frac{2p}{1-p}$
$[n(x P_i z P_i y) + n(z P_i y P_i x)] + \frac{p}{1-p} n(z P_i y I_i x) + \frac{3p-1}{1-p} n(x I_i z P_i y) +$
$\frac{2p-1}{1-p}[n(y P_i x P_i z) + n(z P_i x P_i y)]$           (T6.4-13)

$\frac{1}{1-p}[n(y P_i z P_i x) + n(z P_i x P_i y)] + \frac{p}{1-p}[n(y I_i z P_i x) + n(z P_i x I_i y)] > \frac{1}{1-p}[n(x P_i$
$z P_i y) + n(y P_i x P_i z)] + n(x P_i z I_i y) + n(y I_i x P_i z) + \frac{2p-1}{1-p}$
$[n(x I_i z P_i y) + n(y P_i x I_i z)]$           (T6.4-14)

We have $\frac{2p-1}{1-p} \geq 0$ as $\frac{1}{2} \leq p < 1$. Therefore, (T6.4-8)–(T6.4-10) imply, respectively:

$2n(x P_i y P_i z) + n(x P_i y I_i z) + n(x I_i y P_i z) > 0$           (T6.4-15)

$\frac{1}{1-p} n(y P_i z P_i x) + n(y P_i z I_i x) + \frac{p}{1-p} n(y I_i z P_i x) > 0$           (T6.4-16)

$\frac{1}{1-p} n(z P_i x P_i y) + \frac{p}{1-p} n(z P_i x I_i y) + n(z I_i x P_i y) > 0$           (T6.4-17)

(T6.4-15)–(T6.4-17) imply, respectively:

$(\exists$ concerned $R_i|\{x, y, z\})(x R_i y R_i z)$           (T6.4-18)

$(\exists$ concerned $R_j|\{x, y, z\})(y R_j z R_j x)$           (T6.4-19)

$(\exists$ concerned $R_k|\{x, y, z\})(z R_k x R_k y)$           (T6.4-20)

(T6.4-18)–(T6.4-20) imply that $\mathscr{D}|\{x, y, z\}$ contains $LS(xyzx)$.           (T6.4-21)

(T6.4-12)–(T6.4-14) imply, respectively:

$$n(xP_iyP_iz) + n(zP_ixP_iy) + n(xP_iyI_iz) + n(zI_ixP_iy) > 0 \qquad \text{(T6.4-22)}$$

$$n(xP_iyP_iz) + n(yP_izP_ix) + n(xI_iyP_iz) + n(yP_izI_ix) > 0 \qquad \text{(T6.4-23)}$$

$$n(yP_izP_ix) + n(zP_ixP_iy) + n(yI_izP_ix) + n(zP_ixI_iy) > 0 \qquad \text{(T6.4-24)}$$

$$\text{(T6.4-22)} \rightarrow (\exists R_i)[R_i | \{x, y, z\} \in T[LS(xyzx)] \cap \mathscr{D}| \{x, y, z\} \wedge xP_iy] \qquad \text{(T6.4-25)}$$

$$\text{(T6.4-23)} \rightarrow (\exists R_j)[R_j | \{x, y, z\} \in T[LS(xyzx)] \cap \mathscr{D}| \{x, y, z\} \wedge yP_jz] \qquad \text{(T6.4-26)}$$

$$\text{(T6.4-24)} \rightarrow (\exists R_k)[R_k | \{x, y, z\} \in T[LS(xyzx)] \cap \mathscr{D}| \{x, y, z\} \wedge zP_kx] \qquad \text{(T6.4-27)}$$

$$\text{(T6.4-11)} \rightarrow (\exists l \in N)[xP_lyP_lz \vee yP_lzP_lx \vee zP_lxP_ly] \qquad \text{(T6.4-28)}$$

(T6.4-21), (T6.4-28), (T6.4-25), (T6.4-26), and (T6.4-27) imply that LSPA is violated; establishing the theorem. $\qquad\square$

**Theorem 6.5** *Let $f : \mathscr{T}^n \mapsto \mathscr{C}$ be a p-majority rule, $p \in (\frac{1}{2}, 1)$. Let $\mathscr{D} \subseteq \mathscr{T}$. Then there exists a positive integer $n$ such that $f$ yields quasi-transitive social $R$, $R = f(R_1, \ldots, R_n)$, for every $(R_1, \ldots, R_n) \in \mathscr{D}^n$ only if $\mathscr{D}$ satisfies the condition of Latin Square partial agreement.*

*Proof* Let $f : \mathscr{T}^n \mapsto \mathscr{C}$ be a $p$-majority rule, $p \in (\frac{1}{2}, 1)$; and let $\mathscr{D}$ violate LSPA. Then $\mathscr{D}$ violates LSPA over some triple $A \subseteq S$. From the definition of LSPA it follows that $\mathscr{D}|A$ must contain one of the following six sets, except for a formal interchange of alternatives.

| | | | |
|---|---|---|---|
| $(i)$ | $xP^syP^sz$ | $(ii)$ | $xP^syP^sz$ |
| | $yP^tzP^tx$ | | $yP^tzP^tx$ |
| | $zP^uxP^uy$ | | $zP^uxI^uy$ |
| $(iii)$ | $xP^syP^sz$ | $(iv)$ | $xP^syP^sz$ |
| | $yP^tzP^tx$ | | $yP^tzI^tx$ |
| | $zI^uxP^uy$ | | $zP^uxI^uy$ |
| $(v)$ | $xP^syP^sz$ | $(vi)$ | $xP^syP^sz$ |
| | $yI^tzP^tx$ | | $yI^tzP^tx$ |
| | $zP^uxI^uy$ | | $zI^uxP^uy$ |

Choose a positive integer $m > \frac{p}{1-p}$. Choose a positive integer $n > max\{\frac{m+2}{(1-p)^2}, \frac{1}{2p-1}\}$. Let $p(1 - p)n = \lfloor p(1 - p)n \rfloor + \varepsilon_1, 0 \le \varepsilon_1 < 1; (1 - p)n = \lfloor (1 - p)n \rfloor + \varepsilon_2, 0 \le \varepsilon_2 < 1;$ and $(1 - p)n = \lceil (1 - p)n \rceil - \varepsilon_3, 0 \le \varepsilon_3 < 1$.
We now show that for such a choice of $n$, for every $\mathscr{D}$ containing one of the six sets (i)–(vi) there exists a profile $(R_1, \ldots, R_n) \in \mathscr{D}^n$ which yields non-quasi-transitive $R = f(R_1, \ldots, R_n)$ under the $p$-majority rule.

Set (i): If $\mathscr{D}$ contains set (i) then consider any $(R_1, \ldots, R_n) \in \mathscr{D}^n$ such that $n(x P^s y P^s z) = \lfloor pn \rfloor, n(y P^t z P^t x) = n - 1 - \lfloor pn \rfloor$, and $n(z P^u x P^u y) = 1$. We have: $n(x P_i y) = \lfloor pn \rfloor + 1, n(y P_i x) = n - 1 - \lfloor pn \rfloor, n(y P_i z) = n - 1, n(z P_i y) = 1, n(x P_i z) = \lfloor pn \rfloor, n(z P_i x) = n - \lfloor pn \rfloor$.
$n(x P_i y) = \lfloor pn \rfloor + 1 > pn \rightarrow x P y$.
$n(y P_i z) = n - 1 \rightarrow y P z$, as $n - 1 > pn$ in view of $n > \frac{1}{1-p}$.
$n(x P_i z) = \lfloor pn \rfloor \leq pn \rightarrow \sim x P z$.

Set (ii): If $\mathscr{D}$ contains set (ii) then consider any $(R_1, \ldots, R_n) \in \mathscr{D}^n$ such that $n(x P^s y P^s z) = n + 1 - \lfloor p(1-p)n \rfloor - \lfloor (1-p)n \rfloor, n(y P^t z P^t x) = \lfloor p(1-p)n \rfloor$, and $n(z P^u x I^u y) = \lfloor (1-p)n \rfloor - 1$. We have: $n(x P_i y) = n + 1 - \lfloor p(1-p)n \rfloor - \lfloor (1-p)n \rfloor, n(y P_i x) = \lfloor p(1-p)n \rfloor, n(y P_i z) = n + 1 - \lfloor (1-p)n \rfloor, n(z P_i y) = \lfloor (1-p)n \rfloor - 1, n(x P_i z) = n + 1 - \lfloor p(1-p)n \rfloor - \lfloor (1-p)n \rfloor, n(z P_i x) = \lfloor p(1-p)n \rfloor + \lfloor (1-p)n \rfloor - 1$.
$n(x P_i y) = n + 1 - \lfloor p(1-p)n \rfloor - \lfloor (1-p)n \rfloor = n + 1 - p(1-p)n - (1-p)n + \varepsilon_1 + \varepsilon_2 = p^2 n + 1 + \varepsilon_1 + \varepsilon_2 > p^2 n + p + p\varepsilon_2 = p[n + 1 - \lfloor (1-p)n \rfloor] = p[n(x P_i y) + n(y P_i x)]$. Therefore $x P y$.
$n(y P_i z) = n + 1 - \lfloor (1-p)n \rfloor > pn \rightarrow y P z$.
$n(x P_i z) = n + 1 - \lfloor p(1-p)n \rfloor - \lfloor (1-p)n \rfloor < n - p(1-p)n - (1-p)n + 3 = p^2 n + 3$. Now, $n > \frac{m+2}{(1-p)^2} > \frac{m+1}{(1-p)^2} > \frac{m+1}{p(1-p)} \rightarrow n > \frac{3}{p(1-p)}$, as $m > \frac{p}{1-p} \rightarrow m \geq 2$. Therefore $pn > \frac{3}{(1-p)}$ implying $pn > p^2 n + 3$. Thus $\sim x P z$.

Set (iii): If $\mathscr{D}$ contains set (iii) then consider any $(R_1, \ldots, R_n) \in \mathscr{D}^n$ such that $n(x P^s y P^s z) = \lfloor p(1-p)n \rfloor, n(y P^t z P^t x) = n + 1 - \lfloor p(1-p)n \rfloor - \lfloor (1-p)n \rfloor$, and $n(z I^u x P^u y) = \lfloor (1-p)n \rfloor - 1$. We have: $n(x P_i y) = \lfloor p(1-p)n \rfloor + \lfloor (1-p)n \rfloor - 1, n(y P_i x) = n + 1 - \lfloor p(1-p)n \rfloor - \lfloor (1-p)n \rfloor, n(y P_i z) = n + 1 - \lfloor (1-p)n \rfloor, n(z P_i y) = \lfloor (1-p)n \rfloor - 1, n(x P_i z) = \lfloor p(1-p)n \rfloor, n(z P_i x) = n + 1 - \lfloor p(1-p)n \rfloor - \lfloor (1-p)n \rfloor$.
$n(y P_i z) = n + 1 - \lfloor (1-p)n \rfloor > pn \rightarrow y P z$.
$n(z P_i x) = n + 1 - \lfloor p(1-p)n \rfloor - \lfloor (1-p)n \rfloor > p[n + 1 - \lfloor (1-p)n \rfloor] = p[n(x P_i z) + n(z P_i x)]$. Therefore $z P x$.
$n(y P_i x) = n + 1 - \lfloor p(1-p)n \rfloor - \lfloor (1-p)n \rfloor < pn$. Therefore $\sim y P x$.

Set (iv): If $\mathscr{D}$ contains set (iv) then consider any $(R_1, \ldots, R_n) \in \mathscr{D}^n$ such that $n(x P^s y P^s z) = n - m - \lceil (1-p)n \rceil, n(y P^t z I^t x) = m + 1$, and $n(z P^u x I^u y) = \lceil (1-p)n \rceil - 1$. We have: $n(x P_i y) = n - m - \lceil (1-p)n \rceil, n(y P_i x) = m + 1, n(y P_i z) = n + 1 - \lceil (1-p)n \rceil, n(z P_i y) = \lceil (1-p)n \rceil - 1, n(x P_i z) = n - m - \lceil (1-p)n \rceil, n(z P_i x) = \lceil (1-p)n \rceil - 1$.
$n > \frac{m+2}{(1-p)^2} > \frac{m+1}{(1-p)^2} > \frac{m+1}{p(1-p)} \rightarrow p(1-p)n = pn - p^2 n > m + 1 = m + \varepsilon_3 + (1-\varepsilon_3) > m + \varepsilon_3 + p(1-\varepsilon_3)$
Now, $n(x P_i y) = n - m - \lceil (1-p)n \rceil = n - m - (1-p)n - \varepsilon_3 = pn - m - \varepsilon_3 > p^2 n + p(1-\varepsilon_3) = p[pn + (1-\varepsilon_3)] = p[n + 1 - (1-p)n - \varepsilon_3)] = p[n + 1 - \lceil (1-p)n \rceil] = p[n(x P_i y) + n(y P_i x)]$. Thus we have $x P y$.
$n(y P_i z) = n + 1 - \lceil (1-p)n \rceil > pn \rightarrow y P z$.
$m > \frac{p}{1-p} \rightarrow m - pm > p$
$\rightarrow m - pm > p - \varepsilon_3$

$\rightarrow -m - \varepsilon_3 < -p(m+1)$

$n(xP_iz) = n - m - \lceil(1-p)n\rceil = n - m - (1-p)n - \varepsilon_3 = pn - m - \varepsilon_3 < pn - p(m+1) = p[n - m - 1] = p[n(xP_iz) + n(zP_ix)]$. Therefore $\sim xPz$.

Set (v): If $\mathscr{D}$ contains set (v) then consider any $(R_1, \ldots, R_n) \in \mathscr{D}^n$ such that $n(xP^sy P^sz) = \lceil(1-p)n\rceil - 1, n(yI^tzP^tx) = m + 1$, and $n(zP^uxI^uy) = n - m - \lceil(1-p)n\rceil$. We have: $n(xP_iy) = \lceil(1-p)n\rceil - 1, n(yP_ix) = m + 1, n(yP_iz) = \lceil(1-p)n\rceil - 1, n(zP_iy) = n - m - \lceil(1-p)n\rceil, n(xP_iz) = \lceil(1-p)n\rceil - 1, n(zP_ix) = n + 1 - \lceil(1-p)n\rceil$.

$n(zP_ix) = n + 1 - \lceil(1-p)n\rceil > pn \rightarrow zPx$.

$n > \frac{m+1}{(1-p)^2} \rightarrow n(1-p)^2 > m + 1 > pm + 1$

$n(xP_iy) - p[n(xP_iy) + n(yP_ix)] = \lceil(1-p)n\rceil - 1 - p[\lceil(1-p)n\rceil - 1 + m + 1] = (1-p)\lceil(1-p)n\rceil - 1 - pm \geq (1-p)^2n - 1 - pm > 0$. Therefore $xPy$ obtains.

$n(zP_iy) = n - m - \lceil(1-p)n\rceil < p[n - m - 1] = p[n(yP_iz) + n(zP_iy)]$. Therefore $\sim zPy$.

Set (vi): If $\mathscr{D}$ contains set (vi) then consider any $(R_1, \ldots, R_n) \in \mathscr{D}^n$ such that $n(xP^sy P^sz) = n - m - \lceil(1-p)n\rceil, n(yI^tzP^tx) = \lceil(1-p)n\rceil - 1$, and $n(zI^uxP^uy) = m + 1$. We have: $n(xP_iy) = n + 1 - \lceil(1-p)n\rceil, n(yP_ix) = \lceil(1-p)n\rceil - 1, n(yP_iz) = n - m - \lceil(1-p)n\rceil, n(zP_iy) = m + 1, n(xP_iz) = n - m - \lceil(1-p)n\rceil, n(zP_ix) = \lceil(1-p)n\rceil - 1$.

$n(xP_iy) = n + 1 - \lceil(1-p)n\rceil > pn \rightarrow xPy$.

$n(yP_iz) = n - m - \lceil(1-p)n\rceil > p[n + 1 - \lceil(1-p)n\rceil] = p[n(yP_iz) + n(zP_iy)]$. Therefore $yPz$

$n(xP_iz) = n - m - \lceil(1-p)n\rceil < p[n - m - 1] = p[n(xP_iz) + n(zP_ix)]$. Therefore $\sim xPz$. $\qquad \square$

*Remark 6.2* Let positive integer $m$ be greater than $\frac{p}{1-p}$. If positive integer $n'$ is such that $n' > max\{\frac{m+2}{(1-p)^2}, \frac{1}{2p-1}\}$ then for any positive integer $n > n'$ we have: $n > max\{\frac{m+2}{(1-p)^2}, \frac{1}{2p-1}\}$. Consequently in view of Theorems 6.4 and 6.5 the following theorem holds. $\qquad\qquad \diamond$

**Theorem 6.6** *Let* $f : \mathscr{T}^n \mapsto \mathscr{C}$ *be a p-majority rule,* $p \in (\frac{1}{2}, 1)$. *Let n be a positive integer such that* $n > max\{\frac{m+2}{(1-p)^2}, \frac{1}{2p-1}\}$, *where m is a positive integer greater than* $\frac{p}{1-p}$. *Let* $\mathscr{D} \subseteq \mathscr{T}$. *Then f yields quasi-transitive social R, R = f(R_1, \ldots, R_n)$, *for every* $(R_1, \ldots, R_n) \in \mathscr{D}^n$ *iff* $\mathscr{D}$ *satisfies the condition of Latin Square partial agreement.*

## 6.4  Two-Thirds Majority Rule

**Theorem 6.7** *Let* $f : \mathscr{T}^n \mapsto \mathscr{C}$ *be the two-thirds majority rule defined for n individuals,* $n \geq 10$. *Let* $\mathscr{D} \subseteq \mathscr{T}$. *Then f yields transitive social R, R = f(R_1, \ldots, R_n)$, *for every* $(R_1, \ldots, R_n) \in \mathscr{D}^n$ *iff* $\mathscr{D}$ *satisfies the condition of placement restriction.*

*Proof* Let $\mathscr{D} \subseteq \mathscr{T}$. If $\mathscr{D}$ satisfies PR then $f$ yields transitive social $R$ for every $(R_1, \ldots, R_n) \in \mathscr{D}^n$ follows from Theorem 6.1.

Suppose $\mathscr{D} \subseteq \mathscr{T}$ violates PR. Then, by Lemma 6.1, it must contain one of the 10 sets (i)–(x) listed in the statement of the lemma, except for a formal interchange of alternatives. In what follows we consider each of these 10 cases.

If $\mathscr{D}$ contains set (i) then consider any $(R_1, \ldots, R_n) \in \mathscr{D}^n$ such that $n(x P^s y P^s z) = n(y P^t z P^t x) = k$, if $n = 2k, k \geq 5$; and any $(R_1, \ldots, R_n) \in \mathscr{D}^n$ such that $n(x P^s y P^s z) = k + 1$ and $n(y P^t z P^t x) = k$ if $n = 2k + 1, k \geq 5$. In each case we obtain: $x I y \wedge y P z \wedge x I z$.

If $\mathscr{D}$ contains set (ii) then consider any $(R_1, \ldots, R_n) \in \mathscr{D}^n$ such that $n(x P^s y P^s z) = n(z P^t x I^t y) = k$, if $n = 2k, k \geq 5$; and any $(R_1, \ldots, R_n) \in \mathscr{D}^n$ such that $n(x P^s y P^s z) = k + 1$ and $n(z P^t x I^t y) = k$ if $n = 2k + 1, k \geq 5$. In each case we obtain: $x P y \wedge y I z \wedge x I z$.

If $\mathscr{D}$ contains set (iii) then consider any $(R_1, \ldots, R_n) \in \mathscr{D}^n$ such that $n(x P^s y P^s z) = n(y I^t z P^t x) = k$, if $n = 2k, k \geq 5$; and any $(R_1, \ldots, R_n) \in \mathscr{D}^n$ such that $n(x P^s y P^s z) = k + 1$ and $n(y I^t z P^t x) = k$ if $n = 2k + 1, k \geq 5$. In each case we obtain: $x I y \wedge y P z \wedge x I z$.

If $\mathscr{D}$ contains set (iv) then consider any $(R_1, \ldots, R_n) \in \mathscr{D}^n$ such that $n(x P^s y P^s z) = 2k - 2, n(z P^t y P^t x) = k - 1, n(y P^u z I^u x) = 3$, if $n = 3k, k \geq 4$; any $(R_1, \ldots, R_n) \in \mathscr{D}^n$ such that $n(x P^s y P^s z) = 2k - 2, n(z P^t y P^t x) = k, n(y P^u z I^u x) = 3$, if $n = 3k + 1, k \geq 3$; and any $(R_1, \ldots, R_n) \in \mathscr{D}^n$ such that $n(x P^s y P^s z) = 2k - 1, n(z P^t y P^t x) = k, n(y P^u z I^u x) = 3$, if $n = 3k + 2, k \geq 3$. In each case it results in $(x I y \wedge y P z \wedge x I z)$.

If $\mathscr{D}$ contains set (v) then consider any $(R_1, \ldots, R_n) \in \mathscr{D}^n$ such that $n(x P^s y P^s z) = 2k - 2, n(z P^t y P^t x) = k - 1, n(z I^u x P^u y) = 3$, if $n = 3k, k \geq 4$; any $(R_1, \ldots, R_n) \in \mathscr{D}^n$ such that $n(x P^s y P^s z) = 2k - 2, n(z P^t y P^t x) = k, n(z I^u x P^u y) = 3$, if $n = 3k + 1, k \geq 3$; and any $(R_1, \ldots, R_n) \in \mathscr{D}^n$ such that $n(x P^s y P^s z) = 2k - 1, n(z P^t y P^t x) = k, n(z I^u x P^u y) = 3$, if $n = 3k + 2, k \geq 3$. In each case it results in $(x P y \wedge y I z \wedge x I z)$.

If $\mathscr{D}$ contains set (vi) then consider any $(R_1, \ldots, R_n) \in \mathscr{D}^n$ such that $n(x P^s y P^s z) = n(y P^t z I^t x) = n(z I^u x P^u y) = k$, if $n = 3k, k \geq 4$; any $(R_1, \ldots, R_n) \in \mathscr{D}^n$ such that $n(x P^s y P^s z) = k - 1, n(y P^t z I^t x) = k + 1, n(z I^u x P^u y) = k + 1$, if $n = 3k + 1, k \geq 3$; and any $(R_1, \ldots, R_n) \in \mathscr{D}^n$ such that $n(x P^s y P^s z) = k, n(y P^t z I^t x) = k + 1, n(z I^u x P^u y) = k + 1$, if $n = 3k + 2, k \geq 3$. In each case it results in $(x I y \wedge y I z \wedge x P z)$.

If $\mathscr{D}$ contains set (vii) then consider any $(R_1, \ldots, R_n) \in \mathscr{D}^n$ such that $n(x P^s y I^s z) = n(y P^t z I^t x) = n(z I^u x P^u y) = k$, if $n = 3k, k \geq 4$; any $(R_1, \ldots, R_n) \in \mathscr{D}^n$ such that $n(x P^s y I^s z) = k - 1, n(y P^t z I^t x) = k + 1, n(z I^u x P^u y) = k + 1$, if $n = 3k + 1, k \geq 3$; and any $(R_1, \ldots, R_n) \in \mathscr{D}^n$ such that $n(x P^s y I^s z) = k, n(y P^t z I^t x) = k + 1, n(z I^u x P^u y) = k + 1$, if $n = 3k + 2, k \geq 3$. In each case it results in $(x I y \wedge y I z \wedge x P z)$.

If $\mathscr{D}$ contains set (viii) then consider any $(R_1, \ldots, R_n) \in \mathscr{D}^n$ such that $n(xI^s y P^s z) = n(yP^t zI^t x) = n(zI^u x P^u y) = k$, if $n = 3k, k \geq 4$; any $(R_1, \ldots, R_n) \in \mathscr{D}^n$ such that $n(xI^s y P^s z) = k - 1, n(yP^t zI^t x) = k + 1, n(zI^u x P^u y) = k + 1$, if $n = 3k + 1, k \geq 3$; and any $(R_1, \ldots, R_n) \in \mathscr{D}^n$ such that $n(xI^s y P^s z) = k, n(yP^t zI^t x) = k + 1, n(zI^u x P^u y) = k + 1$, if $n = 3k + 2, k \geq 3$. In each case it results in $(xIy \wedge yIz \wedge xPz)$.

If $\mathscr{D}$ contains set (ix) then consider any $(R_1, \ldots, R_n) \in \mathscr{D}^n$ such that $n(xP^s yI^s z) = 2k + 1, n(yP^t zI^t x) = k + 1, n(zP^u xI^u y) = k$, if $n = 4k + 2, k \geq 2$; any $(R_1, \ldots, R_n) \in \mathscr{D}^n$ such that $n(xP^s yI^s z) = 2k + 2, n(yP^t zI^t x) = k + 1, n(zP^u xI^u y) = k$, if $n = 4k + 3, k \geq 2$; any $(R_1, \ldots, R_n) \in \mathscr{D}^n$ such that $n(xP^s yI^s z) = 2k + 2, n(yP^t zI^t x) = k + 2, n(zP^u xI^u y) = k$, if $n = 4k + 4, k \geq 2$; and any $(R_1, \ldots, R_n) \in \mathscr{D}^n$ such that $n(xP^s yI^s z) = 2k + 3, n(yP^t zI^t x) = k + 2, n(zP^u xI^u y) = k$, if $n = 4k + 5, k \geq 2$. In each case it results in $(xIy \wedge yIz \wedge xPz)$.

If $\mathscr{D}$ contains set (x) then consider any $(R_1, \ldots, R_n) \in \mathscr{D}^n$ such that $n(xI^s y P^s z) = 2k + 1, n(yI^t z P^t x) = k + 1, n(zI^u x P^u y) = k$, if $n = 4k + 2, k \geq 2$; any $(R_1, \ldots, R_n) \in \mathscr{D}^n$ such that $n(xI^s y P^s z) = 2k + 2, n(yI^t z P^t x) = k + 1, n(zI^u x P^u y) = k$, if $n = 4k + 3, k \geq 2$; any $(R_1, \ldots, R_n) \in \mathscr{D}^n$ such that $n(xI^s y P^s z) = 2k + 2, n(yI^t z P^t x) = k + 2, n(zI^u x P^u y) = k$, if $n = 4k + 4, k \geq 2$; and any $(R_1, \ldots, R_n) \in \mathscr{D}^n$ such that $n(xI^s y P^s z) = 2k + 3, n(yI^t z P^t x) = k + 2, n(zI^u x P^u y) = k$, if $n = 4k + 5, k \geq 2$. In each case it results in $(xIy \wedge yPz \wedge xIz)$. $\qquad\square$

**Theorem 6.8** *Let $f : \mathscr{T}^n \mapsto \mathscr{C}$ be the two-thirds majority rule defined for n individuals, $n \in \mathbb{N}, n \geq 10$. Let $\mathscr{D} \subseteq \mathscr{T}$. Then $f$ yields quasi-transitive social $R$, $R = f(R_1, \ldots, R_n)$, for every $(R_1, \ldots, R_n) \in \mathscr{D}^n$ iff $\mathscr{D}$ satisfies the condition of Latin Square partial agreement.*

*Proof* Let $\mathscr{D} \subseteq \mathscr{T}$. If $\mathscr{D}$ satisfies LSPA then $f$ yields quasi-transitive social $R$ for every $(R_1, \ldots, R_n) \in \mathscr{D}^n$ follows from Theorem 6.4.

Suppose $\mathscr{D} \subseteq \mathscr{T}$ violates LSPA. Then, it must contain one of the six sets (i)–(vi) listed in the proof of Theorem 6.5, except for a formal interchange of alternatives. In what follows we consider each of these six cases.

If $\mathscr{D}$ contains set (i) then consider any $(R_1, \ldots, R_n) \in \mathscr{D}^n$ such that $n(xP^s y P^s z) = 2k, n(yP^t z P^t x) = k - 1, n(zP^u x P^u y) = 1$, if $n = 3k, k \geq 4$; any $(R_1, \ldots, R_n) \in \mathscr{D}^n$ such that $n(xP^s y P^s z) = 2k, n(yP^t z P^t x) = k, n(zP^u x P^u y) = 1$, if $n = 3k + 1, k \geq 3$; and any $(R_1, \ldots, R_n) \in \mathscr{D}^n$ such that $n(xP^s y P^s z) = 2k + 1, n(yP^t z P^t x) = k, n(zP^u x P^u y) = 1$, if $n = 3k + 2, k \geq 3$. In each case we obtain: $xPy \wedge yPz \wedge \sim xPz$.

If $\mathscr{D}$ contains set (ii) then consider any $(R_1, \ldots, R_n) \in \mathscr{D}^n$ such that $n(xP^s y P^s z) = 2k, n(yP^t z P^t x) = k - 1, n(zP^u xI^u y) = 1$, if $n = 3k, k \geq 4$; any $(R_1, \ldots, R_n) \in \mathscr{D}^n$ such that $n(xP^s y P^s z) = 2k, n(yP^t z P^t x) = 1, n(zP^u xI^u y) = k$, if $n = 3k + 1, k \geq 3$; and any $(R_1, \ldots, R_n) \in \mathscr{D}^n$ such that $n(xP^s y P^s z) = 2k + 1, n(yP^t z P^t x)$

$= k, n(z P^u x I^u y) = 1$, if $n = 3k + 2, k \geq 3$. In each case we obtain: $x P y \wedge y P z \wedge \sim x P z$.

If $\mathscr{D}$ contains set (iii) then consider any $(R_1, \ldots, R_n) \in \mathscr{D}^n$ such that $n(x P^s y P^s z) = k - 1, n(y P^t z P^t x) = 2k, n(z I^u x P^u y) = 1$, if $n = 3k, k \geq 4$; any $(R_1, \ldots, R_n) \in \mathscr{D}^n$ such that $n(x P^s y P^s z) = 1, n(y P^t z P^t x) = 2k, n(z I^u x P^u y) = k$, if $n = 3k + 1, k \geq 3$; and any $(R_1, \ldots, R_n) \in \mathscr{D}^n$ such that $n(x P^s y P^s z) = k, n(y P^t z P^t x) = 2k + 1, n(z I^u x P^u y) = 1$, if $n = 3k + 2, k \geq 3$. In each case we obtain: $y P z \wedge z P x \wedge \sim y P x$.

If $\mathscr{D}$ contains set (iv) then consider any $(R_1, \ldots, R_n) \in \mathscr{D}^n$ such that $n(x P^s y P^s z) = 2k - 2, n(y P^t z I^t x) = 3, n(z P^u x I^u y) = k - 1$, if $n = 3k, k \geq 5$; any $(R_1, \ldots, R_n) \in \mathscr{D}^n$ such that $n(x P^s y P^s z) = 2k - 2, n(y P^t z I^t x) = 3, n(z P^u x I^u y) = k$, if $n = 3k + 1, k \geq 5$; any $(R_1, \ldots, R_n) \in \mathscr{D}^n$ such that $n(x P^s y P^s z) = 2k - 1, n(y P^t z I^t x) = 3, n(z P^u x I^u y) = k$, if $n = 3k + 2, k \geq 4$; any $(R_1, \ldots, R_n) \in \mathscr{D}^n$ such that $n(x P^s y P^s z) = 5, n(y P^t z I^t x) = 2, n(z P^u x I^u y) = 3$, if $n = 10$; and any $(R_1, \ldots, R_n) \in \mathscr{D}^n$ such that $n(x P^s y P^s z) = 7, n(y P^t z I^t x) = 2, n(z P^u x I^u y) = 4$, if $n = 13$. In each case we obtain: $x P y \wedge y P z \wedge \sim x P z$. If $n = 11$ then consider any $(R_1, \ldots, R_n) \in \mathscr{D}^n$ such that $n(x P^s y P^s z) = 3, n(y P^t z I^t x) = 1, n(z P^u x I^u y) = 7$; and if $n = 12$ then consider any $(R_1, \ldots, R_n) \in \mathscr{D}^n$ such that $n(x P^s y P^s z) = 3, n(y P^t z I^t x) = 1, n(z P^u x I^u y) = 8$. In both cases we obtain: $z P x \wedge x P y \wedge \sim z P y$.

If $\mathscr{D}$ contains set (v) then consider any $(R_1, \ldots, R_n) \in \mathscr{D}^n$ such that $n(x P^s y P^s z) = 2k - 1, n(y I^t z P^t x) = 3, n(z P^u x I^u y) = k - 2$, if $n = 3k, k \geq 4$; any $(R_1, \ldots, R_n) \in \mathscr{D}^n$ such that $n(x P^s y P^s z) = 2k, n(y I^t z P^t x) = 3, n(z P^u x I^u y) = k - 2$, if $n = 3k + 1, k \geq 4$; and any $(R_1, \ldots, R_n) \in \mathscr{D}^n$ such that $n(x P^s y P^s z) = 2k + 1, n(y I^t z P^t x) = 3, n(z P^u x I^u y) = k - 2$, if $n = 3k + 2, k \geq 3$. In each case we obtain: $x P y \wedge y P z \wedge \sim x P z$. If $n = 10$ then consider any $(R_1, \ldots, R_n) \in \mathscr{D}^n$ such that $n(x P^s y P^s z) = 3, n(y I^t z P^t x) = 1, n(z P^u x I^u y) = 6$; which results in $z P x \wedge x P y \wedge \sim z P y$.

If $\mathscr{D}$ contains set (vi) then consider any $(R_1, \ldots, R_n) \in \mathscr{D}^n$ such that $n(x P^s y P^s z) = 2k - 2, n(y I^t z P^t x) = k - 1, n(z I^u x P^u y) = 3$, if $n = 3k, k \geq 5$; any $(R_1, \ldots, R_n) \in \mathscr{D}^n$ such that $n(x P^s y P^s z) = 2k - 2, n(y I^t z P^t x) = k, n(z I^u x P^u y) = 3$, if $n = 3k + 1, k \geq 5$; any $(R_1, \ldots, R_n) \in \mathscr{D}^n$ such that $n(x P^s y P^s z) = 2k - 1, n(y I^t z P^t x) = k, n(z I^u x P^u y) = 3$, if $n = 3k + 2, k \geq 4$; any $(R_1, \ldots, R_n) \in \mathscr{D}^n$ such that $n(x P^s y P^s z) = 5, n(y I^t z P^t x) = 3, n(z I^u x P^u y) = 2$, if $n = 10$; and any $(R_1, \ldots, R_n) \in \mathscr{D}^n$ such that $n(x P^s y P^s z) = 7, n(y I^t z P^t x) = 4, n(z I^u x P^u y) = 2$, if $n = 13$. In each case we obtain: $x P y \wedge y P z \wedge \sim x P z$. If $n = 11$ then consider any $(R_1, \ldots, R_n) \in \mathscr{D}^n$ such that $n(x P^s y P^s z) = 3, n(y I^t z P^t x) = 7, n(z I^u x P^u y) = 1$; and if $n = 12$ then consider any $(R_1, \ldots, R_n) \in \mathscr{D}^n$ such that $n(x P^s y P^s z) = 3, n(y I^t z P^t x) = 8, n(z I^u x P^u y) = 1$. In both cases we obtain: $y P z \wedge z P x \wedge \sim y P x$. $\qquad \square$

# Reference

Jain, Satish K. 1983. Necessary and sufficient conditions for quasi-transitivity and transitivity of special majority rules. *Keio Economic Studies* 20: 55–63.

# The Class of Strict Majority Rules

The strict majority rule discussed in Chap. 4 belongs to the class of strict majority rules. Let $\frac{1}{2} \leq p < 1$. Under $p$-strict majority rule, an alternative $x$ is considered to be socially at least as good as some other alternative $y$ iff more than $p$ fraction of total number of individuals do not prefer $y$ to $x$. Thus, under the $p$-strict majority rule, between two alternatives social strict preference prevails iff one of the alternatives of the pair is strictly preferred by more than $p$ fraction of the total number of individuals over the other alternative of the pair; otherwise the social indifference prevails. If $p$ is $\frac{1}{2}$ then the rule is the strict majority rule studied in Chap. 4; if $\frac{1}{2} < p < 1$ then the rule will be called a strict special majority rule. The class of strict majority rules is characterized by the following six conditions: (i) independence of irrelevant alternatives, (ii) neutrality, (iii) monotonicity, (iv) weak Pareto-criterion, (v) anonymity and (vi) a set of individuals is a blocking coalition iff it is a strictly blocking coalition (Theorem 7.1). In Chap. 4 it was shown that a social decision rule is the strict majority rule iff it satisfies these six conditions and has the property that every proper superset of a blocking coalition is a winning coalition. Thus this last property separates out the strict majority rule from other rules belonging to the class of strict majority rules.

For the purpose of derivation of conditions for transitivity under strict majority rules, we partition the set of rules defined for all possible number of individuals $\#N = n \geq 2$ into two subsets: (i) The rules which are such that there exists a partition $(N_1, N_2)$ of $N$ such that $\#N_1 \leq pn \wedge \#N_2 \leq pn$; (ii) the rules which are such that there does not exist a partition $(N_1, N_2)$ of $N$ such that $\#N_1 \leq pn \wedge \#N_2 \leq pn$. Let $\mathscr{D}$ be a set of orderings of the set of alternatives $S$. We show that: (i) If the $p$-strict majority rule is such that there exists a partition $(N_1, N_2)$ of $N$ such that $\#N_1 \leq pn \wedge \#N_2 \leq pn$, then the rule yields transitive social weak preference relation for every profile of individual orderings belonging to $\mathscr{D}^n$ iff $\mathscr{D}$ satisfies the condition of Latin Square extremal value restriction over every triple of alternatives (Theorem 7.4); and (ii) if the $p$-strict majority rule is such that there does not exist a partition $(N_1, N_2)$ of $N$ such that $\#N_1 \leq pn \wedge \#N_2 \leq pn$, then the rule yields

© Springer Nature Singapore Pte Ltd. 2019
S. K. Jain, *Domain Conditions and Social Rationality*,
https://doi.org/10.1007/978-981-13-9672-4_7

transitive social weak preference relation for every profile of individual orderings belonging to $\mathcal{D}^n$ iff $\mathcal{D}$ satisfies the condition of weak Latin Square extremal value restriction over every triple of alternatives (Theorem 7.7). If the number of individuals is even then there will always exist a partition $(N_1, N_2)$ of $N$ such that $\#N_1 \leq pn \wedge \#N_2 \leq pn$. Consequently, when the number of individuals is even then under every $p$-strict majority rule every profile of individual orderings belonging to $\mathcal{D}^n$ yields transitive social weak preference relation iff $\mathcal{D}$ satisfies the condition of Latin Square extremal value restriction over every triple of alternatives. Theorem 4.2 follows as a corollary of this general result. If the number of individuals is odd then there can be no partition $(N_1, N_2)$ of $N$ such that neither $N_1$ nor $N_2$ has a majority of individuals. Consequently, it follows from Theorem 7.7 that when the number of individuals is odd, under the strict majority rule every profile of individual orderings belonging to $\mathcal{D}^n$ yields transitive social weak preference relation iff $\mathcal{D}$ satisfies the condition of weak Latin Square extremal value restriction over every triple of alternatives. Thus Theorem 4.3 can be derived as a corollary of Theorem 7.7.

If the number of individuals $n = \lfloor pn \rfloor + 1$ then the $p$-strict majority rule coincides with the weak Pareto-rule. The weak Pareto-rule yields quasi-transitive social weak preference relation for every profile of individual orderings. Therefore when $n = \lfloor pn \rfloor + 1$, $p$-strict majority rule yields quasi-transitive social weak preference relation for every profile of individual orderings. If $n > \lfloor pn \rfloor + 1$ then it is shown in Theorem 7.8 that under every $p$-strict majority rule every profile of individual orderings belonging to $\mathcal{D}^n$ yields quasi-transitive social weak preference relation iff $\mathcal{D}$ satisfies the condition of Latin Square unique value restriction over every triple of alternatives.

The chapter is divided into three sections. The first section contains a characterization of the class of strict majority rules. Sections 7.2 and 7.3 discuss, respectively, conditions for transitivity and quasi-transitivity under the class of strict majority rules.[1]

## 7.1 Characterization of the Class of Strict Majority Rules

The Class of Strict Majority Rules: Let $\frac{1}{2} \leq p < 1$. A social decision rule $f : \mathcal{T}^n \mapsto \mathcal{C}$ is the $p$-strict majority rule iff $(\forall (R_1, \ldots, R_n) \in \mathcal{T}^n)(\forall x, y \in S)[xPy \leftrightarrow n(xP_iy) > pn]$.

**Theorem 7.1** *A social decision rule* $f : \mathcal{T}^n \mapsto \mathcal{C}$ *is a $p$-strict majority rule, $\frac{1}{2} \leq p < 1$, iff it satisfies the conditions of (i) independence of irrelevant alternatives, (ii) neutrality, (iii) monotonicity, (iv) weak Pareto-criterion, (v) anonymity and (vi) its structure is such that a coalition is blocking iff it is strictly blocking.*

---

[1] This chapter is based on Jain (1994, 1984).

*Proof* Let $f$ be a $p$-strict majority rule, $\frac{1}{2} \leq p < 1$. Then from the definition of $p$-strict majority rule, $\frac{1}{2} \leq p < 1$, it is clear that $f$ satisfies conditions (i)–(v). Suppose $V, V \subseteq N$, is a blocking coalition. Then from the definitions of a blocking coalition and $p$-strict majority rule, $\frac{1}{2} \leq p < 1$, it follows that $\#(N - V) \leq pn$. Therefore, for any $(R_1, \ldots, R_n) \in \mathscr{T}^n$ and any $x, y \in S$, $(\forall i \in V)(x R_i y)$ implies that $n(y P_i x) \leq pn$, which in turn implies $\sim y P x$, i.e., $x R y$. This establishes that $V$ is a strictly blocking coalition, thus proving that (vi) holds.

Next let social decision rule $f : \mathscr{T}^n \mapsto \mathscr{C}$ satisfy conditions (i)–(vi). As $f$ satisfies condition I we conclude that for any $x, y \in S$ the social $R$ over $\{x, y\}$ is completely determined by individual preferences over $\{x, y\}$. By neutrality the rule for determining social $R$ from individual preferences is the same for all ordered pairs of alternatives. Consider any profile $(R_1, \ldots, R_n) \in \mathscr{T}^n$ and any $x, y \in S$ such that $x P y$. Let $N_1, N_2, N_3$ designate the sets $\{i \in N \mid x P_i y\}, \{i \in N \mid x I_i y\}, \{i \in N \mid y P_i x\}$, respectively. Now consider any profile $(R_1', \ldots, R_n') \in \mathscr{T}^n$ such that $[(\forall i \in N_1)(x P_i' y) \wedge (\forall i \in N_2 \cup N_3)(y P_i' x)]$. Suppose $y R' x$. Then $N_2 \cup N_3$ is a blocking coalition as a consequence of conditions I, M and N. As every blocking coalition is strictly blocking we conclude that $N_2 \cup N_3$ is strictly blocking. But then in $(R_1, \ldots, R_n)$ situation we must have $y R x$, as we have $(\forall i \in N_2 \cup N_3)(y R_i x)$. As this contradicts $x P y$, we conclude that in $(R_1', \ldots, R_n')$ situation $y R' x$ is impossible, i.e., we must have $x P' y$. $x P' y$ in turn implies, in view of conditions I, M and N, that $N_1$ is a winning coalition. Thus we have shown that $(\forall (R_1, \ldots, R_n) \in \mathscr{T}^n)(\forall x, y \in S)[x P y \rightarrow (\exists V \in W)(\forall i \in V)(x P_i y)]$. If $V \in W$ then $(\forall (R_1, \ldots, R_n) \in \mathscr{T}^n)(\forall x, y \in S)[(\forall i \in V)(x P_i y) \rightarrow x P y]$, by the definition of a winning coalition. Thus, $(\forall (R_1, \ldots, R_n) \in \mathscr{T}^n)(\forall x, y \in S)[x P y \leftrightarrow (\exists V \in W)(\forall i \in V)(x P_i y)]$. Now, by the weak Pareto-criterion, $N$ is winning and thus $W$ is nonempty. If $V \in W$ and $\#V = k$, then by anonymity and the definition of a winning coalition we conclude that $(\forall V \subseteq N)(\#V \geq k \rightarrow V \in W)$. Next we note that $(\forall V \subseteq N)(V \in W \rightarrow \#V > \frac{n}{2})$, otherwise as a consequence of anonymity, there will exist two nonempty disjoint winning coalitions leading to a contradiction (see Remark 4.1). Let $\bar{k} = min\{k \mid (\exists V \in W)(\#V = k)\}$. As $\bar{k} > \frac{n}{2}$, we obtain $\bar{k} = \lfloor pn \rfloor + 1$ for some $p \in [\frac{1}{2}, 1)$. Therefore, we conclude that $(\exists p \in [\frac{1}{2}, 1))(\forall V \subseteq N)[V \in W \leftrightarrow \#V > pn]$. This coupled with the earlier inference that $(\forall (R_1, \ldots, R_n) \in \mathscr{T}^n)(\forall x, y \in S)[x P y \leftrightarrow (\exists V \in W)(\forall i \in V)(x P_i y)]$ implies that $(\exists p \in [\frac{1}{2}, 1))(\forall (R_1, \ldots, R_n) \in \mathscr{T}^n)(\forall x, y \in S)[x P y \leftrightarrow n(x P_i y) > pn]$. This establishes that $f$ is a $p$-strict majority rule, $\frac{1}{2} \leq p < 1$. $\square$

## 7.2 Transitivity Under the Strict Majority Rules

**Theorem 7.2** *Let $f : \mathscr{T}^n \mapsto \mathscr{C}$ be a $p$-strict majority rule; $\frac{1}{2} \leq p < 1$. Let $\mathscr{D} \subseteq \mathscr{T}$. Then $f$ yields transitive social $R$, $R = f(R_1, \ldots, R_n)$, for every $(R_1, \ldots, R_n) \in \mathscr{D}^n$ if $\mathscr{D}$ satisfies the condition of Latin Square extremal value restriction.*

*Proof* Let $f : \mathcal{T}^n \mapsto \mathcal{C}$ be a $p$-strict majority rule, $p \in [\frac{1}{2}, 1)$, and let $\mathcal{D} \subseteq \mathcal{T}$. Suppose $f$ does not yield transitive social $R$ for every $(R_1, \ldots, R_n) \in \mathcal{D}^n$. Then:

$$(\exists (R_1, \ldots, R_n) \in \mathcal{D}^n)(\exists x, y, z \in S)(xRy \land yRz \land zPx). \tag{T7.2-1}$$

$$zPx \rightarrow n(zP_ix) > pn \tag{T7.2-2}$$

$$xRy \rightarrow n(xR_iy) \geq (1 - p)n, \tag{T7.2-3}$$

as $n(xR_iy) < (1 - p)n$ would imply $n(yP_ix) > pn$, which in turn would imply $yPx$, contradicting $xRy$

$$yRz \rightarrow n(yR_iz) \geq (1 - p)n, \tag{T7.2-4}$$

as $n(yR_iz) < (1 - p)n$ would imply $n(zP_iy) > pn$, which in turn would imply $zPy$, contradicting $yRz$

$$(T7.2\text{-}2) \land (T7.2\text{-}3) \rightarrow (\exists j \in N)(zP_jxR_jy) \tag{T7.2-5}$$

$$(T7.2\text{-}2) \land (T7.2\text{-}4) \rightarrow (\exists k \in N)(yR_kzP_kx) \tag{T7.2-6}$$

(T7.2-1) implies that $x$, $y$, $z$ are distinct alternatives. $zP_jxR_jy$ and $yR_kzP_kx$ belong to $T[LS(xyzx)]$. In the triple $\{x, y, z\}$, $z$ is uniquely best according to $zP_jxR_jy$, and medium according to $yR_kzP_kx$ without being worst; furthermore $x$ is uniquely worst according to $yR_kzP_kx$ and medium according to $zP_jxR_jy$ without being best. Therefore LSEVR is violated over the triple $\{x, y, z\}$. Thus $\mathcal{D}$ violates LSEVR. It has been shown that violation of transitivity by $R = f(R_1, \ldots, R_n)$, $(R_1, \ldots, R_n) \in \mathcal{D}^n$, implies violation of LSEVR by $\mathcal{D}$, i.e., if $\mathcal{D}$ satisfies LSEVR then every $(R_1, \ldots, R_n) \in \mathcal{D}^n$ yields transitive social $R$. This completes the proof of the theorem. $\qquad \square$

**Theorem 7.3** *Let $f : \mathcal{T}^n \mapsto \mathcal{C}$ be a $p$-strict majority rule; $\frac{1}{2} \leq p < 1$. Let $n = \#N$ be such that there exists a partition $(N_1, N_2)$ of $N$ such that $\#N_1 \leq pn \land \#N_2 \leq pn$. Let $\mathcal{D} \subseteq \mathcal{T}$. Then $f$ yields transitive social $R$, $R = f(R_1, \ldots, R_n)$, for every $(R_1, \ldots, R_n) \in \mathcal{D}^n$ only if $\mathcal{D}$ satisfies the condition of Latin Square extremal value restriction.*

*Proof* Suppose $\mathcal{D} \subseteq \mathcal{T}$ violates LSEVR. Then there exist distinct $a, b, c \in S$ such that $\mathcal{D}$ violates LSEVR over $\{a, b, c\}$. Violation of LSEVR by $\mathcal{D}$ over $\{a, b, c\}$ implies $(\exists$ distinct $x, y, z \in \{a, b, c\})(\exists R^s, R^t \in \mathcal{D})[xP^syR^sz \land zR^txP^ty]$. Partition $N$ into $(N_1, N_2)$ such that $\#N_1 \leq pn \land \#N_2 \leq pn$. Now consider any $(R_1, \ldots, R_n) \in \mathcal{D}^n$ such that the restriction of $(R_1, \ldots, R_n)$ to $\{a, b, c\} = \{x, y, z\}$, $(R_1|\{a, b, c\}, \ldots, R_n|\{a, b, c\})$, is given by: $[(\forall i \in N_1)(xP_iyR_iz) \land (\forall i \in N_2)(zR_ixP_iy)]$. From $n(xP_iy) = n \land n(zP_iy) \leq pn \land n(yP_iz) \leq pn \land n(xP_iz) \leq pn \land n(zP_ix) \leq pn$, we conclude that $(xPy \land yIz \land xIz)$ holds, which violates transitivity. We have shown that if $\mathcal{D} \subseteq \mathcal{T}$ violates LSEVR then there exists a $(R_1, \ldots, R_n) \in \mathcal{D}^n$ such that $R = f(R_1, \ldots, R_n)$ is intransitive, i.e., if $f$ yields transitive $R$ for every $(R_1, \ldots, R_n) \in \mathcal{D}^n$ then $\mathcal{D}$ must satisfy LSEVR. This establishes the theorem. $\qquad \square$

Combining Theorems 7.2 and 7.3 we obtain:

**Theorem 7.4** *Let $f : \mathscr{T}^n \mapsto \mathscr{C}$ be a p-strict majority rule; $\frac{1}{2} \leq p < 1$. Let $n = \#N$ be such that there exists a partition $(N_1, N_2)$ of $N$ such that $\#N_1 \leq pn \wedge \#N_2 \leq pn$. Let $\mathscr{D} \subseteq \mathscr{T}$. Then $f$ yields transitive social $R$, $R = f(R_1, \ldots, R_n)$, for every $(R_1, \ldots, R_n) \in \mathscr{D}^n$ iff $\mathscr{D}$ satisfies the condition of Latin Square extremal value restriction.*

*Remark 7.1* If $\#N$ is even and greater than or equal to 2, i.e., if $\#N = 2k, k \geq 1$, then $(N_1, N_2)$, where $\#N_1 = \#N_2 = k = \frac{n}{2}$, is a partition of $N$ such that $\#N_1 \leq pn \wedge \#N_2 \leq pn$ for any $p \in [\frac{1}{2}, 1)$. Thus when $\#N$ is even and greater than or equal to 2, there always exists a partition $(N_1, N_2)$ of $N$ such that $\#N_1 \leq pn \wedge \#N_2 \leq pn$. Consequently, in view of Theorem 7.4, the following theorem holds. $\Diamond$

**Theorem 7.5** *Let $f : \mathscr{T}^n \mapsto \mathscr{C}$ be a p-strict majority rule; $\frac{1}{2} \leq p < 1$. Let $\#N = 2k, k \geq 1$. Let $\mathscr{D} \subseteq \mathscr{T}$. Then $f$ yields transitive social $R$, $R = f(R_1, \ldots, R_n)$, for every $(R_1, \ldots, R_n) \in \mathscr{D}^n$ iff $\mathscr{D}$ satisfies the condition of Latin Square extremal value restriction.*

*Remark 7.2* Let $\frac{1}{2} < p < 1$. For $n \geq \frac{1}{2p-1}$, we have: $\lfloor pn \rfloor \leq pn \wedge n - \lfloor pn \rfloor \leq pn$. Therefore if $\frac{1}{2} < p < 1$ and $n \geq \frac{1}{2p-1}$ then there exists a partition $(N_1, N_2)$ of $N$, where $\#N_1 = \lfloor pn \rfloor \wedge \#N_2 = n - \lfloor pn \rfloor$, such that the cardinality of neither of the two sets exceeds $pn$. Consequently, in view of Theorem 7.3 the following theorem holds. $\Diamond$

**Theorem 7.6** *Let $f : \mathscr{T}^n \mapsto \mathscr{C}$ be a p-strict majority rule; $\frac{1}{2} < p < 1$. Let $\#N = n \geq \frac{1}{2p-1}$. Let $\mathscr{D} \subseteq \mathscr{T}$. Then $f$ yields transitive social $R$, $R = f(R_1, \ldots, R_n)$, for every $(R_1, \ldots, R_n) \in \mathscr{D}^n$ iff $\mathscr{D}$ satisfies the condition of Latin Square extremal value restriction.*

**Theorem 7.7** *Let $f : \mathscr{T}^n \mapsto \mathscr{C}$ be a p-strict majority rule; $\frac{1}{2} \leq p < 1$. Let $\#N \geq 3$. Let $N$ be such that there does not exist any partition $(N_1, N_2)$ of it such that $\#N_1 \leq pn \wedge \#N_2 \leq pn$. Let $\mathscr{D} \subseteq \mathscr{T}$. Then $f$ yields transitive social $R$, $R = f(R_1, \ldots, R_n)$, for every $(R_1, \ldots, R_n) \in \mathscr{D}^n$ iff $\mathscr{D}$ satisfies the condition of weak Latin Square extremal value restriction.*

*Proof* Let $f : \mathscr{T}^n \mapsto \mathscr{C}$ be a p-strict majority rule; $\frac{1}{2} \leq p < 1$. Let $\#N \geq 3$. Let $N$ be such that there does not exist any partition $(N_1, N_2)$ of it such that $\#N_1 \leq pn \wedge \#N_2 \leq pn$. Let $\mathscr{D} \subseteq \mathscr{T}$. Suppose $f$ does not yield transitive social $R$ for every $(R_1, \ldots, R_n) \in \mathscr{D}^n$. Then:

$$(\exists(R_1, \ldots, R_n) \in \mathscr{D}^n)(\exists x, y, z \in S)(xRy \wedge yRz \wedge zPx). \tag{T7.7-1}$$

$$xRy \rightarrow \{i \in N \mid yP_ix\} \leq pn$$

$$\rightarrow \{i \in N \mid xR_iy\} > pn, \text{ as there does not exist any partition } (N_1, N_2) \text{ of } N \text{ such that } \#N_1 \leq pn \wedge \#N_2 \leq pn \tag{T7.7-2}$$

$yRz \rightarrow \{i \in N \mid zP_iy\} \leq pn$

$\rightarrow \{i \in N \mid yR_iz\} > pn$, as there does not exist any partition $(N_1, N_2)$ of $N$ such

that $\#N_1 \leq pn \wedge \#N_2 \leq pn$                                                                (T7.7-3)

$zPx \rightarrow \{i \in N \mid zP_ix\} > pn$                                                                   (T7.7-4)

$(T7.7\text{-}2) \wedge (T7.7\text{-}3) \rightarrow (\exists i \in N)(xR_iyR_iz)$, as $\frac{1}{2} \leq p < 1$                        (T7.7-5)

$(T7.7\text{-}3) \wedge (T7.7\text{-}4) \rightarrow (\exists j \in N)(yR_jzP_jx)$, as $\frac{1}{2} \leq p < 1$                       (T7.7-6)

$(T7.7\text{-}4) \wedge (T7.7\text{-}2) \rightarrow (\exists k \in N)(zP_kxR_ky)$, as $\frac{1}{2} \leq p < 1$.                      (T7.7-7)

(T7.7-1) implies that $x, y, z$ are distinct alternatives. $xR_iyR_iz$, $yR_jzP_jx$ and $zP_k$ $xR_ky$ form $WLS(xyzx)$, and belong to $T[WLS(xyzx)]$. In the triple $\{x, y, z\}$, $z$ is uniquely best according to $zP_kxR_ky$, and medium according to $yR_jzP_jx$ without being worst; furthermore $x$ is uniquely worst according to $yR_jzP_jx$ and medium according to $zP_kxR_ky$ without being best. Therefore WLSEVR is violated over the triple $\{x, y, z\}$. Thus $\mathscr{D}$ violates WLSEVR. It has been shown that violation of transitivity by $R = f(R_1, \ldots, R_n)$, $(R_1, \ldots, R_n) \in \mathscr{D}^n$, implies violation of WLSEVR by $\mathscr{D}$, i.e., if $\mathscr{D}$ satisfies WLSEVR then every $(R_1, \ldots, R_n) \in \mathscr{D}^n$ yields transitive social $R$.

Suppose $\mathscr{D} \subseteq \mathscr{T}$ violates WLSEVR. Then there exist distinct $a, b, c \in S$ such that $\mathscr{D}$ violates WLSEVR over $\{a, b, c\}$. Violation of WLSEVR by $\mathscr{D}$ over $\{a, b, c\}$ implies $(\exists$ distinct $x, y, z \in \{a, b, c\})(\exists R^s, R^t, R^u \in \mathscr{D})[yR^uzR^ux \wedge zR^txP^ty \wedge xP^syR^sz]$. As $\frac{1}{2} \leq p < 1$ and $n \geq 3$, it follows that $\lfloor pn \rfloor \geq 1$. As $\lfloor pn \rfloor \leq pn$, we obtain $n - \lfloor pn \rfloor > pn$, in view of the fact that there does not exist any partition $(N_1, N_2)$ of $N$ such that $\#N_1 \leq pn \wedge \#N_2 \leq pn$. As $\lfloor pn \rfloor + 1 > pn$, it follows that $n - \lfloor pn \rfloor - 1 \leq pn$. Now consider any $(R_1, \ldots, R_n) \in \mathscr{D}^n$ such that the restriction of $(R_1, \ldots, R_n)$ to $\{x, y, z\} = \{a, b, c\}$, $(R_1|\{x, y, z\}, \ldots, R_n|\{x, y, z\})$, is such that: $[n(yR_izR_ix) = 1 \wedge n(zR_ixP_iy) = \lfloor pn \rfloor \wedge n(xP_iyR_iz) = n - \lfloor pn \rfloor - 1]$. We have: $n(xP_iy) = n - 1, n(yR_ix) = 1; n(zP_iy) = \lfloor pn \rfloor, n(yR_iz) = n - \lfloor pn \rfloor; n(xP_iz) = n - \lfloor pn \rfloor - 1, n(zR_ix) = \lfloor pn \rfloor + 1$. Therefore we conclude that $[xPy \wedge yRz \wedge zRx]$ holds, which violates transitivity. We have shown that if $\mathscr{D} \subseteq \mathscr{T}$ violates WLSEVR then there exists a $(R_1, \ldots, R_n) \in \mathscr{D}^n$ such that $R = f(R_1, \ldots, R_n)$ is intransitive, i.e., if $f$ yields transitive $R$ for every $(R_1, \ldots, R_n) \in \mathscr{D}^n$ then $\mathscr{D}$ must satisfy WLSEVR. This establishes the theorem. $\qquad\square$

## 7.3  Quasi-transitivity Under the Strict Majority Rules

*Remark 7.3* If $\lfloor pn \rfloor = n - 1$, then the $p$-strict majority rule coincides with the weak Pareto-rule. Thus, if $n = \lfloor pn \rfloor + 1$ then the $p$-strict majority rule yields quasi-transitive social $R$ for every $(R_1, \ldots, R_n) \in \mathscr{T}^n$.                                                      $\Diamond$

**Theorem 7.8** *Let $f : \mathscr{T}^n \mapsto \mathscr{C}$ be a p-strict majority rule; $p \in [\frac{1}{2}, 1)$. Let $\#N = n > \lfloor pn \rfloor + 1$. Let $\mathscr{D} \subseteq \mathscr{T}$. Then $f$ yields quasi-transitive social $R$, $R = f(R_1, \ldots, R_n)$, for every $(R_1, \ldots, R_n) \in \mathscr{D}^n$ iff $\mathscr{D}$ satisfies the condition of Latin Square unique value restriction.*

*Proof* Let $f : \mathcal{T}^n \mapsto \mathcal{C}$ be a $p$-strict majority rule, $p \in [\frac{1}{2}, 1)$, $\#N = n > \lfloor pn \rfloor + 1$; and let $\mathcal{D} \subseteq \mathcal{T}$. Suppose $f$ does not yield quasi-transitive social $R$ for every $(R_1, \dots, R_n) \in \mathcal{D}^n$. Then:

$$(\exists(R_1, \dots, R_n) \in \mathcal{D}^n)(\exists x, y, z \in S)(xPy \wedge yPz \wedge zRx). \tag{T7.8-1}$$

$$xPy \rightarrow n(xP_iy) > pn \tag{T7.8-2}$$

$$yPz \rightarrow n(yP_iz) > pn \tag{T7.8-3}$$

$$zRx \rightarrow n(zR_ix) \geq (1-p)n, \tag{T7.8-4}$$

as $n(zR_ix) < (1-p)n$ would imply $n(xP_iz) > pn$, which in turn would imply $xPz$, contradicting $zRx$

$$(\text{T7.8-2}) \wedge (\text{T7.8-3}) \rightarrow (\exists i \in N)(xP_iyP_iz) \tag{T7.8-5}$$

$$(\text{T7.8-3}) \wedge (\text{T7.8-4}) \rightarrow (\exists j \in N)(yP_jzR_jx) \tag{T7.8-6}$$

$$(\text{T7.8-4}) \wedge (\text{T7.8-2}) \rightarrow (\exists k \in N)(zR_kxP_ky) \tag{T7.8-7}$$

(T7.8-1) implies that $x, y, z$ are distinct alternatives. $xP_iyP_iz$, $yP_jzR_jx$, $zR_kxP_ky$ belong to $T[LS(xyzx)]$ and form $LS(xyzx)$. In the triple $\{x, y, z\}$, $y$ is uniquely medium according to $xP_iyP_iz$, uniquely best according to $yP_jzR_jx$, and uniquely worst according to $zR_kxP_ky$. Therefore LSUVR is violated over the triple $\{x, y, z\}$. Thus $\mathcal{D}$ violates LSUVR. It has been shown that violation of quasi-transitivity by $R = f(R_1, \dots, R_n)$, $(R_1, \dots, R_n) \in \mathcal{D}^n$, implies violation of LSUVR by $\mathcal{D}$, i.e., if $\mathcal{D}$ satisfies LSUVR then every $(R_1, \dots, R_n) \in \mathcal{D}^n$ yields quasi-transitive social $R$.

Suppose $\mathcal{D} \subseteq \mathcal{T}$ violates LSUVR. Then there exist distinct $a, b, c \in S$ such that $\mathcal{D}$ violates LSUVR over $\{a, b, c\}$. Violation of LSUVR by $\mathcal{D}$ over $\{a, b, c\}$ implies $(\exists \text{ distinct } x, y, z \in \{a, b, c\})(\exists R^s, R^t, R^u \in \mathcal{D})[xP^syP^sz \wedge yP^tzR^tx \wedge zR^uxP^u y]$. As $n > \lfloor pn \rfloor + 1$, there exists a partition of $N$, $(N_1, N_2, N_3)$, such that $\#N_1 = \lfloor pn \rfloor \wedge \#N_2 = n - \lfloor pn \rfloor - 1 \wedge \#N_3 = 1$. Now consider any $(R_1, \dots, R_n) \in \mathcal{D}^n$ such that the restriction of $(R_1, \dots, R_n)$ to $\{x, y, z\} = \{a, b, c\}$, $(R_1|\{x, y, z\}, \dots, R_n|\{x, y, z\})$, is given by: $[(\forall i \in N_1)(xP_iyP_iz) \wedge (\forall i \in N_2)(yP_izR_ix) \wedge (\forall i \in N_3)(zR_ixP_iz)]$. From $n(xP_iy) = \lfloor pn \rfloor + 1 \wedge n(yP_iz) = n - 1 > \lfloor pn \rfloor \wedge n(xP_iz) = \lfloor pn \rfloor$ we conclude that $(xPy \wedge yPz \wedge zRx)$ holds, which violates quasi-transitivity. We have shown that if $\mathcal{D} \subseteq \mathcal{T}$ violates LSUVR then there exists a $(R_1, \dots, R_n) \in \mathcal{D}^n$ such that $R = f(R_1, \dots, R_n)$ is not quasi-transitive, i.e., if $f$ yields quasi-transitive $R$ for every $(R_1, \dots, R_n) \in \mathcal{D}^n$ then $\mathcal{D}$ must satisfy LSUVR. This establishes the theorem. $\square$

# References

Jain, Satish K. 1994. Characterization of non-minority rules. DSA working paper 12/94. Centre for Economic Studies and Planning, Jawaharlal Nehru University.

Jain, Satish K. 1984. Non-minority rules: Characterization of configurations with rational social preferences. *Keio Economic Studies* 21: 45–54.

# The Class of Pareto-Inclusive Strict Majority Rules

A major drawback of strict majority rules is that they fail to satisfy the Pareto-criterion. The class of Pareto-inclusive strict majority rules, while retaining the 'non-minority' aspect of strict majority rules, by explicitly incorporating the Pareto-criterion remedies this problematic aspect of strict majority rules. Under a Pareto-inclusive $p$-strict majority rule an alternative $x$ is better than another alternative $y$ iff more than $p$ fraction ($\frac{1}{2} \leq p < 1$) of the total number of individuals prefer $x$ to $y$ or $x$ is Pareto-superior to $y$.

When the number of individuals is two, Pareto-inclusive $p$-strict majority rule, $\frac{1}{2} \leq p < 1$, becomes identical with the method of majority decision. Let $\mathscr{D}$ be a set of orderings of the set of social alternatives. From Theorem 3.2 we know that, when the number of individuals is even and greater than or equal to 2, the method of majority decision yields transitive social weak preference relation for every profile of individual orderings belonging to $\mathscr{D}^n$ iff $\mathscr{D}$ satisfies extremal restriction. Therefore it follows that when the number of individuals is two, for every Pareto-inclusive $p$-strict majority rule, extremal restriction completely characterizes all $\mathscr{D}$ such that every profile of individual orderings belonging to $\mathscr{D}^n$ gives rise to transitive social weak preference relation.

If the smallest integer greater than $pn$ is $n$, the total number of individuals, then the Pareto-inclusive $p$-strict majority rule coincides with the Pareto-rule. We establish that under the Pareto-rule every profile of individual orderings belonging to $\mathscr{D}^n$ gives rise to transitive social weak preference relation iff $\mathscr{D}$ satisfies over every triple of alternatives at least one of the three conditions of placement restriction, absence of unique extremal value and strongly antagonistic preferences (2).

It turns out that for every Pareto-inclusive $p$-strict majority rule satisfaction over every triple of alternatives of placement restriction or absence of unique extremal value by $\mathscr{D}$ is sufficient to ensure transitivity of social weak preference relation generated by every profile belonging to $\mathscr{D}^n$, regardless of the value of $n$. Furthermore, if $n$ is greater than the smallest integer greater than $pn$ and is such that there exists a threefold partition of the set of individuals $N$ such that union of no two of them has more than $pn$ individuals then every profile of individual orderings belonging to $\mathscr{D}^n$

© Springer Nature Singapore Pte Ltd. 2019
S. K. Jain, *Domain Conditions and Social Rationality*,
https://doi.org/10.1007/978-981-13-9672-4_8

gives rise to transitive social weak preference relation iff $\mathscr{D}$ satisfies over every triple of alternatives placement restriction or absence of unique extremal value. Therefore, it follows that for the class of Pareto-inclusive strict majority rules satisfaction over every triple of alternatives of placement restriction or absence of unique extremal value is maximally sufficient for transitivity.

As noted above, if the smallest integer greater than $pn$ is $n$, then the Pareto-inclusive $p$-strict majority rule coincides with the Pareto-rule. As Pareto-rule yields quasi-transitive social weak preference relation for every logically possible profile of individual orderings, it follows that when the smallest integer greater than $pn$ is $n$, every $p$-strict majority rule yields quasi-transitive social weak preference relation for every logically possible profile of individual orderings.

It turns out that for every Pareto-inclusive $p$-strict majority rule satisfaction over every triple of alternatives of Latin Square unique value restriction or limited agreement by $\mathscr{D}$ is sufficient to ensure quasi-transitivity of social weak preference relation generated by every profile of individual orderings belonging to $\mathscr{D}^n$, regardless of the value of $n$. If $n$ is such that the smallest integer greater than $pn$ is $n - 1$ then every profile of individual orderings belonging to $\mathscr{D}^n$ yields quasi-transitive social weak preference relation iff $\mathscr{D}$ satisfies the condition of Latin Square linear ordering restriction. And, in case $n$ is such that the smallest integer greater than $pn$ is less than $n - 1$ then every profile of individual orderings belonging to $\mathscr{D}^n$ yields quasi-transitive social weak preference relation iff $\mathscr{D}$ satisfies over every triple of alternatives Latin Square unique value restriction or limited agreement. From these propositions it follows that from the perspective of quasi-transitivity, the class of all Pareto-inclusive strict majority rules can be partitioned into three subclasses: (i) The subclass of rules which are such that the smallest integer greater than $pn$ is $n$. These rules yield quasi-transitive social weak preference relation for every profile of individual orderings belonging to $\mathscr{T}^n$. (ii) The subclass of rules which are such that the smallest integer greater than $pn$ is $n - 1$. These rules yield quasi-transitive social weak preference relation for every profile of individual orderings belonging to $\mathscr{D}^n$ iff $\mathscr{D}$ satisfies Latin Square linear ordering restriction. (iii) The subclass of rules which are such that the smallest integer greater than $pn$ is less than $n - 1$. These rules yield quasi-transitive social weak preference relation for every profile of individual orderings belonging to $\mathscr{D}^n$ iff $\mathscr{D}$ satisfies over every triple of alternatives Latin Square unique value restriction or limited agreement.

## 8.1  Definitions of Pareto-Rule and the Class of Pareto-Inclusive Strict Majority Rules

A social decision rule $f : \mathscr{T}^n \mapsto \mathscr{C}$ is a Pareto-inclusive $p$-strict majority rule, $\frac{1}{2} \leq p < 1$, iff $(\forall (R_1, \ldots, R_n) \in \mathscr{T}^n)(\forall x, y \in S)[x P y \leftrightarrow [(\forall i \in N)(x R_i y) \wedge (\exists i \in N)(x P_i y)] \vee n(x P_i y) > pn]$.

A social decision rule $f : \mathcal{T}^n \mapsto \mathcal{C}$ is the Pareto-rule iff $(\forall (R_1, \ldots, R_n) \in \mathcal{T}^n)(\forall x, y \in S)[x P y \leftrightarrow (\forall i \in N)(x R_i y) \wedge (\exists i \in N)(x P_i y)]$. Suppose $x P y \wedge y P z, x, y, z \in S$, under the Pareto-rule. $x P y$ and $y P z$ imply $[(\forall i \in N)(x R_i y) \wedge (\exists i \in N)(x P_i y) \wedge (\forall i \in N)(y R_i z) \wedge (\exists i \in N)(y P_i z)]$. As individual preferences are transitive, from $(\forall i \in N)(x R_i y \wedge y R_i z)$ we obtain $(\forall i \in N)(x R_i z)$, and from $(\forall i \in N)(x R_i y) \wedge (\exists i \in N)(y P_i z)$ we obtain $(\exists i \in N)(x P_i z)$. $(\forall i \in N)(x R_i z) \wedge (\exists i \in N)(x P_i z)$ under the Pareto-rule imply $x P z$. Therefore it follows that the Pareto-rule yields quasi-transitive social $R$ for every $(R_1, \ldots, R_n) \in \mathcal{T}^n$.

## 8.2 Transitivity Under the Pareto-Inclusive Strict Majority Rules

**Theorem 8.1** *Let $f : \mathcal{T}^n \mapsto \mathcal{C}$ be a Pareto-inclusive p-strict majority rule; $p \in [\frac{1}{2}, 1)$. Let $\#N = 2$. Let $\mathcal{D} \subseteq \mathcal{T}$. Then $f$ yields transitive social $R$, $R = f(R_1, R_2)$, for every $(R_1, R_2) \in \mathcal{D}^2$ iff $\mathcal{D}$ satisfies the condition of extremal restriction.*

*Proof* When there are only two individuals, a Pareto-inclusive $p$-strict majority rule, $p \in [\frac{1}{2}, 1)$, becomes identical to the simple majority rule. Therefore the theorem follows from Theorem 3.2. ☐

**Theorem 8.2** *Let $f : \mathcal{T}^n \mapsto \mathcal{C}$ be a Pareto-inclusive p-strict majority rule; $p \in [\frac{1}{2}, 1)$. Let $\mathcal{D} \subseteq \mathcal{T}$. Then $f$ yields transitive social $R$, $R = f(R_1, \ldots, R_n)$, for every $(R_1, \ldots, R_n) \in \mathcal{D}^n$ if $\mathcal{D}$ satisfies over every triple of alternatives the placement restriction or the condition of absence of unique extremal value.*

*Proof* Let $f : \mathcal{T}^n \mapsto \mathcal{C}$ be a Pareto-inclusive $p$-strict majority rule, $p \in [\frac{1}{2}, 1)$, and let $\mathcal{D} \subseteq \mathcal{T}$. Suppose $f$ does not yield transitive social $R$ for every $(R_1, \ldots, R_n) \in \mathcal{D}^n$. Then:

$(\exists (R_1, \ldots, R_n) \in \mathcal{D}^n)(\exists x, y, z \in S)(x R y \wedge y R z \wedge z P x).$

$x R y \rightarrow \sim y P x$

$\rightarrow \sim [n(y P_i x) > pn \vee [(\forall i \in N)(y R_i x) \wedge (\exists i \in N)(y P_i x)]]$

$\rightarrow n(y P_i x) \leq pn \wedge [(\exists i \in N)(x P_i y) \vee (\forall i \in N)(x R_i y)]$

$\rightarrow [n(y P_i x) \leq pn \wedge (\exists i \in N)(x P_i y)] \vee (\forall i \in N)(x R_i y) \qquad \text{(T8.2-1)}$

Similarly,

$y R z \rightarrow [n(z P_i y) \leq pn \wedge (\exists i \in N)(y P_i z)] \vee (\forall i \in N)(y R_i z) \qquad \text{(T8.2-2)}$

$z P x \rightarrow [n(z P_i x) > pn \vee [(\forall i \in N)(z R_i x) \wedge (\exists i \in N)(z P_i x)]] \qquad \text{(T8.2-3)}$

(T8.2-1) $\rightarrow n(xR_iy) \geq (1-p)n$                                                      (T8.2-4)

(T8.2-2) $\rightarrow n(yR_iz) \geq (1-p)n$                                                      (T8.2-5)

(T8.2-3) $\rightarrow (\exists i \in N)(zP_ix)$                                                      (T8.2-6)

Suppose $(\forall i \in N)(xR_iy)$. By (T8.2-6) then, there exists an individual $i$ for whom $zP_iy$ holds, which in turn, by (T8.2-2), implies that $(\exists i \in N)(yP_iz)$. From $(\forall i \in N)(xR_iy)$ and $(\exists i \in N)(yP_iz)$, we conclude that $(\exists i \in N)(xP_iz)$. Therefore, by (T8.2-3), $n(zP_ix) > pn$. But (T8.2-5), together with $n(zP_ix) > pn$, implies that $(\exists i \in N)(yR_izP_ix)$, entailing that $(\exists i \in N)(yP_ix)$, contradicting the starting assumption that $(\forall i \in N)(xR_iy)$. This contradiction establishes that $(\forall i \in N)(xR_iy)$ is false. Consequently, we conclude:

$(\exists i \in N)(yP_ix)$                                                                  (T8.2-7)

Next suppose $(\forall i \in N)(yR_iz)$. By (T8.2-6) then, there exists an individual $i$ for whom $yP_ix$ holds, which in turn, by (T8.2-1), implies that $(\exists i \in N)(xP_iy)$. From $(\forall i \in N)(yR_iz) \wedge (\exists i \in N)(xP_iy)$, we conclude that $(\exists i \in N)(xP_iz)$. Therefore, by (T8.2-3), $n(zP_ix) > pn$. But (T8.2-4), together with $n(zP_ix) > pn$, implies that $(\exists i \in N)(zP_ixR_iy)$, entailing that $(\exists i \in N)(zP_iy)$, contradicting the supposition that $(\forall i \in N)(yR_iz)$. This contradiction establishes that $(\forall i \in N)(yR_iz)$ is false. Consequently, we conclude:

$(\exists i \in N)(zP_iy)$                                                                  (T8.2-8)

(T8.2-7) $\wedge$ (T8.2-1) $\rightarrow (\exists i \in N)(xP_iy)$                                          (T8.2-9)

(T8.2-8) $\wedge$ (T8.2-2) $\rightarrow (\exists i \in N)(yP_iz)$                                          (T8.2-10)

First suppose that $n(zP_ix) > pn$. Then by (T8.2-4) and (T8.2-5), $(\exists i \in N)(zP_ixR_iy)$ $\wedge (\exists i \in N)(yR_izP_ix)$. This, coupled with (T8.2-9) and (T8.2-10), implies that both PR and AUEV are violated.                                                      (T8.2-11)

Next suppose that $[(\forall i \in N)(zR_ix) \wedge (\exists i \in N)(zP_ix)]$. Then, by (T8.2-9) and (T8.2-10), we conclude that $[(\exists i \in N)(zR_ixP_iy) \wedge (\exists i \in N)(yP_izR_ix)]$. This, together with (T8.2-6), implies that both PR and AUEV are violated.          (T8.2-12)

(T8.2-11) and (T8.2-12) establish the theorem.                                          $\square$

*Remark 8.1* From the proof of Theorem 8.2 it is clear that if a Pareto-inclusive $p$-strict majority rule, $p \in [\frac{1}{2}, 1)$, violates transitivity then we must have for some distinct $x, y, z \in S$:

$[(\exists i \in N)(xP_iy) \wedge (\exists i \in N)(yP_ix) \wedge (\exists i \in N)(yP_iz) \wedge (\exists i \in N)(zP_iy) \wedge (\exists i$

$\in N)(zP_ix)]$ and

$(\exists i, j \in N)[(zR_ixP_iy \wedge yP_jzR_jx) \vee (zP_ixR_iy \wedge yR_jzP_jx)].$          $\Diamond$

**Lemma 8.1** *Let $\mathscr{D} \subseteq \mathscr{T}$. Let $A = \{x, y, z\} \subseteq S$ be a triple of alternatives. Then $\mathscr{D}$ violates both the conditions of placement restriction and absence of unique extremal* *value over A iff $\mathscr{D}|A$ contains one of the following eight sets of orderings of A, except for a formal interchange of alternatives.*

$(A)$  $xP^syP^sz$
      $yP^tzP^tx$

$(B)$  $xP^syP^sz$
      $zP^txI^ty$

$(C)$  $xP^syP^sz$
      $yI^tzP^tx$

$(D)$  $xP^syP^sz$
      $zP^tyP^tx$
      $yP^uzI^ux$

$(E)$  $xP^syP^sz$
      $zP^tyP^tx$
      $zI^uxP^uy$

$(F)$  $xP^syP^sz$
      $yP^tzI^tx$
      $zI^uxP^uy$

$(G)$  $xP^syI^sz$
      $yP^tzI^tx$
      $zI^uxP^uy$

$(H)$  $xI^syP^sz$
      $yP^tzI^tx$
      $zI^uxP^uy$

*Proof* By Lemma 6.1, $\mathscr{D} \subseteq \mathscr{T}$ violates PR over the triple $A = \{x, y, z\}$ iff $\mathscr{D} \mid A$ contains one of the following 10 sets, except for a formal interchange of alternatives.

$(i)$   $xP^syP^sz$
      $yP^tzP^tx$

$(ii)$  $xP^syP^sz$
      $zP^txI^ty$

$(iii)$ $xP^syP^sz$
      $yI^tzP^tx$

$(iv)$  $xP^syP^sz$
      $zP^tyP^tx$
      $yP^uzI^ux$

$(v)$   $xP^syP^sz$
      $zP^tyP^tx$
      $zI^uxP^uy$

$(vi)$  $xP^syP^sz$
      $yP^tzI^tx$
      $zI^uxP^uy$

$(vii)$ $xP^syI^sz$
      $yP^tzI^tx$
      $zI^uxP^uy$

$(viii)$ $xI^syP^sz$
      $yP^tzI^tx$
      $zI^uxP^uy$

$(ix)$  $xP^syI^sz$
      $yP^tzI^tx$
      $zP^uxI^uy$

$(x)$   $xI^syP^sz$
      $yI^tzP^tx$
      $zI^uxP^uy$

The sets (i)–(viii) are the same as the sets (A)–(H), respectively. The sets (ix) and (x) do not violate AEUV. $\mathscr{D} \mid A$ containing (ix) or (x) would violate AUEV iff a concerned ordering $R^v$ of A other than the ones in the set is included in it. With the inclusion of a concerned ordering, not already contained in (ix) or (x) as the case may be, one of the sets (A)–(H) would be contained in $\mathscr{D} \mid A$. This establishes the Lemma.  $\square$

Strongly Antagonistic Preferences (2) (SAP(2)): $\mathscr{D} \subseteq \mathscr{T}$ satisfies SAP(2) over the triple $A \subseteq S$ iff ($\exists$ distinct $x, y, z \in A$)[($\forall R \in \mathscr{D}|A)(xPyPz \vee zPyPx \vee yPx$ $Iz \vee xIyIz)$] $\vee$ ($\exists$ distinct $x, y, z \in A$)[($\forall R \in \mathscr{D}|A)(xPyPz \vee zPyPx \vee xIzPy \vee$

$x I y I z)]$. $\mathscr{D}$ satisfies SAP(2) iff it satisfies SAP(2) over every triple contained in $S$.

**Lemma 8.2** *Let $\mathscr{D} \subseteq \mathscr{T}$. Let $A = \{x, y, z\} \subseteq S$ be a triple of alternatives. Then $\mathscr{D}$ violates all three conditions of placement restriction, absence of unique extremal value and strongly antagonistic preferences (2) over A iff $\mathscr{D}\backslash A$ contains one of the following six sets of orderings of A, except for a formal interchange of alternatives.*

$$(A)\ \ x P^s y P^s z \qquad\qquad (B)\ \ x P^s y P^s z$$
$$y P^t z P^t x \qquad\qquad\qquad z P^t x I^t y$$

$$(C)\ \ x P^s y P^s z \qquad\qquad (F)\ \ x P^s y P^s z$$
$$y I^t z P^t x \qquad\qquad\qquad y P^t z I^t x$$
$$\qquad\qquad\qquad\qquad\qquad z I^u x P^u y$$

$$(G)\ \ x P^s y I^s z \qquad\qquad (H)\ \ x I^s y P^s z$$
$$y P^t z I^t x \qquad\qquad\qquad y P^t z I^t x$$
$$z I^u x P^u y \qquad\qquad\qquad z I^u x P^u y$$

*Proof* By Lemma 8.1, $\mathscr{D} \subseteq \mathscr{T}$ violates both PR and AUEV over the triple $A = \{x, y, z\}$ iff $\mathscr{D} \mid A$ contains one of the eight sets (A)–(H) listed in the statement of the Lemma, except for a formal interchange of alternatives. Excepting sets (D) and (E), all other sets violate SAP(2) as well. $\mathscr{D} \mid A$ containing (D) or (E) would violate SAP(2) iff a concerned ordering of $A$ not belonging to (D) or (E), as the case may be, is included in it. With the inclusion of the required ordering $\mathscr{D} \mid A$ would contain one of the six sets A, B, C, F, G, H. This establishes the lemma. $\qquad\square$

*Remark 8.2* If $n = \lfloor pn \rfloor + 1$, then Pareto-inclusive $p$-strict majority rule, $p \in [\frac{1}{2}, 1)$, becomes identical with the Pareto-rule. $\qquad\diamond$

**Theorem 8.3** *Let $f : \mathscr{T}^n \mapsto \mathscr{C}$ be the Pareto-rule; and let $n \geq 3$. Let $\mathscr{D} \subseteq \mathscr{T}$. Then $f$ yields transitive social $R$, $R = f(R_1, \ldots, R_n)$, for every $(R_1, \ldots, R_n) \in \mathscr{D}^n$ iff $\mathscr{D}$ satisfies over every triple of alternatives the placement restriction or the condition of absence of unique extremal value or strongly antagonistic preferences (2).*

*Proof* Let $f : \mathscr{T}^n \mapsto \mathscr{C}$ be the Pareto-rule. Let $\mathscr{D} \subseteq \mathscr{T}$. Suppose $f$ does not yield transitive social $R$ for every $(R_1, \ldots, R_n) \in \mathscr{D}^n$. Then $(\exists(R_1, \ldots, R_n) \in \mathscr{D}^n)(\exists x, y, z \in S)(x R y \wedge y R z \wedge z P x)$.

As the Pareto-rule can be viewed as a Pareto-inclusive $p$-strict majority rule for some $p \in [\frac{1}{2}, 1)$, in view of Theorem 8.2 it follows that:

$\mathscr{D} \mid A$ violates PR and AUEV. $\qquad\qquad\qquad$ (T8.3-1)

As in the proof of Theorem 8.2, we can conclude that:

$[(\exists i \in N)(x P_i y) \wedge (\exists i \in N)(y P_i z)]$ $\qquad\qquad\qquad$ (T8.3-2)

$z P x \rightarrow [(\forall i \in N)(z R_i x) \wedge (\exists i \in N)(z P_i x)]$ $\qquad\qquad$ (T8.3-3)

(T8.3-3) $\wedge$ (T8.3-2) $\rightarrow$ $(\exists i, j \in N)(z R_i x P_i y \wedge y P_j z R_j x)$                  (T8.3-4)

(T8.3-4) implies that $\mathscr{D} \mid A$ violates SAP(2).                                    (T8.3-5)

(T8.3-1) and (T8.3-5) establish the sufficiency part of the theorem.

Let $A = \{x, y, z\} \subseteq S$ be a triple; and let $\mathscr{D} \mid A$ violate all three conditions PR, AUEV and SAP(2). Then by Lemma 8.2, $\mathscr{D} \mid A$ must contain one of the six sets listed in the statement of the Lemma, except for a formal interchange of alternatives. If $\mathscr{D} \mid A$ contains (A) or (B) or (C), then consider any $(R_1, \ldots, R_n) \in \mathscr{D}^n$ such that $[\#\{i \in N \mid R_i \mid A = R^s \mid A\} = n - 1 \wedge \#\{i \in N \mid R_i \mid A = R^t \mid A\} = 1]$; and if $\mathscr{D} \mid A$ contains any of (F) or (G) or (H), then consider any $(R_1, \ldots, R_n) \in \mathscr{D}^n$ such that $[\#\{i \in N \mid R_i \mid A = R^s \mid A\} = n - 2 \wedge \#\{i \in N \mid R_i \mid A = R^t \mid A\} = 1 \wedge \#\{i \in N \mid R_i \mid A = R^u \mid A\} = 1]$. Then we obtain: $(x I y \wedge y P z \wedge x I z)$ for sets (A) and (C); $(x P y \wedge y I z \wedge x I z)$ for set (B); and $(x I y \wedge y I z \wedge x P z)$ for sets (F), (G) and (H). Thus transitivity is violated in each case. This establishes the necessity part of theorem.                                                                                        $\square$

**Theorem 8.4** *Let $f : \mathscr{T}^n \mapsto \mathscr{C}$ be a Pareto-inclusive p-strict majority rule; $p \in [\frac{1}{2}, 1)$. Let $n \geq 3 \wedge n > \lfloor pn \rfloor + 1$. Let N be such that there is a partition $(N_1, N_2, N_3)$ of N such that $(\#N_1 + \#N_2 \leq pn \wedge \#N_1 + \#N_3 \leq pn \wedge \#N_2 + \#N_3 \leq pn)$. Let $\mathscr{D} \subseteq \mathscr{T}$. Then f yields transitive social $R, R = f(R_1, \ldots, R_n)$, for every $(R_1, \ldots, R_n) \in \mathscr{D}^n$ iff $\mathscr{D}$ satisfies over every triple of alternatives the placement restriction or the condition of absence of unique extremal value.*

*Proof* The sufficiency part follows from Theorem 8.2.

Let there be a partition $(N_1, N_2, N_3)$ of $N$ such that $(\#N_1 + \#N_2 \leq pn \wedge \#N_1 + \#N_3 \leq pn \wedge \#N_2 + \#N_3 \leq pn)$. Existence of such a partition implies the existence of a twofold partition of $N$ such that the cardinality of neither set exceeds $pn$. From the existence of a twofold partition of $N$ such that the cardinality of neither of the two sets exceeds $pn$, it follows that $n - \lfloor pn \rfloor \leq pn$.

Let $A = \{x, y, z\} \subseteq S$ be a triple; and let $\mathscr{D} \mid A$ violate both PR and AUEV. Then by Lemma 8.1, $\mathscr{D} \mid A$ must contain one of the sets (A)–(H), except for a formal interchange of alternatives. If $\mathscr{D} \mid A$ contains (A) or (B) or (C), then consider any $(R_1, \ldots, R_n) \in \mathscr{D}^n$ such that $[(\forall i \in N_1 \cup N_2)(R_i \mid A = R^s \mid A) \wedge (\forall i \in N_3)(R_i \mid A = R^t \mid A)]$; if $\mathscr{D} \mid A$ contains (D) or (E), then consider any $(R_1, \ldots, R_n) \in \mathscr{D}^n$ such that $[(\#\{i \in N \mid R_i \mid A = R^s \mid A\} = \lfloor pn \rfloor) \wedge (\#\{i \in N \mid R_i \mid A = R^t \mid A\} = n - \lfloor pn \rfloor - 1)) \wedge (\#\{i \in N \mid R_i \mid A = R^u \mid A\} = 1]$; and if $\mathscr{D} \mid A$ contains (F) or (G) or (H), then consider any $(R_1, \ldots, R_n) \in \mathscr{D}^n$ such that $[(\forall i \in N_1)(R_i \mid A = R^s \mid A) \wedge (\forall i \in N_2)(R_i \mid A = R^t \mid A) \wedge (\forall i \in N_3)(R_i \mid A = R^u \mid A)]$. Then we obtain: $(x I y \wedge y P z \wedge x I z)$ for sets (A), (C) and (D); $(x P y \wedge y I z \wedge x I z)$ for sets (B) and (E); and $(x I y \wedge y I z \wedge x P z)$ for sets (F), (G) and (H). Thus transitivity is violated in each case. This establishes the theorem.                                                       $\square$

*Remark 8.3* An example of a rule for which the above theorem is applicable is provided by the Pareto-inclusive two-thirds strict majority rule defined for nine individuals. We have: $n \geq 3$ and $n > \lfloor pn \rfloor + 1 = 7$. $(N_1, N_2, N_3)$, where $\#N_1 =$

$\#N_2 = \#N_3 = 3$, is a partition of $N$ such that $(\#N_1 + \#N_2 \leq pn \wedge \#N_1 + \#N_3 \leq pn \wedge \#N_2 + \#N_3 \leq pn)$. ◇

## 8.3   Quasi-transitivity Under the Pareto-Inclusive Strict Majority Rules

**Theorem 8.5** *Let* $f : \mathcal{T}^n \mapsto \mathcal{C}$ *be a Pareto-inclusive p-strict majority rule;* $p \in [\frac{1}{2}, 1)$. *Let* $\mathcal{D} \subseteq \mathcal{T}$. *Then* $f$ *yields quasi-transitive social R,* $R = f(R_1, \ldots, R_n)$, *for every* $(R_1, \ldots, R_n) \in \mathcal{D}^n$ *if* $\mathcal{D}$ *satisfies over every triple of alternatives the condition of Latin Square unique value restriction or the condition of limited agreement.*

*Proof* Let $f : \mathcal{T}^n \mapsto \mathcal{C}$ be a Pareto-inclusive $p$-strict majority rule, $p \in [\frac{1}{2}, 1)$, and let $\mathcal{D} \subseteq \mathcal{T}$. Suppose $f$ does not yield quasi-transitive social $R$ for every $(R_1, \ldots, R_n) \in \mathcal{D}^n$. Then:

$$(\exists(R_1, \ldots, R_n) \in \mathcal{D}^n)(\exists x, y, z \in S)(xPy \wedge yPz \wedge zRx). \tag{T8.5-1}$$

$$xPy \to n(xP_iy) > pn \vee [(\forall i \in N)(xR_iy) \wedge (\exists i \in N)(xP_iy)] \tag{T8.5-2}$$

$$yPz \to n(yP_iz) > pn \vee [(\forall i \in N)(yR_iz) \wedge (\exists i \in N)(yP_iz)] \tag{T8.5-3}$$

$$zRx \to n(xP_iz) \leq pn \wedge \sim [(\forall i \in N)(xR_iz) \wedge (\exists i \in N)(xP_iz)]$$

$$\to n(xP_iz) \leq pn \wedge [(\exists i \in N)(zP_ix) \vee (\forall i \in N)(zR_ix)]$$

$$\to [n(xP_iz) \leq pn \wedge (\exists i \in N)(zP_ix)] \vee (\forall i \in N)(zR_ix) \tag{T8.5-4}$$

$$(\text{T8.5-2}) \to (\exists i \in N)(xP_iy) \tag{T8.5-5}$$

$$(\text{T8.5-3}) \to (\exists i \in N)(yP_iz) \tag{T8.5-6}$$

$$(\text{T8.5-4}) \to n(zR_ix) \geq (1 - p)n \tag{T8.5-7}$$

Suppose $(\forall i \in N)(xR_iy)$. By (T8.5-6) then, there exists an individual for whom $xP_iz$ holds. Then (T8.5-4) implies that there exists an individual for whom $zP_ix$ holds. This in turn implies that $(\exists i \in N)(zP_iy)$ in view of $(\forall i \in N)(xR_iy)$. $(\exists i \in N)(zP_iy)$, by (T8.5-3), implies that $n(yP_iz) > pn$. (T8.5-7), together with $n(yP_iz) > pn$, implies that $(\exists i \in N)(yP_izR_ix)$, which in turn implies that $(\exists i \in N)(yP_ix)$, contradicting the supposition that $(\forall i \in N)(xR_iy)$. Therefore we conclude that:

$$(\exists i \in N)(yP_ix) \tag{T8.5-8}$$

$$(\text{T8.5-8}) \wedge (\text{T8.5-2}) \to n(xP_iy) > pn \tag{T8.5-9}$$

$$(\text{T8.5-9}) \wedge (\text{T8.5-7}) \to (\exists k \in N)(zR_kxP_ky) \tag{T8.5-10}$$

$$(\text{T8.5-10}) \to (\exists i \in N)(zP_iy) \tag{T8.5-11}$$

$$(\text{T8.5-3}) \wedge (\text{T8.5-11}) \to n(yP_iz) > pn \tag{T8.5-12}$$

$$(\text{T8.5-9}) \wedge (\text{T8.5-12}) \to (\exists i \in N)(xP_iyP_iz), \text{ as } \tfrac{1}{2} \leq p < 1 \tag{T8.5-13}$$

$$(\text{T8.5-13}) \to (\exists i \in N)(xP_iz) \tag{T8.5-14}$$

(T8.5-4) $\wedge$ (T8.5-14) $\rightarrow$ ($\exists i \in N$)($z P_i x$)                    (T8.5-15)

(T8.5-12) $\wedge$ (T8.5-7) $\rightarrow$ ($\exists j \in N$)($y P_j z R_j x$)                    (T8.5-16)

(T8.5-1) implies that $x$, $y$, $z$ are distinct alternatives. (T8.5-13), (T8.5-16) and (T8.5-10) establish that: ($\exists i, j, k \in N$)$[x P_i y P_i z \wedge y P_j z R_j x \wedge z R_k x P_k y]$. $x P_i y P_i z$, $y P_j z R_j x$, $z R_k x P_k y$ belong to $T[LS(xyzx)]$ and form $LS(xyzx)$. In the triple $\{x, y, z\}$, $y$ is uniquely medium according to $x P_i y P_i z$, uniquely best according to $y P_j z R_j x$ and uniquely worst according to $z R_k x P_k y$. Therefore LSUVR is violated over the triple $\{x, y, z\}$. From (T8.5-5), (T8.5-8), (T8.5-6), (T8.5-11), (T8.5-14) and (T8.5-15), it follows that LA is violated over the triple $\{x, y, z\}$. Thus $\mathscr{D} \mid \{x, y, z\}$ violates both LSUVR and LA. It has been shown that violation of quasi-transitivity by $R = f(R_1, \ldots, R_n)$, $(R_1, \ldots, R_n) \in \mathscr{D}^n$, implies violation of both LSUVR and LA by $\mathscr{D}$ over a triple, i.e., if $\mathscr{D}$ satisfies over every triple of alternatives LSUVR or LA then every $(R_1, \ldots, R_n) \in \mathscr{D}^n$ yields quasi-transitive social $R$.                    □

*Remark 8.4* Let $\frac{1}{2} \leq p < 1$. If $\lfloor pn \rfloor = n - 1$, then the Pareto-inclusive $p$-strict majority rule coincides with the Pareto-rule. Thus, if $n = \lfloor pn \rfloor + 1$ then the Pareto-inclusive $p$-strict majority rule yields quasi-transitive social $R$ for every $(R_1, \ldots, R_n) \in \mathscr{T}^n$.                    ◇

**Theorem 8.6** *Let* $f : \mathscr{T}^n \mapsto \mathscr{C}$ *be a Pareto-inclusive $p$-strict majority rule; $p \in [\frac{1}{2}, 1)$. Let $n = \lfloor pn \rfloor + 2$. Let $\mathscr{D} \subseteq \mathscr{T}$. Then $f$ yields quasi-transitive social $R$, $R = f(R_1, \ldots, R_n)$, for every $(R_1, \ldots, R_n) \in \mathscr{D}^n$ iff $\mathscr{D}$ satisfies the condition of Latin Square linear ordering restriction.*

*Proof* Let $f : \mathscr{T}^n \mapsto \mathscr{C}$ be a Pareto-inclusive $p$-strict majority rule, $p \in [\frac{1}{2}, 1)$, and let $n = \lfloor pn \rfloor + 2$. Let $\mathscr{D} \subseteq \mathscr{T}$. Suppose $f$ does not yield quasi-transitive social $R$ for every $(R_1, \ldots, R_n) \in \mathscr{D}^n$. Then ($\exists (R_1, \ldots, R_n) \in \mathscr{D}^n$)($\exists x, y, z \in S$)($x P y \wedge y P z \wedge z R x$). Then, as in the proof of Theorem 8.5, we can infer that ($\exists i, j, k \in N$)($x P_i y P_i z \wedge y P_j z R_j x \wedge z R_k x P_k y$).                    (T8.6-1)

$x P y \rightarrow n(x P_i y) > pn$

$\rightarrow n(x P_i y) \geq \lfloor pn \rfloor + 1$                    (T8.6-2)

Similarly, $y P z \rightarrow n(y P_i z) \geq \lfloor pn \rfloor + 1$                    (T8.6-3)

(T8.6-2) $\wedge$ (T8.6-3) $\rightarrow n(x P_i y P_i z) \geq \lfloor pn \rfloor$, as $n = \lfloor pn \rfloor + 2$                    (T8.6-4)

(T8.6-1) $\wedge$ (T8.6-4) $\rightarrow n(x P_i y P_i z) = \lfloor pn \rfloor \wedge n(y P_j z R_j x) = 1 \wedge n(z R_k x P_k y) = 1$                    (T8.6-5)

By Theorem 8.5, LA must be violated over $\{x, y, z\}$. Thus, in view of (T8.6-5) it follows that we must have:

($\exists i, j, k \in N$)$[(x P_i y P_i z \wedge y P_j z P_j x \wedge z R_k x P_k y) \vee (x P_i y P_i z \wedge y P_j z R_j x \wedge z P_k x P_k y)]$                    (T8.6-6)

(T8.6-6) implies that LSLOR is violated. Thus we have shown that if $n = \lfloor pn \rfloor +$ 2 then violation of quasi-transitivity by $R = f(R_1, \ldots, R_n)$, $(R_1, \ldots, R_n) \in \mathscr{D}^n$, implies violation of LSLOR by $\mathscr{D}$, establishing the sufficiency part of the Theorem.

Suppose $\mathscr{D} \subseteq \mathscr{T}$ violates LSLOR. Then there exists a triple $A \subseteq S$ such that $\mathscr{D}$ violates LSLOR over $A$. Violation of LSLOR by $\mathscr{D}$ over $A$ implies ($\exists$ distinct $x, y, z \in A$)($\exists R^s, R^t, R^u \in \mathscr{D}$)[$(x P^t y P^t z \wedge y P^s z P^s x \wedge z P^u x R^u y) \vee (x P^s y P^s z \wedge y P^t z P^t x \wedge z R^u x P^u y)$]. Consider any $(R_1, \ldots, R_n) \in \mathscr{D}^n$ such that the restriction of $(R_1, \ldots, R_n)$ to $\{x, y, z\} = A$, $(R_1|\{x, y, z\}, \ldots, R_n|\{x, y, z\})$, is given by: [#$\{i \in N \mid R_i \mid A = R^s \mid A\} = \lfloor pn \rfloor \wedge$ #$\{i \in N \mid R_i \mid A = R^t \mid A\} = 1 \wedge$ #$\{i \in N \mid R_i \mid A = R^u \mid A\} = 1$]. If $(R^t \mid A = x P^t y P^t z \wedge R^s \mid A = y P^s z P^s x \wedge R^u \mid A = z P^u x R^u y)$ then we have: $n(y P_i z) = \lfloor pn \rfloor + 1 \wedge n(z P_i x) = \lfloor pn \rfloor + 1 \wedge n(y P_i x) = \lfloor pn \rfloor$ giving rise to $(y P z \wedge z P x \wedge x R y)$. And, if $(R^s \mid A = x P^s y P^s z \wedge R^t \mid A = y P^t z P^t x \wedge R^u \mid A = z R^u x P^u y)$ then we have: $n(x P_i y) = \lfloor pn \rfloor + 1 \wedge n(y P_i z) = \lfloor pn \rfloor + 1 \wedge n(x P_i z) = \lfloor pn \rfloor$ giving rise to $(x P y \wedge y P z \wedge z R x)$. In either case quasi-transitivity is violated. We have shown that if $\mathscr{D} \subseteq \mathscr{T}$ violates LSLOR then there exists a $(R_1, \ldots, R_n) \in \mathscr{D}^n$ such that $R = f(R_1, \ldots, R_n)$ is not quasi-transitive, i.e., if $f$ yields quasi-transitive $R$ for every $(R_1, \ldots, R_n) \in \mathscr{D}^n$ then $\mathscr{D}$ must satisfy LSLOR, establishing the necessity part of the Theorem.     □

**Lemma 8.3** *Let $\mathscr{D} \subseteq \mathscr{T}$. Let $A \subseteq S$ be a triple of alternatives. Then $\mathscr{D}$ violates both Latin Square unique value restriction and limited agreement over $A$ iff $\mathscr{D}|A$ contains one of the following six sets of orderings of $A$, except for a formal interchange of alternatives.*

| | | | |
|---|---|---|---|
| (i) | $x P^s y P^s z$ | (ii) | $x P^s y P^s z$ |
| | $y P^t z P^t x$ | | $y P^t z P^t x$ |
| | $z P^u x P^u y$ | | $z I^u x P^u y$ |
| (iii) | $x P^s y P^s z$ | (iv) | $x P^s y P^s z$ |
| | $y P^t z I^t x$ | | $y P^t z I^t x$ |
| | $z P^u x P^u y$ | | $z I^u x P^u y$ |
| | | | $z P^v y P^v x$ |
| (v) | $x P^s y P^s z$ | (vi) | $x P^s y P^s z$ |
| | $y P^t z I^t x$ | | $y P^t z I^t x$ |
| | $z I^u x P^u y$ | | $z I^u x P^u y$ |
| | $z P^v y I^v x$ | | $z I^v y P^v x$ |

*Proof* By the definition of LSUVR, it is violated over the triple $A$ iff ($\exists x, y, z \in A$)($\exists R^s, R^t, R^u \in \mathscr{D}|A$)[$x P^s y P^s z \wedge y P^t z R^t x \wedge z R^u x P^u y$]. A set of orderings over $A$ containing $\{x P^s y P^s z, y P^t z R^t x, z R^u x P^u y\}$ would violate LA iff it contains an ordering in which $z P^v x$ holds, i.e., iff it contains $(y P^v z P^v x \vee y I^v z P^v x \vee z P^v y P^v x \vee z P^v y I^v x \vee z P^v x P^v y)$. Therefore, a set of orderings over $A$ containing $\{x P^s y P^s z, y P^t z R^t x, z R^u x P^u y\}$ would violate LA iff $\mathscr{D}|A$ contains [$(x P^s y P^s z \wedge y P^t z P^t x \wedge z R^u x P^u y) \vee (x P^s y P^s z \wedge y P^t z R^t x \wedge z P^u x P^u y) \vee (x P^s y P^s z \wedge y P^t z I^t x \wedge z I^u x P^u y \wedge (z P^v y P^v x \vee z P^v y I^v x \vee z I^v y P^v x))$], i.e., iff it contains one of the six sets listed in the statement of the Lemma, except for a formal interchange of alternatives.                                                                                  □

**Theorem 8.7** *Let* $f : \mathscr{T}^n \mapsto \mathscr{C}$ *be a Pareto-inclusive p-strict majority rule; $p \in [\frac{1}{2}, 1)$. Let $n > \lfloor pn \rfloor + 2$. Let $\mathscr{D} \subseteq \mathscr{T}$. Then $f$ yields quasi-transitive social $R$, $R = f(R_1, \ldots, R_n)$, for every $(R_1, \ldots, R_n) \in \mathscr{D}^n$ only if $\mathscr{D}$ satisfies Latin Square unique value restriction or limited agreement over every triple of alternatives.*

*Proof* Let $f : \mathscr{T}^n \mapsto \mathscr{C}$ be a Pareto-inclusive $p$-strict majority rule, $p \in [\frac{1}{2}, 1)$, and let $n > \lfloor pn \rfloor + 2$. Suppose $\mathscr{D} \subseteq \mathscr{T}$ violates both Latin Square unique value restriction and limited agreement over some triple $A$. Let $A = \{a, b, c\}$. Then, by Lemma 8.3 restriction of $\mathscr{D}$ over $A$ must contain one of the six sets listed in the statement of the Lemma, except for a formal interchange of alternatives. If $\mathscr{D} \mid A$ contains (i) or (ii) or (iii), then consider any $(R_1, \ldots, R_n) \in \mathscr{D}^n$ such that $\#\{i \in N \mid R_i \mid A = R^s \mid A\} = \lfloor pn \rfloor \wedge \#\{i \in N \mid R_i \mid A = R^t \mid A\} = n - \lfloor pn \rfloor - 1 \wedge \#\{i \in N \mid R_i \mid A = R^u \mid A\} = 1$; and if $\mathscr{D} \mid A$ contains (iv) or (v) or (vi), then consider any $(R_1, \ldots, R_n) \in \mathscr{D}^n$ such that $\#\{i \in N \mid R_i \mid A = R^s \mid A\} = \lfloor pn \rfloor \wedge \#\{i \in N \mid R_i \mid A = R^t \mid A\} = n - \lfloor pn \rfloor - 2 \wedge \#\{i \in N \mid R_i \mid A = R^u \mid A\} = 1 \wedge \#\{i \in N \mid R_i \mid A = R^v \mid A\} = 1$; $(x, y, z)$ being a permutation of $(a, b, c)$. In each case we obtain: $\#\{i \in N \mid x P_i y\} > pn \wedge \#\{i \in N \mid y P_i z\} > pn \wedge \#\{i \in N \mid x P_i z\} = \lfloor pn \rfloor$ yielding $(x P y \wedge y P z \wedge \sim x P z)$, violating quasi-transitivity. We have shown that if $\mathscr{D} \subseteq \mathscr{T}$ violates both LSUVR and LA over a triple then there exists a $(R_1, \ldots, R_n) \in \mathscr{D}^n$ such that $R = f(R_1, \ldots, R_n)$ is not quasi-transitive, i.e., if $f$ yields quasi-transitive $R$ for every $(R_1, \ldots, R_n) \in \mathscr{D}^n$ then $\mathscr{D}$ must satisfy LSUVR or LA over every triple of alternatives, establishing the theorem.                                           □

In view of Theorems 8.5 and 8.7, it follows that the following theorem holds.

**Theorem 8.8** *Let* $f : \mathscr{T}^n \mapsto \mathscr{C}$ *be a Pareto-inclusive p-strict majority rule; $p \in [\frac{1}{2}, 1)$. Let $n > \lfloor pn \rfloor + 2$. Let $\mathscr{D} \subseteq \mathscr{T}$. Then $f$ yields quasi-transitive social $R$, $R = f(R_1, \ldots, R_n)$, for every $(R_1, \ldots, R_n) \in \mathscr{D}^n$ iff $\mathscr{D}$ satisfies Latin Square unique value restriction or limited agreement over every triple of alternatives.*

A simple game social decision rule is defined by the condition that under it an alternative $x$ is socially preferred to another alternative $y$ iff all individuals belonging to some winning coalition unanimously prefer $x$ to $y$. In Theorem 9.1 of this chapter it is shown that a social decision rule is a simple game iff it satisfies the following four properties: (i) independence of irrelevant alternatives, (ii) neutrality, (iii) monotonicity and (iv) a set of individuals is a blocking coalition iff it is a strictly blocking coalition. These four properties, along with the two other conditions, namely, anonymity and weak Pareto-criterion, figured in the characterization of the class of strict majority rules. Thus the additional conditions of anonymity and weak Pareto-criterion separate the set of strict majority rules from other simple game social decision rules. An important subclass of the simple game social decision rules is that of strong simple game social decision rules. A simple game is a strong simple game iff it is the case that whenever a subset of individuals is not winning, its complement is winning. In Theorem 9.2 of this chapter a characterization for this subclass is given. It is shown that a social decision rule is a strong simple game iff it satisfies the properties of (i) independence of irrelevant alternatives, (ii) neutrality, (iii) monotonicity and (iv) the three sets, the set of blocking coalitions, the set of strictly blocking coalitions and the set of winning coalitions, being identical to each other.

For discussing transitivity under simple game social decision rules we partition the set of all simple game social decision rules into three subsets: (i) the rules that are null or dictatorial, (ii) the rules that are non-null non-strong simple games and (iii) the rules that are non-dictatorial strong simple games. Let $\mathscr{D}$ be a set of orderings of the set of alternatives $S$. We show that: (i) a simple game social decision rule yields transitive social weak preference relation for every profile of individual orderings iff it is null or dictatorial (Proposition 9.1); (ii) a non-null non-strong simple game social decision rule yields transitive social weak preference relation for every profile

© Springer Nature Singapore Pte Ltd. 2019
S. K. Jain, *Domain Conditions and Social Rationality*,
https://doi.org/10.1007/978-981-13-9672-4_9

of individual orderings belonging to $\mathscr{D}^n$ iff $\mathscr{D}$ satisfies the condition of Latin Square extremal value restriction over every triple of alternatives (Theorem 9.3); (iii) a non-dictatorial strong simple game social decision rule yields transitive social weak preference relation for every profile of individual orderings belonging to $\mathscr{D}^n$ iff $\mathscr{D}$ satisfies the condition of weak Latin Square extremal value restriction over every triple of alternatives (Theorem 9.4).

Under every $p$-strict majority rule, $\frac{1}{2} \le p < 1$, the set of all individuals $N$ is a winning coalition. Thus every $p$-strict majority rule is non-null. If there exists a partition $(N_1, N_2)$ of $N$ such that $\#N_1 \le pn \wedge \#N_2 \le pn$ then clearly the $p$-strict majority rule under consideration is not a strong simple game. Thus Theorem 7.4 can be derived as a corollary of Theorem 9.3. If the number of individuals is at least two then a $p$-strict majority rule is non-dictatorial. If the $p$-strict majority rule is such that there does not exist a partition $(N_1, N_2)$ of $N$ such that $\#N_1 \le pn \wedge \#N_2 \le pn$ then clearly it is the case that for every partition $(N_1, N_2)$ of $N$ we must have $\#N_1 > pn \vee \#N_2 > pn$, i.e., for every partition $(N_1, N_2)$ of $N$ either $N_1$ or $N_2$ is winning. In other words the $p$-strict majority rule in question is a strong simple game. Thus Theorem 7.7 can be derived as a corollary of Theorem 9.4.

For the purpose of discussing quasi-transitivity under simple game social decision rules we partition the set of all simple game social decision rules into two subsets: (i) the simple game social decision rules that are null or are such that there is a unique minimal winning coalition (i.e., are oligarchic) and (ii) the simple game social decision rules that are non-null and are such that there are at least two minimal winning coalitions (i.e., are non-oligarchic). We show that: (i) a simple game social decision rule yields quasi-transitive social weak preference relation for every profile of individual orderings iff it is null or is such that there is a unique minimal winning coalition (Proposition 9.3); (ii) a non-null simple game social decision rule such that there are at least two minimal winning coalitions yields quasi-transitive social weak preference relation for every profile of individual orderings belonging to $\mathscr{D}^n$ iff $\mathscr{D}$ satisfies the condition of Latin Square unique value restriction over every triple of alternatives (Theorem 9.5).

If $p$-strict majority rule, $\frac{1}{2} \le p < 1$, is such that $n = \#N = \lfloor pn \rfloor + 1$ then clearly the only winning coalition is that which consists of all individuals, hence there is a unique minimal winning coalition. On the other hand if $n = \#N > \lfloor pn \rfloor + 1$ then every set of individuals consisting of $\lfloor pn \rfloor + 1$ individuals is a minimal winning coalition. Thus Theorem 7.8 can be derived as a corollary of Theorem 9.5.

The chapter is divided into three sections. Section 1 provides characterizations of simple game SDRs and strong simple game SDRs. Sections 2 and 3 discuss, respectively, conditions for transitivity and quasi-transitivity under simple game SDRS.[1]

---

[1] This chapter relies on Jain (1989).

## 9.1 Characterization of Social Decision Rules Which Are Simple Games

A social decision rule $f : \mathcal{T}^n \mapsto \mathcal{C}$ is a simple game iff $(\forall (R_1, \ldots, R_n) \in \mathcal{T}^n)(\forall x, y \in S)[x P y \leftrightarrow (\exists V \in W)(\forall i \in V)(x P_i y)]$. A social decision rule $f : \mathcal{T}^n \mapsto \mathcal{C}$ is a strong simple game iff it is a simple game and $(\forall V \subseteq N)[V \notin W \rightarrow N - V \in W]$.

**Theorem 9.1** *A social decision rule $f : \mathcal{T}^n \mapsto \mathcal{C}$ is a simple game iff it satisfies the conditions of (i) independence of irrelevant alternatives, (ii) neutrality and (iii) monotonicity, and (iv) its structure is such that a coalition is blocking iff it is strictly blocking.*

*Proof* Let SDR $f : \mathcal{T}^n \mapsto \mathcal{C}$ be a simple game.
Consider any $(R_1, \ldots, R_n), (R'_1, \ldots, R'_n) \in \mathcal{T}^n$ and any $x, y \in S$ such that $(\forall i \in N)[(x R_i y \leftrightarrow x R'_i y) \wedge (y R_i x \leftrightarrow y R'_i x)]$.
$x P y \rightarrow (\exists V \in W)[(\forall i \in V)(x P_i y)]$
$\rightarrow (\forall i \in V)(x P'_i y)$
$\rightarrow x P' y$
$x P' y \rightarrow (\exists V' \in W)[(\forall i \in V')(x P'_i y)]$
$\rightarrow (\forall i \in V')(x P_i y)$
$\rightarrow x P y$
Thus we have: $x P y \leftrightarrow x P' y$. Analogously we obtain: $y P x \leftrightarrow y P' x$. $[(x P y \leftrightarrow x P' y) \wedge (y P x \leftrightarrow y P' x)]$ implies $(x I y \leftrightarrow x I' y)$, in view of connectedness of $R$ and $R'$. Thus $f$ satisfies condition I.
Consider any $(R_1, \ldots, R_n), (R'_1, \ldots, R'_n) \in \mathcal{T}^n$ and any $x, y, z, w \in S$ such that $(\forall i \in N)[(x R_i y \leftrightarrow z R'_i w) \wedge (y R_i x \leftrightarrow w R'_i z)]$.
Let $N_1 = \{i \in N \mid x P_i y \wedge z P'_i w\}$, $N_2 = \{i \in N \mid x I_i y \wedge z I'_i w\}$, $N_3 = \{i \in N \mid y P_i x \wedge w P'_i z\}$.
$x P y \rightarrow (\exists V \subseteq N_1)[V \in W]$
$\rightarrow z P' w$.
Similarly, $z P' w \rightarrow x P y$. Thus we have: $x P y \leftrightarrow z P' w$. Analogously we obtain: $y P x \leftrightarrow w P' z$. $[(x P y \leftrightarrow z P' w) \wedge (y P x \leftrightarrow w P' z)]$ implies $(x I y \leftrightarrow z I' w)$, in view of connectedness of $R$ and $R'$. Thus $f$ satisfies neutrality.
Now consider any $(R_1, \ldots, R_n), (R'_1, \ldots, R'_n) \in \mathcal{T}^n$ and any $x, y \in S$ such that $(\forall i \in N)[(x P_i y \rightarrow x P'_i y) \wedge (x I_i y \rightarrow x R'_i y)]$.
Designate by $N_1, N_2, N_3$ the sets $\{i \in N \mid x P_i y\}$, $\{i \in N \mid x I_i y\}$, $\{i \in N \mid y P_i x\}$, respectively; and by $N'_1, N'_2, N'_3$ the sets $\{i \in N \mid x P'_i y\}$, $\{i \in N \mid x I'_i y\}$, $\{i \in N \mid y P'_i x\}$, respectively.
$x P y \rightarrow (\exists V \subseteq N_1)[V \in W]$
$\rightarrow (\forall i \in V)(x P' y)$, as $N_1 \subseteq N'_1$
$\rightarrow x P' y$.
$y P' x \rightarrow (\exists V' \subseteq N'_3)[V' \in W]$
$\rightarrow (\forall i \in V')(y P x)$, as $N'_3 \subseteq N_3$
$\rightarrow y P x$.

$(yP'x \rightarrow yPx)$ is equivalent to $(xRy \rightarrow xR'y)$ in view of connectedness of $R$ and $R'$. $(xPy \rightarrow xP'y)$ and $(xRy \rightarrow xR'y)$ establish that $f$ is monotonic.

If $V$ is strictly blocking then $V$ is blocking.

Suppose $V$ is not strictly blocking

$\rightarrow (\exists (R_1, \ldots, R_n) \in \mathscr{T}^n)(\exists x, y \in S)[(\forall i \in V)(xR_iy) \wedge yPx]$

$\rightarrow (\exists V' \subseteq N - V)(V' \in W)$

$\rightarrow (\forall (R_1, \ldots, R_n) \in \mathscr{T}^n)[(\forall i \in V)(xP_iy) \wedge (\forall i \in N - V)(yP_ix) \rightarrow yPx]$

$\rightarrow V$ is not blocking.

Thus a coalition is blocking iff it is strictly blocking.

Thus it has been shown that if $f : \mathscr{T}^n \mapsto \mathscr{C}$ is a simple game then $f$ satisfies I, M, N and its structure is such that a coalition is blocking iff it is strictly blocking.

Now let $f : \mathscr{T}^n \mapsto \mathscr{C}$ be a social decision rule such that it satisfies the conditions of (i) independence of irrelevant alternatives, (ii) neutrality and (iii) monotonicity, and (iv) its structure is such that a coalition is blocking iff it is strictly blocking.

Consider any profile $(R'_1, \ldots, R'_n) \in \mathscr{T}^n$ and any $x, y \in S$ such that $xP'y$. By condition I, $xP'y$ is a consequence solely of individual preferences over $\{x, y\}$ in profile $(R'_1, \ldots, R'_n)$. Designate by $N_1, N_2, N_3$ the sets $\{i \in N \mid xP'_iy\}$, $\{i \in N \mid xI'_iy\}$, $\{i \in N \mid yP'_ix\}$, respectively. Now consider any profile $(R''_1, \ldots, R''_n) \in \mathscr{T}^n$ such that $(\forall i \in N_1)(xP''_iy) \wedge (\forall i \in N_2 \cup N_3)(yP''_ix)$. Suppose $yR''x$. Then by conditions I, N and M we conclude: $(\forall (R_1, \ldots, R_n) \in \mathscr{T}^n)(\forall a, b \in S)[(\forall i \in N_2 \cup N_3)(aP_ib) \rightarrow aRb]$. In other words, $N_2 \cup N_3$ is a blocking coalition. As every blocking coalition is strictly blocking, it follows that $N_2 \cup N_3$ is strictly blocking. Consequently we must have $yR'x$ as $(\forall i \in N_2 \cup N_3)(yR'_ix)$. This, however, contradicts the hypothesis $xP'y$. Therefore, we conclude that $(\forall i \in N_1)(xP''_iy) \wedge (\forall i \in N_2 \cup N_3)(yP''_ix)$ must result in $xP''y$. Then it follows by conditions I, N and M that $(\forall (R_1, \ldots, R_n) \in \mathscr{T}^n)(\forall a, b \in S)[(\forall i \in N_1)(xP_iy) \rightarrow xPy]$. That is to say, $N_1$ is a winning coalition. We have shown that $(\forall (R_1, \ldots, R_n) \in \mathscr{T}^n)(\forall a, b \in S)[aPb \rightarrow (\exists V \in W)(\forall i \in V)(aP_ib)]$. This coupled with the fact that if $V \in W$ then $(\forall (R_1, \ldots, R_n) \in \mathscr{T}^n)(\forall a, b \in S)[(\forall i \in V)(aP_ib) \rightarrow aPb]$ establishes that $f$ is a simple game.                                                                                   $\square$

**Theorem 9.2** *A social decision rule $f : \mathscr{T}^n \mapsto \mathscr{C}$ is a strong simple game iff it satisfies the conditions of (i) independence of irrelevant alternatives, (ii) neutrality and (iii) monotonicity, and its structure is such that (iv) a coalition is blocking iff it is strictly blocking, and (v) a coalition is blocking iff it is winning.*

*Proof* Let SDR $f : \mathscr{T}^n \mapsto \mathscr{C}$ be a strong simple game.

By Theorem 9.1 it follows that $f$ satisfies conditions (i)–(iv). Suppose $V$ is a blocking coalition. Then $N - V$ cannot be a winning coalition. By the definition of a strong simple game then it follows that the complement of $N - V$, i.e., $V$ must be winning. This coupled with the fact that every winning coalition is blocking establishes that (v) holds.

Let SDR $f : \mathscr{T}^n \mapsto \mathscr{C}$ satisfy (i)–(v).

From Theorem 9.1 we know that (i)–(iv) imply that $f$ is a simple game.

Suppose $V$ is not winning. Then $(\exists(R'_1, \ldots, R'_n) \in \mathscr{T}^n)(\exists x, y \in S)[(\forall i \in V)$ $(x P'_i y) \wedge y R' x]$. By conditions I, N and M, then it follows that $(\forall(R_1, \ldots, R_n) \in \mathscr{T}^n)(\forall a, b \in S)[(\forall i \in N - V)(a P_i b) \rightarrow a R b]$. In other words, $N - V$ is a blocking coalition. As every blocking coalition is winning, it follows that $N - V$ is winning. Thus we have shown that $(\forall V \subseteq N)(V \notin W \rightarrow N - V \in W)$, which establishes that f is a strong simple game. $\qquad\square$

## 9.2 Transitivity Under Simple Games

**Proposition 9.1** *Let social decision rule* $f : \mathscr{T}^n \mapsto \mathscr{C}$ *be a simple game. Then,* $f$ *yields transitive social weak preference relation for every* $(R_1, \ldots, R_n) \in \mathscr{T}^n$ *iff it is null or dictatorial.*

*Proof* If $f$ is null then obviously $R = f(R_1, \ldots, R_n)$ is transitive for every $(R_1, \ldots, R_n) \in \mathscr{T}^n$.

If $f$ is dictatorial then there is a minimal winning coalition consisting of a single individual, say individual $j$. As every winning coalition is blocking, it follows that $\{j\}$ is a blocking coalition. By Theorem 9.1, if $f$ is a simple game then every blocking coalition is strictly blocking. From the fact that $\{j\}$ is both winning and strictly blocking we conclude that for every $(R_1, \ldots, R_n) \in \mathscr{T}^n$, $R = f(R_1, \ldots, R_n)$ coincides with $R_j$. Transitivity of $R$ follows from the fact that $R_j$ is an ordering.

Now suppose $f$ yields transitive $R$ for every $(R_1, \ldots, R_n) \in \mathscr{T}^n$. As $f$ is a simple game, it satisfies conditions I, N and M, by Theorem 9.1. If $f$ satisfies the weak Pareto-criterion then from Arrow Impossibility Theorem it follows that $f$ must be dictatorial.

If the weak Pareto-criterion is violated then it must be the case that $(\exists(R_1, \ldots, R_n) \in \mathscr{T}^n)(\exists x, y \in S)[(\forall i \in N)(x P_i y) \wedge y R x]$. In view of conditions I and N then it follows that we must have:

$$(\forall(R_1, \ldots, R_n) \in \mathscr{T}^n)(\forall x, y \in S)[(\forall i \in N)(x P_i y) \rightarrow y R x]. \tag{P9.1-1}$$

Now, a consequence of conditions I and N is that $(\forall(R_1, \ldots, R_n) \in \mathscr{T}^n)(\forall x, y \in S)[(\forall i \in N)(x I_i y) \rightarrow x I y]$ holds. In view of condition M then it follows that it must be the case that:

$$(\forall(R_1, \ldots, R_n) \in \mathscr{T}^n)(\forall x, y \in S)[(\forall i \in N)(x P_i y) \rightarrow x R y]. \tag{P9.1-2}$$

$$(P9.1\text{-}1) \wedge (P9.1\text{-}2) \rightarrow (\forall(R_1, \ldots, R_n) \in \mathscr{T}^n)(\forall x, y \in S)[(\forall i \in N)(x P_i y) \rightarrow x I y]. \tag{P9.1-3}$$

In view of condition M we conclude from (P9.1-3) that it must be the case that $(\forall(R_1, \ldots, R_n) \in \mathscr{T}^n)[f(R_1, \ldots, R_n) = R = (\forall x, y \in S)(x I y)]$, i.e., $f$ is null. This establishes the proposition. $\qquad\square$

**Proposition 9.2** *Let social decision rule* $f : \mathscr{T}^n \mapsto \mathscr{C}$ *be a strong simple game. Then,* $f$ *yields transitive social weak preference relation for every* $(R_1, \ldots, R_n) \in \mathscr{T}^n$ *iff it is dictatorial.*

*Proof* As $f$ is a strong simple game, the set of all individuals $N$ is winning. Therefore $f$ cannot be null. The proposition now follows directly from Proposition 9.1.  □

**Theorem 9.3** *Let social decision rule $f : \mathcal{T}^n \mapsto \mathscr{C}$ be a non-null non-strong simple game. Let $\mathscr{D} \subseteq \mathcal{T}$. Then $f$ yields transitive social $R$, $R = f(R_1, \ldots, R_n)$, for every $(R_1, \ldots, R_n) \in \mathscr{D}^n$ iff $\mathscr{D}$ satisfies the condition of Latin Square extremal value restriction.*

*Proof* Let $f : \mathcal{T}^n \mapsto \mathscr{C}$ be a non-null non-strong simple game; and let $\mathscr{D} \subseteq \mathcal{T}$. Suppose $f$ does not yield transitive social $R$ for every $(R_1, \ldots, R_n) \in \mathscr{D}^n$. Then:

$$(\exists (R_1, \ldots, R_n) \in \mathscr{D}^n)(\exists x, y, z \in S)(xRy \wedge yRz \wedge zPx). \tag{T9.3-1}$$

$$zPx \rightarrow (\exists V \in W)(\forall i \in V)(zP_ix), \tag{T9.3-2}$$

by the definition of a simple game

$$xRy \rightarrow (\exists j \in V)(xR_jy), \tag{T9.3-3}$$
as $(\forall i \in V)(yP_ix)$ would imply $yPx$

$$yRz \rightarrow (\exists k \in V)(yR_kz), \tag{T9.3-4}$$
as $(\forall i \in V)(zP_iy)$ would imply $zPy$

$$\text{(T9.3-2)} \wedge \text{(T9.3-3)} \rightarrow (\exists j \in V)(zP_jxR_jy) \tag{T9.3-5}$$

$$\text{(T9.3-2)} \wedge \text{(T9.3-4)} \rightarrow (\exists k \in V)(yR_kzP_kx) \tag{T9.3-6}$$

(T9.3-1) implies that $x, y, z$ are distinct alternatives. $zP_jxR_jy$ and $yR_kzP_kx$ belong to $T[LS(xyzx)]$. In the triple $\{x, y, z\}$, $z$ is uniquely best according to $zP_jxR_jy$, and medium according to $yR_kzP_kx$ without being worst; furthermore $x$ is uniquely worst according to $yR_kzP_kx$ and medium according to $zP_jxR_jy$ without being best. Therefore LSEVR is violated over the triple $\{x, y, z\}$. Thus $\mathscr{D}$ violates LSEVR. It has been shown that violation of transitivity by $R = f(R_1, \ldots, R_n)$, $(R_1, \ldots, R_n) \in \mathscr{D}^n$, implies violation of LSEVR by $\mathscr{D}$, i.e., if $\mathscr{D}$ satisfies LSEVR then every $(R_1, \ldots, R_n) \in \mathscr{D}^n$ yields transitive social $R$.

Suppose $\mathscr{D} \subseteq \mathcal{T}$ violates LSEVR. Then there exist distinct $x, y, z \in S$ such that $\mathscr{D}$ violates LSEVR over $\{x, y, z\}$. Violation of LSEVR by $\mathscr{D}$ over $\{x, y, z\}$ implies $(\exists \text{ distinct } a, b, c \in \{x, y, z\})(\exists R^s, R^t \in \mathscr{D})[aP^sbR^sc \wedge cR^taP^tb]$. As $f$ is non-null we conclude that $N$ is a winning coalition. Because $f$ is not a strong simple game, there exists a partition of $N$, $(V, N - V)$, such that neither $V$ nor $N - V$ is a winning coalition. Now consider any $(R_1, \ldots, R_n) \in \mathscr{D}^n$ such that the restriction of $(R_1, \ldots, R_n)$ to $\{x, y, z\} = \{a, b, c\}$, $(R_1|\{x, y, z\}, \ldots, R_n|\{x, y, z\})$, is given by: $[(\forall i \in V)(aP_ibR_ic) \wedge (\forall i \in N - V)(cR_iaP_ib)]$. In view of the fact that $N$ is winning but neither $V$ nor $N - V$ is winning we conclude that $(aPb \wedge bIc \wedge aIc)$ holds, which violates transitivity. We have shown that if $\mathscr{D} \subseteq \mathcal{T}$ violates LSEVR then there exists a $(R_1, \ldots, R_n) \in \mathscr{D}^n$ such that $R = f(R_1, \ldots, R_n)$ is intransitive, i.e., if $f$ yields transitive $R$ for every $(R_1, \ldots, R_n) \in \mathscr{D}^n$ then $\mathscr{D}$ must satisfy LSEVR. This establishes the theorem.  □

**Theorem 9.4** *Let social decision rule* $f : \mathscr{T}^n \mapsto \mathscr{C}$ *be a non-dictatorial strong simple game. Let* $\mathscr{D} \subseteq \mathscr{T}$. *Then* $f$ *yields transitive social* $R$, $R = f(R_1, \ldots, R_n)$, *for every* $(R_1, \ldots, R_n) \in \mathscr{D}^n$ *iff* $\mathscr{D}$ *satisfies the condition of weak Latin Square extremal value restriction.*

*Proof* Let $f : \mathscr{T}^n \mapsto \mathscr{C}$ be a non-dictatorial strong simple game; and let $\mathscr{D} \subseteq \mathscr{T}$. Suppose $f$ does not yield transitive social $R$ for every $(R_1, \ldots, R_n) \in \mathscr{D}^n$. Then:

$$(\exists (R_1, \ldots, R_n) \in \mathscr{D}^n)(\exists x, y, z \in S)(xRy \wedge yRz \wedge zPx). \qquad \text{(T9.4-1)}$$

Designate by $V_1, V_2, V_3$ the sets $\{i \in N \mid xR_iy\}, \{i \in N \mid yR_iz\}, \{i \in N \mid zP_ix\}$, respectively.

$xRy \rightarrow N - V_1$ is not a winning coalition

$\rightarrow V_1$ is a winning coalition by the definition of a strong simple game $\qquad \text{(T9.4-2)}$

$yRz \rightarrow N - V_2$ is not a winning coalition

$\rightarrow V_2$ is a winning coalition by the definition of a strong simple game $\qquad \text{(T9.4-3)}$

$zPx \rightarrow V_3$ is a winning coalition by the definition of a simple game $\qquad \text{(T9.4-4)}$

As intersection of any two winning coalitions is nonempty (Remark 4.1), we conclude that:

$(\exists i \in N)(xR_iyR_iz)$, as $V_1 \cap V_2 \neq \emptyset$

$(\exists j \in N)(yR_jzP_jx)$, as $V_2 \cap V_3 \neq \emptyset$

$(\exists k \in N)(zP_kxR_ky)$, as $V_3 \cap V_1 \neq \emptyset$.

(T9.4-1) implies that $x, y, z$ are distinct alternatives. $xR_iyR_iz$, $yR_jzP_jx$ and $zP_kxR_ky$ form $WLS(xyzx)$, and belong to $T[WLS(xyzx)]$. In the triple $\{x, y, z\}$, $z$ is uniquely best according to $zP_kxR_ky$, and medium according to $yR_jzP_jx$ without being worst; furthermore $x$ is uniquely worst according to $yR_jzP_jx$ and medium according to $zP_kxR_ky$ without being best. Therefore WLSEVR is violated over the triple $\{x, y, z\}$. Thus $\mathscr{D}$ violates WLSEVR. It has been shown that violation of transitivity by $R = f(R_1, \ldots, R_n)$, $(R_1, \ldots, R_n) \in \mathscr{D}^n$, implies violation of WLSEVR by $\mathscr{D}$, i.e., if $\mathscr{D}$ satisfies WLSEVR then every $(R_1, \ldots, R_n) \in \mathscr{D}^n$ yields transitive social $R$.

Suppose $\mathscr{D} \subseteq \mathscr{T}$ violates WLSEVR. Then there exist distinct $x, y, z \in S$ such that $\mathscr{D}$ violates WLSEVR over $\{x, y, z\}$. Violation of WLSEVR by $\mathscr{D}$ over $\{x, y, z\}$ implies $(\exists$ distinct $a, b, c \in \{x, y, z\})(\exists R^s, R^t, R^u \in \mathscr{D})[bR^ucR^ua \wedge cR^taP^tb \wedge aP^sbR^s c]$. As $f$ is a strong simple game, it follows that $N \in W$. Consequently the set of minimal winning coalitions $W_m$ is nonempty. Let $V \in W_m$. As $f$ is non-dictatorial, $V$ must contain at least two individuals. Because $f$ is a strong simple game it follows that $V \neq N$. Let $(V_1, V_2)$ be a partition of $V$ such that both $V_1$ and $V_2$ are nonempty. Now consider any $(R_1, \ldots, R_n) \in \mathscr{D}^n$ such that the restriction of $(R_1, \ldots, R_n)$ to $\{x, y, z\} = \{a, b, c\}$, $(R_1|\{x, y, z\}, \ldots, R_n|\{x, y, z\})$, is given by: $[(\forall i \in V_1)(bR_icR_ia) \wedge (\forall i \in V_2)(cR_iaP_ib) \wedge (\forall i \in N - V)(aP_ibR_ic)]$. As $f$ is a strong simple game, union of any two of the sets, $V_1, V_2, N - V$, is a winning coalition. By Theorem 9.2, if $f$ is a strong simple game then a coalition is winning iff it is strictly blocking. So, $V_1 \cup V_2, V_1 \cup (N - V)$ and $V_2 \cup (N - V)$ are strictly blocking. In view of the fact that $V_1 \cup V_2, V_1 \cup (N - V), V_2 \cup (N - V)$ are winning as well as strictly blocking and that none of the sets, $V_1, V_2, N - V$, is winning or strictly blocking, we conclude that $[aPb \land bRc \land cRa]$ holds, which violates transitivity. We have shown that if $\mathscr{D} \subseteq \mathscr{T}$ violates WLSEVR then there exists a $(R_1, \ldots, R_n) \in \mathscr{D}^n$ such that $R = f(R_1, \ldots, R_n)$ is intransitive, i.e., if $f$ yields transitive $R$ for every $(R_1, \ldots, R_n) \in \mathscr{D}^n$ then $\mathscr{D}$ must satisfy WLSEVR. This establishes the theorem.                                                                                          □

## 9.3  Quasi-transitivity Under Simple Games

**Proposition 9.3** *Let social decision rule* $f : \mathscr{T}^n \mapsto \mathscr{C}$ *be a simple game. Then,* $f$ *yields quasi-transitive social weak preference relation for every* $(R_1, \ldots, R_n) \in \mathscr{T}^n$ *iff it is null or there is a unique minimal winning coalition.*

*Proof* If $f$ is null then $R = f(R_1, \ldots, R_n)$ is transitive for every $(R_1, \ldots, R_n) \in \mathscr{T}^n$. Suppose there is a unique minimal winning coalition $V$. Consider any $(R_1, \ldots, R_n) \in \mathscr{T}^n$ and any $x, y, z \in S$ such that $(xPy \land yPz)$ obtains. $xPy \to (\exists V' \in W)(\forall i \in V')(xP_i y)$, by the definition of a simple game. Now it must be the case that $V \subseteq V'$ otherwise the fact that $V$ is the unique minimal winning coalition will be contradicted. Consequently, $xPy \to (\forall i \in V)(xP_i y)$. By an analogous argument we obtain $yPz \to (\forall i \in V)(yP_i z)$. From $(\forall i \in V)(xP_i y \land yP_i z)$ we obtain $(\forall i \in V)(xP_i z)$, which implies $xPz$. This proves that social weak preference relation is quasi-transitive for every $(R_1, \ldots, R_n) \in \mathscr{T}^n$.

Now suppose $f$ yields quasi-transitive social weak preference relation for every $(R_1, \ldots, R_n) \in \mathscr{T}^n$. As $f$ is a simple game it satisfies conditions I, N and M, by Theorem 9.1. If the weak Pareto-criterion is satisfied then by Gibbard's Theorem Gibbard (1969) it follows that there must be a unique minimal winning coalition. If the weak Pareto-criterion is violated then it must be the case that $(\exists (R_1, \ldots, R_n) \in \mathscr{T}^n)(\exists x, y \in S)[(\forall i \in N)(xP_i y) \land yRx]$. In view of conditions I and N then it follows that we must have:

$$(\forall (R_1, \ldots, R_n) \in \mathscr{T}^n)(\forall x, y \in S)[(\forall i \in N)(xP_i y) \to yRx]. \qquad \text{(P9.3-1)}$$

Now, a consequence of conditions I and N is that $(\forall (R_1, \ldots, R_n) \in \mathscr{T}^n)(\forall x, y \in S)[(\forall i \in N)(xI_i y) \to xIy]$ holds. In view of condition M then it follows that it must be the case that:

$$(\forall (R_1, \ldots, R_n) \in \mathscr{T}^n)(\forall x, y \in S)[(\forall i \in N)(xP_i y) \to xRy]. \qquad \text{(P9.3-2)}$$

$$\text{(P9.3-1)} \ \land \ \text{(P9.3-2)} \ \to (\forall (R_1, \ldots, R_n) \in \mathscr{T}^n)(\forall x, y \in S)[(\forall i \in N)(xP_i y) \to xIy]. \qquad \text{(P9.3-3)}$$

In view of condition M we conclude from (P9.3-3) that it must be the case that $(\forall (R_1, \ldots, R_n) \in \mathscr{T}^n)[f(R_1, \ldots, R_n) = R = (\forall x, y \in S)(xIy)]$, i.e., $f$ is null. This establishes the proposition.                                                                                       □

**Theorem 9.5** *Let social decision rule* $f : \mathcal{T}^n \mapsto \mathcal{C}$ *be a non-null simple game such that there are at least two minimal winning coalitions. Let* $\mathcal{D} \subseteq \mathcal{T}$. *Then* $f$ *yields quasi-transitive social* $R$, $R = f(R_1, \ldots, R_n)$, *for every* $(R_1, \ldots, R_n) \in \mathcal{D}^n$ *iff* $\mathcal{D}$ *satisfies the condition of Latin Square unique value restriction.*

*Proof* Let $f : \mathcal{T}^n \mapsto \mathcal{C}$ be a non-null simple game such that there are at least two minimal winning coalitions; and let $\mathcal{D} \subseteq \mathcal{T}$. Suppose $f$ does not yield quasi-transitive social $R$ for every $(R_1, \ldots, R_n) \in \mathcal{D}^n$. Then:

$$(\exists(R_1, \ldots, R_n) \in \mathcal{D}^n)(\exists x, y, z \in S)(xPy \wedge yPz \wedge zRx). \tag{T9.5-1}$$

$$xPy \rightarrow (\exists V_1 \in W)(\forall i \in V_1)(xP_iy), \tag{T9.5-2}$$

by the definition of a simple game

$$yPz \rightarrow (\exists V_2 \in W)(\forall i \in V_2)(yP_iz), \tag{T9.5-3}$$

by the definition of a simple game
(T9.5-2) $\wedge$ (T9.5-3) $\rightarrow$ ($\exists i \in V_1 \cap V_2$)($xP_iyP_iz$), as $V_1 \cap V_2 \neq \emptyset$, by Remark 4.1
$zRx \rightarrow (\exists j \in V_2)(yP_jzR_jx)$, as ($\forall i \in V_2$)($xP_iz$) would imply $xPz$
$zRx \rightarrow (\exists k \in V_1)(zR_kxP_ky)$, as ($\forall i \in V_1$)($xP_iz$) would imply $xPz$.
(T9.5-1) implies that $x, y, z$ are distinct alternatives. $xP_iyP_iz$, $yP_jzR_jx$, $zR_kxP_ky$ belong to $T[LS(xyzx)]$, and form $LS(xyzx)$. In the triple $\{x, y, z\}$, $y$ is uniquely medium according to $xP_iyP_iz$; is uniquely best according to $yP_jzR_jx$ and is uniquely worst according to $zR_kxP_ky$. Therefore LSUVR is violated over the triple $\{x, y, z\}$. Thus $\mathcal{D}$ violates LSUVR. We have shown that violation of quasi-transitivity by $R = f(R_1, \ldots, R_n)$, $(R_1, \ldots, R_n) \in \mathcal{D}^n$, implies violation of LSUVR by $\mathcal{D}$, i.e., if $\mathcal{D}$ satisfies LSUVR then every $(R_1, \ldots, R_n) \in \mathcal{D}^n$ yields quasi-transitive social $R$.

Suppose $\mathcal{D} \subseteq \mathcal{T}$ violates LSUVR. Then there exist distinct $x, y, z \in S$ such that $\mathcal{D}$ violates LSUVR over the triple $\{x, y, z\}$. Violation of LSUVR by $\mathcal{D}$ over $\{x, y, z\}$ implies ($\exists$ distinct $a, b, c \in \{x, y, z\}$)($\exists R^s, R^t, R^u \in \mathcal{D}$)[$aP^sbP^sc \wedge bP^tcR^ta \wedge cR^u aP^ub$]. It is given that $f$ is non-null and that there are at least two minimal winning coalitions. Let $V_1$ and $V_2$ be distinct minimal winning coalitions. $V_1 \cap V_2 \neq \emptyset$ in view of Remark 4.1; and [$V_1 - V_2 \neq \emptyset \wedge N - V_1 \neq \emptyset$] as $V_1$ and $V_2$ are distinct minimal winning coalitions. From the fact that $V_1$ and $V_2$ are distinct minimal winning coalitions we conclude that $V_1 \cap V_2 \notin W$. Now consider any $(R_1, \ldots, R_n) \in \mathcal{D}^n$ such that the restriction of $(R_1, \ldots, R_n)$ to $\{x, y, z\} = \{a, b, c\}$, $(R_1|\{x, y, z\}, \ldots, R_n|\{x, y, z\})$, is given by: [($\forall i \in V_1 \cap V_2$)($aP_ibP_ic$) $\wedge$ ($\forall i \in V_1 - V_2$)($bP_icR_ia$) $\wedge$ ($\forall i \in N - V_1$)($cR_iaP_ib$)].        [$V_2 \in W \wedge (\forall i \in V_2$) ($aP_ib$) $\rightarrow aPb$] and [$V_1 \in W \wedge (\forall i \in V_1$)($bP_ic$) $\rightarrow bPc$]. $\{i \in N \mid aP_ic\} = V_1 \cap V_2$ and $V_1 \cap V_2 \notin W$ imply $cRa$ as $f$ is a simple game. ($aPb \wedge bPc \wedge cRa$) implies that $R$ violates quasi-transitivity. We have shown that if $\mathcal{D} \subseteq \mathcal{T}$ violates LSUVR then there exists $(R_1, \ldots, R_n) \in \mathcal{D}^n$ such that $R = f(R_1, \ldots, R_n)$ violates quasi-transitivity, i.e., if $f$ yields quasi-transitive $R$ for every $(R_1, \ldots, R_n) \in \mathcal{D}^n$ then $\mathcal{D}$ must satisfy LSUVR. This establishes the theorem. □

# References

Gibbard, Allan. 1969. *Social choice and the Arrow conditions*. Discussion Paper: Department of Philosophy, University of Michigan.

Jain, Satish K. 1989. Characterization theorems for social decision rules which are simple games. Paper presented at the IX World Congress of the International Economic Association, Economics Research Center, Athens School of Economics & Business, Athens, Greece, held on Aug 28 - Sep 1, 1989.

# Neutral and Monotonic Binary Social Decision Rules

All the rules for which conditions ensuring transitivity and quasi-transitivity have been discussed so far in this volume satisfy both the conditions of monotonicity and neutrality. In fact, most of the rules that are used for elections and for taking decisions, whether binary or non-binary, are neutral and monotonic. Among the important exceptions are the United Nations Security Council decision-making rule and the single transferable vote used in elections. The United Nations Security Council rule violates neutrality and the single transferable vote fails to satisfy monotonicity.

In this chapter a characterization of neutrality and monotonicity is provided for an important class of binary social decision rules. In this context a hybrid rationality condition called the Pareto quasi-transitivity turns out to be of crucial importance. Pareto quasi-transitivity requires that for any social alternatives $x$, $y$, $z$; if $x$ is socially better than $y$, and $y$ is Pareto-superior to $z$, then $x$ must be socially better than $z$; and if $x$ is Pareto-superior to $y$, and $y$ is socially better than $z$, then $x$ must be socially better than $z$. It is shown in this chapter that a binary social decision rule with unrestricted domain, i.e., with domain $\mathscr{T}^n$ and satisfying the condition of Pareto-indifference is neutral and monotonic iff Pareto quasi-transitivity holds. As conjunction of Pareto-criterion and quasi-transitivity implies Pareto quasi-transitivity, it follows that every binary social decision rule with domain $\mathscr{T}^n$ and satisfying the Pareto-criterion which yields transitive or quasi-transitive social weak preference relation for every profile of individual orderings is neutral and monotonic.

In the social choice literature the notions of decisiveness and semidecisiveness play important roles, particularly in the context of impossibility theorems. In the process of characterizing neutrality and monotonicity, we also generalize the decisiveness and semidecisiveness notions. These notions as defined in Chap. 2 relate to the entire society; they are generalized in this chapter for any sub-society. Let $A$ be a proper subset of $N$, the set of individuals comprising the society. Let $x$ and $y$ be distinct social alternatives. We define a set of individuals $V \subseteq N - A$ to be almost $(N - A)$-decisive for $x$ against $y$ iff whenever all individuals in $A$ are indifferent between $x$ and $y$, all individuals in $V$ prefer $x$ over $y$, and all individuals in

© Springer Nature Singapore Pte Ltd. 2019
S. K. Jain, *Domain Conditions and Social Rationality*,
https://doi.org/10.1007/978-981-13-9672-4_10

$N - (A \cup V)$ prefer $y$ over $x$, then $x$ is socially preferred over $y$; $(N - A)$-decisive for $x$ against $y$ iff whenever all individuals in $A$ are indifferent between $x$ and $y$, and all individuals in $V$ prefer $x$ over $y$, then $x$ is socially preferred over $y$; and $(N - A)$-decisive iff it is $(N - A)$-decisive for every ordered pair of distinct alternatives. Similarly, we define a set of individuals $V \subseteq N - A$ to be almost $(N - A)$-semidecisive for $x$ against $y$ iff whenever all individuals in $A$ are indifferent between $x$ and $y$, all individuals in $V$ prefer $x$ over $y$ and all individuals in $N - (A \cup V)$ prefer $y$ over $x$, then $x$ is socially at least as good as $y$; $(N - A)$-semidecisive for $x$ against $y$ iff whenever all individuals in $A$ are indifferent between $x$ and $y$, and all individuals in $V$ prefer $x$ over $y$, then $x$ is socially at least as good as $y$; and $(N - A)$-semidecisive iff it is $(N - A)$-semidecisive for every ordered pair of distinct alternatives.

For the class of neutral and monotonic binary social decision rules with unrestricted domain, i.e., with domain $\mathscr{T}^n$, and satisfying the Pareto-criterion, the three rationality conditions of transitivity, quasi-transitivity and acyclicity can be characterized as follows: (i) A social decision rule with unrestricted domain and satisfying independence of irrelevant alternatives and the Pareto-criterion yields transitive social weak preference relation for every profile of individual orderings iff for every sub-society $(N - A)$, where $A \subset N$, there is an individual belonging to $(N - A)$ who is $(N - A)$-decisive. (ii) A social decision rule with unrestricted domain and satisfying independence of irrelevant alternatives and the Pareto-criterion yields quasi-transitive social weak preference relation for every profile of individual orderings iff for every sub-society $(N - A)$, where $A \subset N$, there exists $V \subseteq N - A$ such that it is $(N - A)$-oligarchy. (iii) A social decision rule with unrestricted domain and satisfying independence of irrelevant alternatives, Pareto-criterion, neutrality and monotonicity yields acyclic social weak preference relation for every profile of individual orderings iff for every sub-society $(N - A)$, where $A \subset N$, every collection of $m$, $3 \leq m \leq \#S$, $(N - A)$-decisive sets has nonempty intersection.

For every binary neutral and monotonic social decision rule the condition of strict placement restriction is sufficient to ensure transitivity. That is to say, if $\mathscr{D}$ is a set of orderings of the set of alternatives, then under every binary neutral and monotonic social decision rule the social weak preference relation is transitive for every profile of individual orderings belonging to $\mathscr{D}^n$ if $\mathscr{D}$ satisfies the condition of strict placement restriction over every triple of alternatives. For the class of binary neutral and monotonic social decision rules no condition weaker than this exists that is sufficient for transitivity under every rule belonging to the class. In other words, the condition of strict placement restriction is maximally sufficient for transitivity for the class of binary neutral and monotonic social decision rules. If we consider the subclass of binary neutral and monotonic social decision rules satisfying the Pareto-criterion then the condition of placement restriction, less stringent than the condition of strict placement restriction, turns out to be maximally sufficient for transitivity.

With respect to quasi-transitivity under the class of binary neutral and monotonic social decision rules, value restriction turns out to be a sufficient condition. If in addition to independence of irrelevant alternatives, neutrality and monotonicity, Pareto-criterion is also satisfied then the limited agreement also turns out to be sufficient for quasi-transitivity. Thus, under every binary neutral and monotonic social decision rule satisfying the Pareto-criterion every profile of orderings belonging to $\mathscr{D}^n$ yields quasi-transitive social weak preference relation if $\mathscr{D}$ satisfies value restriction or limited agreement over every triple of alternatives.

For the class of binary neutral and monotonic social decision rules the statements of conditions for transitivity and quasi-transitivity become extremely simple when individual orderings are linear. For quasi-transitivity we have the following: (i) Given that the domain consists of all logically possible profiles of individual linear orderings, a binary social decision rule satisfying neutrality and monotonicity yields quasi-transitive social weak preference relation for every profile of individual linear orderings iff there is at most one minimal decisive set. (ii) A binary social decision rule satisfying neutrality and monotonicity which is such that there are at least two minimal decisive sets yields quasi-transitive social weak preference relation for every profile of individual linear orderings belonging to $\mathscr{D}^n$ iff $\mathscr{D}$ is such that it does not contain a Latin Square over any triple of alternatives. For transitivity we have the following: (i) Given that the domain consists of all logically possible profiles of individual linear orderings, a binary social decision rule satisfying neutrality and monotonicity yields transitive social weak preference relation for every profile of individual linear orderings iff it is null or dictatorial. (ii) A non-null non-dictatorial binary social decision rule satisfying neutrality and monotonicity, which is such that for every partition $(N_1, N_2)$ of $N$ either $N_1$ or $N_2$ is a decisive set, yields transitive social weak preference relation for every profile of individual linear orderings belonging to $\mathscr{D}^n$ iff $\mathscr{D}$ is such that it does not contain a Latin Square over any triple of alternatives. (iii) A non-null non-dictatorial binary social decision rule satisfying neutrality and monotonicity, which is such that there exists a partition $(N_1, N_2)$ of $N$ such that neither $N_1$ nor $N_2$ is a decisive set, yields transitive social weak preference relation for every profile of individual linear orderings belonging to $\mathscr{D}^n$ iff $\mathscr{D}$ is such that it does not contain more than one linear ordering belonging to either of the two Latin Squares over any triple of alternatives.

The chapter is divided into five sections. The first section contains the characterization theorems for the class of neutral and monotonic binary social decision rules. Sections 10.2 and 10.3 are concerned with conditions for transitivity and quasi-transitivity, respectively, for the class of neutral and monotonic binary social decision rules. Section 10.4 is concerned with the conditions for transitivity and quasi-transitivity under the class of neutral and monotonic binary social decision rules when individuals have linear orderings of social alternatives. The final section, which is in the appendix to the chapter, contains some notes on the literature.

## 10.1   Characterization Theorems for Neutral and Monotonic Binary Social Decision Rules

Let binary relation $U$ on $S$ be defined by: $(\forall x, y \in S)[xUy \leftrightarrow (\forall i \in N)(xR_iy)]$. Let $P(U)$ and $I(U)$ denote the asymmetric and symmetric parts of $U$, respectively. Thus we have: $(\forall x, y \in S)[[xP(U)y \leftrightarrow (\forall i \in N)(xR_iy) \wedge (\exists i \in N)(xP_iy)] \wedge [xI(U) y \leftrightarrow (\forall i \in N)(xI_iy)]]$.

Define binary relation $\overline{U}$ on $S$ by: $(\forall x, y \in S)[x\overline{U}y \leftrightarrow (\forall i \in N)(xP_iy)]$.

Now we define some conditions on social decision rules.

Let $\mathbb{D} \subseteq \mathscr{T}^n$. $f : \mathbb{D} \mapsto \mathscr{C}$ satisfies (i) the Weak Pareto Quasi-transitivity (WPQT) iff $(\forall(R_1, \ldots, R_n) \in \mathbb{D})(\forall x, y, z \in S)[(xPy \wedge y\overline{U}z \to xPz) \wedge (x\overline{U}y \wedge yPz \to xPz)]$, and (ii) Pareto Quasi-transitivity (PQT) iff $(\forall(R_1, \ldots, R_n) \in \mathbb{D})(\forall x, y, z \in S)[(xPy \wedge yP(U)z \to xPz) \wedge (xP(U)y \wedge yPz \to xPz)]$.

$f : \mathbb{D} \mapsto \mathscr{C}$ satisfies the condition of weak monotonicity (WM) iff $(\forall(R_1, \ldots, R_n), (R'_1, \ldots, R'_n) \in \mathbb{D})(\forall x \in S)(\forall k \in N)[(\forall i \in N - \{k\})(R'_i = R_i) \wedge (R'_k \neq R_k) \wedge (\forall a, b \in S - \{x\})(aR_kb \leftrightarrow aR'_kb) \wedge (\forall y \in S - \{x\})[(xP_ky \to xP'_ky) \wedge (xI_ky \to xR'_ky)] \to (\forall y \in S - \{x\})[(xPy \to xP'y) \wedge (xIy \to xR'y)]]$. It is clear from the definitions of conditions I and WM that an SDR $f : \mathscr{T}^n \mapsto \mathscr{C}$ satisfying condition I satisfies WM iff $(\forall(R_1, \ldots, R_n), (R'_1, \ldots, R'_n) \in \mathscr{T}^n)(\forall x, y \in S)(\forall k \in N)[(\forall i \in N - \{k\})[(xR_iy \leftrightarrow xR'_iy) \wedge (yR_ix \leftrightarrow yR'_ix)] \wedge [(yP_kx \wedge xR'_ky) \vee (xI_ky \wedge xP'_ky)] \to [(xPy \to xP'y) \wedge (xIy \to xR'y)]]$.

Let SDR be $f : \mathbb{D} \mapsto \mathscr{C}$. Let $A \subset N$ and $V \subseteq N - A$. Let $x, y \in S$ and $x \neq y$. We define the set of individuals $V$ to be (i) almost $(N - A)$-decisive for $(x, y)$ $[D_{N-A}(x, y)]$ iff $(\forall(R_1, \ldots, R_n) \in \mathbb{D})[(\forall i \in A)(xI_iy) \wedge (\forall i \in V)(xP_iy) \wedge (\forall i \in N - (A \cup V))(yP_ix) \to xPy]$, (ii) $(N - A)$-decisive for $(x, y)$ $[\overline{D}_{N-A}(x, y)]$ iff $(\forall(R_1, \ldots, R_n) \in \mathbb{D})[(\forall i \in A)(xI_iy) \wedge (\forall i \in V)(xP_iy) \to xPy]$, (iii) $(N - A)$-decisive iff it is $(N - A)$-decisive for every $(a, b) \in S \times S, a \neq b$, (iv) almost $(N - A)$-semidecisive for $(x, y)$ $[S_{N-A}(x, y)]$ iff $(\forall(R_1, \ldots, R_n) \in \mathbb{D})[(\forall i \in A)(xI_iy) \wedge (\forall i \in V)(xP_iy) \wedge (\forall i \in N - (A \cup V))(yP_ix) \to xRy]$, (v) $(N - A)$-semidecisive for $(x, y)[\overline{S}_{N-A}(x, y)]$ iff $(\forall(R_1, \ldots, R_n) \in \mathbb{D})[(\forall i \in A)(xI_iy) \wedge (\forall i \in V)(xP_iy) \to xRy]$, (vi) $(N - A)$-semidecisive iff it is $(N - A)$-semidecisive for every $(a, b) \in S \times S, a \neq b$.

If $A = \emptyset$ then the prefix $(N - A)$ is dropped. When $A = \emptyset$ (i) through (vi) give definitions of an almost decisive set for $(x, y)$ $[D(x, y)]$, a decisive set for $(x, y)$ $[\overline{D}(x, y)]$, a decisive set, an almost semidecisive set for $(x, y)$ $[S(x, y)]$, a semidecisive set for $(x, y)$ $[\overline{S}(x, y)]$ and a semidecisive set, respectively.

Let $A \subset N$ and $V \subseteq N - A$. $V$ is an $(N - A)$-oligarchy iff $V$ is an $(N - A)$-decisive set and every $i \in V$ is an $(N - A)$-semidecisive set. $V$ is a strict $(N - A)$-oligarchy iff $V$ is an $(N - A)$-decisive set and $(\forall j \in V)(\forall(R_1, \ldots, R_n) \in \mathbb{D})(\forall \text{ distinct } a, b \in S)[(\forall i \in A)(aI_ib) \wedge aR_jb \to aRb]$.

**Lemma 10.1** *Let social decision rule* $f : \mathscr{T}^n \mapsto \mathscr{C}$ *satisfy independence of irrelevant alternatives. If $f$ satisfies Pareto quasi-transitivity then, whenever a group of individuals $V$ is almost $(N - A)$-decisive for some ordered pair of distinct alternatives, it is $(N - A)$-decisive for every ordered pair of distinct alternatives, where $A \subset N$ and $V \subseteq N - A$.*

*Proof* Let $f$ satisfy I and PQT. Let $V$ be almost $(N - A)$-decisive for $(x, y)$; $x \neq y, x, y \in S, A \subset N, V \subseteq N - A$. Let $z$ be an alternative distinct from $x$ and $y$, and consider the following configuration of individual preferences:

$(\forall i \in A)[x I_i y I_i z]$

$(\forall i \in V)[x P_i y P_i z]$

$(\forall i \in N - (A \cup V))[y P_i x \land y P_i z]$.

In view of the almost $(N - A)$-decisiveness of $V$ for $(x, y)$ and the fact that $[(\forall i \in A)(x I_i y) \land (\forall i \in V)(x P_i y) \land (\forall i \in N - (A \cup V))(y P_i x)]$, we obtain $x P y$. From $x P y$ and $[(\forall i \in N)(y R_i z) \land (\forall i \in N - A)(y P_i z) \land (N - A) \neq \emptyset]$ we conclude $x P z$ by PQT. As $(\forall i \in A)(x I_i z)$, $(\forall i \in V)(x P_i z)$, and the preferences of individuals in $N - (A \cup V)$ have not been specified over $\{x, z\}$, it follows, in view of condition I, that $V$ is $(N - A)$-decisive for $(x, z)$. Similarly, by considering the configuration $[(\forall i \in A)(z I_i x I_i y) \land (\forall i \in V)(z P_i x P_i y) \land (\forall i \in N - (A \cup V))(z P_i x \land y P_i x)]$, we can show $[D_{N-A}(x, y) \to \overline{D}_{N-A}(z, y)]$. By appropriate interchanges of alternatives it follows that $D_{N-A}(x, y) \to \overline{D}_{N-A}(a, b)$, for all $(a, b) \in \{x, y, z\} \times \{x, y, z\}$, where $a \neq b$. To prove the assertion for any $(a, b) \in S \times S, a \neq b$, first we note that if $[(a = x \lor a = y) \lor (b = x \lor b = y)]$, the desired conclusion $\overline{D}_{N-A}(a, b)$ can be obtained by considering a triple which includes all of $x, y, a$ and $b$. If both $a$ and $b$ are different from $x$ and $y$, then one first considers the triple $\{x, y, a\}$ and deduces $\overline{D}_{N-A}(x, a)$ and hence $D_{N-A}(x, a)$, and then considers the triple $\{x, a, b\}$ and obtains $\overline{D}_{N-A}(a, b)$.                                              □

**Lemma 10.2** *Let social decision rule* $f : \mathscr{T}^n \mapsto \mathscr{C}$ *satisfy independence of irrelevant alternatives. If* $f$ *satisfies Pareto quasi-transitivity then, whenever a group of individuals $V$ is almost $(N - A)$-semidecisive for some ordered pair of distinct alternatives, it is $(N - A)$-semidecisive for every ordered pair of distinct alternatives, where $A \subset N$ and $V \subseteq N - A$.*

*Proof* Let $f$ satisfy I and PQT. Let $V$ be almost $(N - A)$-semidecisive for $(x, y)$; $x \neq y, x, y \in S, A \subset N, V \subseteq N - A$. Let $z$ be an alternative distinct from $x$ and $y$, and consider the following configuration of individual preferences:

$(\forall i \in A)[x I_i y I_i z]$

$(\forall i \in V)[x P_i y P_i z]$

$(\forall i \in N - (A \cup V))[y P_i x \land y P_i z]$.

From the almost $(N - A)$-semidecisiveness of $V$ for $(x, y)$, we obtain $x R y$. Suppose $z P x . [(\forall i \in N)(y R_i z) \land (\forall i \in N - A)(y P_i z) \land (N - A) \neq \emptyset]$ and $z P x$ imply $y P x$ by PQT, which contradicts $x R y$. Therefore $z P x$ cannot be true, which by connectedness of $R$ implies $x R z$. As $(\forall i \in A)(x I_i z)$, $(\forall i \in V)(x P_i z)$ and the preferences of individuals belonging to $N - (A \cup V)$ have not been specified over $\{x, z\}$, $x R z$ implies that $\overline{S}_{N-A}(x, z)$ holds in view of condition I. Thus $S_{N-A}(x, y) \to \overline{S}_{N-A}(x, z)$. Similarly, by considering the configuration $[(\forall i \in A)(z I_i x I_i y), (\forall i \in V)(z P_i x P_i y), (\forall i \in N - (A \cup V))(z P_i x \land y P_i x)]$, we can show that $S_{N-A}(x, y) \to \overline{S}_{N-A}(z, y)$. By appropriate interchanges of alternatives it follows that $S_{N-A}(x, y)$ implies $\overline{S}_{N-A}(a, b)$ for all $(a, b) \in \{x, y, z\} \times \{x, y, z\}, a \neq b$. To prove the assertion for any $(a, b) \in S \times S, a \neq b$, first we note that if $[(a = x \lor a = y) \lor (b = x \lor b = y)]$, the desired conclusion $\overline{S}_{N-A}(a, b)$ can be obtained by considering a triple which includes all of $x, y, a$ and $b$. If both $a$ and $b$ are different from $x$ and

$y$, then one first considers the triple $\{x, y, a\}$ and deduces $\overline{S}_{N-A}(x, a)$ and hence $S_{N-A}(x, a)$, and then considers the triple $\{x, a, b\}$ and obtains $\overline{S}_{N-A}(a, b)$.                    □

**Lemma 10.3** *Let social decision rule* $f : \mathscr{T}^n \mapsto \mathscr{C}$ *satisfy independence of irrelevant alternatives. If* $f$ *satisfies Pareto quasi-transitivity then* $(\forall x, y \in S)[x P(U)y \rightarrow x R y]$.

*Proof* Suppose $(\exists x, y \in S)[x P(U)y \wedge \sim x R y]$.
$x P(U)y \wedge \sim x R y \rightarrow x P(U)y \wedge y P x$, by the connectedness of $R$.
$y P x \wedge x P(U)y \rightarrow y P y$, by PQT.
$y P y$ is a logical contradiction.
Therefore we conclude that $\sim(\exists x, y \in S)[x P(U)y \wedge \sim x R y]$, i.e., $(\forall x, y \in S)$ $[x P(U)y \rightarrow x R y]$ holds.                    □

**Proposition 10.1** *Let social decision rule* $f : \mathscr{T}^n \mapsto \mathscr{C}$ *satisfy independence of irrelevant alternatives and the Pareto-indifference. If* $f$ *satisfies Pareto quasi-transitivity then it is neutral.*

*Proof* Consider any $(R_1, \ldots, R_n)$, $(R'_1, \ldots, R'_n) \in \mathscr{T}^n$ such that $(\forall i \in N)[(x R_i y \leftrightarrow z R'_i w) \wedge (y R_i x \leftrightarrow w R'_i z)]$, $x, y, z, w \in S$. Designate by $N_1$, $N_2$ and $N_3$ the sets $\{i \in N \mid x P_i y \wedge z P'_i w\}$, $\{i \in N \mid x I_i y \wedge z I'_i w\}$ and $\{i \in N \mid y P_i x \wedge w P'_i z\}$, respectively.
If $N_1 \cup N_3 = \emptyset$, then $x I y$ and $z I' w$ follow from the condition of Pareto-indifference. Now assume that $N_1 \cup N_3 \neq \emptyset$. Nonemptiness of $N_1 \cup N_3$ implies that $x \neq y$ and $z \neq w$. Suppose $x P y$. Then $N_1$ is nonempty by PQT (Lemma 10.3) and almost $(N - N_2)$-decisive for $(x, y)$ by condition I. In view of Lemma 10.1 it follows that $N_1$ is $(N - N_2)$-decisive for every ordered pair of distinct alternatives. Therefore $z P' w$ must hold as $[(\forall i \in N_2)(z I'_i w) \wedge (\forall i \in N_1)(z P'_i w)]$. We have shown that $(x P y \rightarrow z P' w)$. By an analogous argument it can be shown that $(z P' w \rightarrow x P y)$. So we have $(x P y \leftrightarrow z P' w)$. Next suppose $y P x$. Then $N_3$ is nonempty by PQT and an almost $(N - N_2)$-decisive set for $(y, x)$ by condition I, and hence an $(N - N_2)$-decisive set by Lemma 10.1. Therefore, we must have $w P' z$ as $[(\forall i \in N_2)(w I'_i z) \wedge (\forall i \in N_3)(w P'_i z)]$. So $(y P x \rightarrow w P' z)$. By a similar argument one obtains $(w P' z \rightarrow y P x)$. Therefore we have $(y P x \leftrightarrow w P' z)$. As $(x P y \leftrightarrow z P' w)$ and $(y P x \leftrightarrow w P' z)$, by the connectedness of $R$ and $R'$ it follows that $(x I y \leftrightarrow z I' w)$. This establishes that the SDR is neutral.                    □

**Lemma 10.4** *Let social decision rule* $f : \mathscr{T}^n \mapsto \mathscr{C}$ *satisfy the condition of independence of irrelevant alternatives. Then* $f$ *is monotonic iff it is weakly monotonic.*

*Proof* By the definitions of monotonicity and weak monotonicity, if $f$ is monotonic then it is weakly monotonic. Now suppose that $f$ is weakly monotonic. Consider any $(R_1, \ldots, R_n)$, $(R'_1, \ldots, R'_n) \in \mathscr{T}^n$ and any $x, y \in S$ such that $(\forall i \in N)[(x P_i y \rightarrow x P'_i y) \wedge (x I_i y \rightarrow x R'_i y)]$. Let $N' = \{j_1, j_2, \ldots, j_m\}$ be the set of individuals for whom $R_i|\{x, y\} \neq R_i'|\{x, y\}$. Rename $(R_1, \ldots, R_n)$ as $(R_1^0, \ldots, R_n^0)$ and construct $(R_1^t, \ldots, R_n^t), t = 1, \ldots, m$, as follows: $[(\forall i \in N - \{j_t\})[R_i^t = R_i^{t-1}] \wedge [R_{j_t}^t = R_{j_t}']]$. So we have $(R_1^m|\{x, y\}, \ldots, R_n^m|\{x, y\}) = (R_1'|\{x, y\}, \ldots, R_n'|\{x, y\})$. By condition I and weak monotonicity, we obtain $[(xP^{t-1}y \rightarrow xP^ty) \wedge (xI^{t-1}y \rightarrow xR^ty)]$ for $t = 1, \ldots, m$, which by condition I implies $[(xPy \rightarrow xP'y) \wedge (xIy \rightarrow xR'y)]$. This establishes that $f$ is monotonic.                                             □

**Proposition 10.2** *Let social decision rule* $f : \mathscr{T}^n \mapsto \mathscr{C}$ *satisfy independence of irrelevant alternatives and the Pareto-indifference. If $f$ satisfies Pareto quasi-transitivity then it is monotonic.*

*Proof* Let SDR $f : \mathscr{T}^n \mapsto \mathscr{C}$ satisfy conditions I, PI and PQT. In view of Lemma 10.4 it suffices to show that the SDR is weakly monotonic. Consider any $(R_1, \ldots, R_n)$, $(R_1', \ldots, R_n') \in \mathscr{T}^n$, any $x, y \in S$ and any individual $k \in N$ such that $(\forall i \in N - \{k\})[(xR_iy \leftrightarrow xR_i'y) \wedge (yR_ix \leftrightarrow yR_i'x)]$ and $[(xP_kx \wedge xR_k'y) \vee (xI_ky \wedge xP_k'y)]$. Designate by $N_1$, $N_2$ and $N_3$ the sets $\{i \in N \mid xP_iy\}, \{i \in N \mid xI_iy\}, \{i \in N \mid yP_ix\}$, respectively.
If $N_1 \cup N_3 = \emptyset$ then $xIy$ follows from PI; and $xR'y$ follows from PQT.
Now let $N_1 \cup N_3 \neq \emptyset$.
Suppose $xPy$. Then $N_1$ is nonempty by PQT and almost $(N - N_2)$-decisive for $(x, y)$ as a consequence of condition I, and hence an $(N - N_2)$-decisive set in view of Lemma 10.1. If $k \in N_3$ then it follows that we must have $xP'y$. Now suppose $k \in N_2$. If $yR'x$ then $N_3$ is almost $(N - (N_2 - \{k\}))$-semidecisive for $(y, x)$, and hence an $(N - (N_2 - \{k\}))$-semidecisive set by Lemma 10.2. As $[(\forall i \in N_3)(yP_ix) \wedge (\forall i \in N_2 - \{k\})(xI_iy)]$, it follows that we must have $yRx$. This, however, contradicts the hypothesis that $xPy$ holds. So $yR'x$ is impossible and therefore $xP'y$ must obtain. Next suppose $xIy$. Then $N_1$ is an $(N - N_2)$-semidecisive set in view of Lemma 10.2. If $k \in N_3$ then it follows that we must have $xR'y$ as $[(\forall i \in N_1)(xP_i'y) \wedge (\forall i \in N_2)(xI_i'y)]$. Suppose $k \in N_2$ and $yP'x$. $yP'x$ implies that $N_3$ is nonempty by PQT and an $(N - (N_2 - \{k\}))$-decisive set, which in turn implies that $yPx$ must obtain as $[(\forall i \in N_3)(yP_ix) \wedge (\forall i \in N_2 - \{k\})(xI_iy)]$, contradicting the hypothesis of $xIy$. So $yP'x$ is impossible and by the connectedness of $R'$ we conclude that $xR'y$ must hold. Thus we have shown that $[(xPy \rightarrow xP'y) \wedge (xIy \rightarrow xR'y)]$, which establishes that the SDR is weakly monotonic.                                                    □

**Proposition 10.3** *Let social decision rule* $f : \mathscr{T}^n \mapsto \mathscr{C}$ *satisfy independence of irrelevant alternatives, neutrality and monotonicity. Then $f$ satisfies Pareto quasi-transitivity.*

*Proof* Consider any $x, y, z \in S$ and any $(R_1, \ldots, R_n) \in \mathscr{T}^n$ such that $[xPy \wedge (\forall i \in N)(yR_iz) \wedge (\exists i \in N)(yP_iz)]$. Designate by $N_1$, $N_2$, $N_3$ the sets $\{i \in N \mid xP_iy\}, \{i \in N \mid xI_iy\}, \{i \in N \mid yP_ix\}$, respectively, and by $N_1'$, $N_2'$, $N_3'$ the sets $\{i \in N \mid xP_iz\}, \{i \in N \mid xI_iz\}, \{i \in N \mid zP_ix\}$, respectively. As individual weak preference relations are transitive, from $[(\forall i \in N)(yR_iz) \wedge (\exists i \in N)(yP_iz)]$ we conclude that $N_1 \subseteq N_1'$ and $N_3' \subseteq N_3$. Let $(R_1', \ldots, R_n') \in \mathscr{T}^n$ be any configuration such that $[(\forall i \in N_1)(x P_i' z) \wedge (\forall i \in N_2)(x I_i' z) \wedge (\forall i \in N_3')(z P_i' x)]$. As $x P y$, we conclude $x P' z$ by conditions I and $N$. $x P' z$ in turn implies $x P z$ in view of $N_1 \subseteq N_1'$ and $N_3' \subseteq N_3$, as a consequence of conditions I and $M$. Thus we have shown that $x P y$ and $[(\forall i \in N)(y R_i z) \wedge (\exists i \in N)(y P_i z)]$ imply $x P z$. By an analogous argument it can be shown that $[(\forall i \in N)(x R_i y) \wedge (\exists i \in N)(x P_i y)]$ and $y P z$ imply $x P z$. This establishes that PQT holds.  □

Combining Propositions 10.1–10.3 we obtain:

**Theorem 10.1** *A binary social decision rule* $f : \mathscr{T}^n \mapsto \mathscr{C}$ *satisfying Pareto-indifference is neutral and monotonic iff Pareto quasi-transitivity holds.*

### 10.1.1  Characterization of Transitivity

*Remark 10.1* If a binary social decision rule $f : \mathscr{T}^n \mapsto \mathscr{C}$ satisfying the Pareto-criterion yields quasi-transitive social weak preference relation for every $(R_1, \ldots, R_n) \in \mathscr{T}^n$ then $f$ satisfies PQT.

*Proof* Immediate.  ◊

**Theorem 10.2** *Let social decision rule* $f : \mathscr{T}^n \mapsto \mathscr{C}$ *satisfy independence of irrelevant alternatives and the the Pareto-criterion. Then, $f$ yields transitive social weak preference relation for every* $(R_1, \ldots, R_n) \in \mathscr{T}^n$ *iff for every* $A \subset N$*, there is an individual belonging to* $N - A$ *who is* $(N - A)$*-decisive.*

*Proof* Let $f : \mathscr{T}^n \mapsto \mathscr{C}$ satisfy I and $\overline{P}$. Suppose $f$ yields transitive social weak preference relation for every $(R_1, \ldots, R_n) \in T^n$. Consider any $A \subset N$. By condition $\overline{P}$, $N - A$ is an $(N - A)$-decisive set. As $N$ is finite there exists $V \subseteq N - A$ such that it is a minimal $(N - A)$-decisive set. By condition $\overline{P}$, $V$ is nonempty. Suppose $\#V \geq 2$. Let $(V_1, V_2)$ be a partition of $V$ [*i.e.*, $V_1 \neq \emptyset, V_2 \neq \emptyset, V_1 \cap V_2 = \emptyset, V_1 \cup V_2 = V$]. Consider the following configuration of individual preferences:
$(\forall i \in A)[x I_i y I_i z]$
$(\forall i \in V_1)[x P_i y P_i z]$
$(\forall i \in V_2)[y P_i z P_i x]$
$(\forall i \in N - (A \cup V))[z P_i x P_i y]$,
where $x, y, z$ are all distinct and belong to $S$.
From $[(\forall i \in A)(y I_i z) \wedge (\forall i \in V)(y P_i z)]$ we obtain $y P z$, by the $(N - A)$-decisiveness of $V$.
$y P x \vee x R y$, as $R$ is connected.
$y P x \rightarrow V_2$ is almost $(N - A)$-decisive for $(y, x)$, by condition I
$\rightarrow V_2$ is an $(N - A)$-decisive set, by Lemma 10.1 and Remark 10.1.
This contradicts the minimality of $V$.

$xRy \to xPz$, by transitivity of $R$

$\to V_1$ is almost $(N - A)$-decisive for $(x, z)$, by condition I

$\to V_1$ is an $(N - A)$-decisive set, by Lemma 10.1 and Remark 10.1.

This contradicts the minimality of $V$.

As each of $yPx$ and $xRy$ leads to a contradiction, we conclude that $V$ consists of a single individual.

Now, let $f$ be such that, for every $A \subset N$, there is an individual belonging to $N - A$ who is $(N - A)$-decisive. Take any $(R_1, \ldots, R_n) \in \mathscr{T}^n$ and any $x, y, z \in S$. Suppose $xRy \wedge yRz$. Let $A = \{i \in N \mid xI_iyI_iz\}$. If $A = N$ then $xIz$, and thus $xRz$, follows from $\overline{P}$. Let $A \subset N$. As $A \subset N$, there is an individual $j \in N - A$ who is $(N - A)$-decisive. For any $a, b \in S$, denote by $V_{ab}$, $V_{(ab)}$ and $V_{ba}$ the sets $\{i \in N - A \mid aP_ib\}, \{i \in N - A \mid aI_ib\}, \{i \in N - A \mid bP_ia\}$, respectively. As $j \in V_{yx}$ would imply $yPx$ in contradiction to the hypothesis $xRy$, we conclude $j \in V_{xy} \cup V_{(xy)}$. Similarly, as $j \in V_{zy}$ would imply $zPy$ in contradiction to the hypothesis $yRz$, we conclude $j \in V_{yz} \cup V_{(yz)}$. From $j \in V_{xy} \cup V_{(xy)}$ and $j \in V_{yz} \cup V_{(yz)}$, we conclude $xR_jz$. Now, as $xI_jz$ would imply $xI_jyI_jz$, which in turn would imply that $j \in A$ contradicting the fact that $j \in N - A$, we conclude $xP_jz$. $xP_jz$ implies $xPz$. Thus $xRz$ holds. $\qquad\square$

## 10.1.2  Characterization of Quasi-transitivity

**Theorem 10.3** *Let binary social decision rule* $f : \mathscr{T}^n \mapsto \mathscr{C}$ *satisfy* $\overline{P}$. $f$ *yields quasi-transitive social weak preference relation for every* $(R_1, \ldots, R_n) \in \mathscr{T}^n$ *iff for every* $A \subset N$, *there is an* $(N - A)$-*oligarchy* $V \subseteq N - A$.

*Proof* Let social decision rule $f : \mathscr{T}^n \mapsto \mathscr{C}$ satisfy I and $\overline{P}$. Suppose $f$ yields quasi-transitive social weak preference relation for every $(R_1, \ldots, R_n) \in \mathscr{T}^n$. Take any $A \subset N$. $N - A$ is $(N - A)$-decisive by condition $\overline{P}$. As a consequence of $\overline{P}$ and nonemptiness and finiteness of $N$, there exists a nonempty set $V \subseteq N - A$ which is minimally $(N - A)$-decisive. Suppose $V'$ is a minimal $(N - A)$-decisive set and $V' \neq V$. Let $x, y, z \in S$ be distinct. Consider the following configuration of individual preferences:

$(\forall i \in A)(xI_iyI_iz)$

$(\forall i \in V \cap V')(xP_iyP_iz)$

$(\forall i \in V - V')(zP_ixP_iy)$

$(\forall i \in N - (A \cup V))(yP_izP_ix)$.

As $[(\forall i \in A)(xI_iy) \wedge (\forall i \in V)(xP_iy)]$ and $[(\forall i \in A)(yI_iz) \wedge (\forall i \in V')(yP_iz)]$, we obtain $xPy$ and $yPz$, which imply $xPz$ as social $R$ is quasi-transitive. $xPz$, in view of condition I, Remark 10.1 and Lemma 10.1, implies that $V \cap V'$ is an $(N - A)$-decisive set. This, however, leads to a contradiction as the hypothesis of $V$ and $V'$ being distinct minimal $(N - A)$-decisive sets implies that $V \cap V'$ is not an $(N - A)$-decisive set. This establishes that there is a unique minimal $(N - A)$-decisive set. Let $V$ be the unique minimal $(N - A)$-decisive set.

Let $j \in V$. We will show that $(\forall (R_1, \ldots, R_n) \in \mathscr{T}^n)(\forall x, y \in S)[(\forall i \in A)(x I_i y) \wedge x P_j y \rightarrow x R y]$. Suppose not. Then $(\exists (R_1, \ldots, R_n) \in \mathscr{T}^n)(\exists x, y \in S)[(\forall i \in A)(x I_i y) \wedge x P_j y \wedge y P x]$. Then by condition I, and by $M$ and $N$ which hold in view of Remark 10.1 and Theorem 10.1, we conclude that: $(\forall (R_1, \ldots, R_n) \in \mathscr{T}^n)(\forall x, y \in S)[(\forall i \in A)(x I_i y) \wedge x P_j y \wedge (\forall i \in N - (A \cup \{j\}))(y P_i x) \rightarrow y P x]$. This implies that $N - (A \cup \{j\})$ is an $(N - A)$-decisive set. Consequently, there exists a nonempty $V' \subseteq N - (A \cup \{j\})$, which is minimally $(N - A)$-decisive. As $j \in V$ and $j \notin V'$, it follows that $V' \neq V$. This however contradicts that $V$ is the unique minimal $(N - A)$-decisive set. This contradiction establishes that we must have: $(\forall (R_1, \ldots, R_n) \in \mathscr{T}^n)(\forall x, y \in S)[(\forall i \in A)(x I_i y) \wedge x P_j y \rightarrow x R y]$. Thus $V$ is an $(N - A)$-oligarchy.

Now, let $f$ be such that, for every $A \subset N$, there is a $V \subseteq N - A$ which is an $(N - A)$-oligarchy. Take any $(R_1, \ldots, R_n) \in \mathscr{T}^n$ and any $x, y, z \in S$. Suppose $x P y \wedge y P z$. Let $A = \{i \in N \mid x I_i y I_i z\}$. $x P y$ and $y P z$ imply that $A \neq N$, by condition $\overline{P}$. Thus $A \subset N$. As $A \subset N$, there is a $V \subseteq N - A$ which is an $(N - A)$-oligarchy. For any $a, b \in S$, denote by $V_{ab}$, $V_{(ab)}$ and $V_{ba}$ the sets $\{i \in N - A \mid a P_i b\}$, $\{i \in N - A \mid a I_i b\}$, $\{i \in N - A \mid b P_i a\}$, respectively. As $V \cap V_{yx} \neq \emptyset$ would imply $y R x$ in contradiction to the hypothesis $x P y$, we conclude $V \subseteq V_{xy} \cup V_{(xy)}$. Similarly, as $V \cap V_{zy} \neq \emptyset$ would imply $z R y$ in contradiction to the hypothesis $y P z$, we conclude $V \subseteq V_{yz} \cup V_{(yz)}$. From $V \subseteq V_{xy} \cup V_{(xy)}$ and $V \subseteq V_{yz} \cup V_{(yz)}$, we conclude $(\forall i \in V)(x R_i z)$. Now, as $(\exists j \in V)(x I_j z)$ would imply $(\exists j \in V)(x I_j y I_j z)$, which in turn would imply that there exists an individual $j \in V \cap A$ contradicting the fact that $V \subseteq N - A$, we conclude $(\forall i \in V)(x P_i z)$. $(\forall i \in V)(x P_i z)$ implies $x P z$. This establishes the theorem.    □

### 10.1.3  Characterization of Acyclicity

**Theorem 10.4** *Let $f : \mathscr{T}^n \mapsto \mathscr{C}$ be a neutral and monotonic binary social decision rule satisfying $\overline{P}$. Then, $f$ yields acyclic social weak preference relation for every $(R_1, \ldots, R_n) \in \mathscr{T}^n$ iff for every $A \subset N$, every nonempty collection $\{V_1, \ldots, V_m\}$, $3 \leq m \leq \#S$, of $(N - A)$-decisive sets has nonempty intersection.*

*Proof* Suppose $f$ does not yield acyclic social weak preference relation for every $(R_1, \ldots, R_n) \in \mathscr{T}^n$. Then for some $(R_1, \ldots, R_n) \in \mathscr{T}^n$ and some distinct $x_1, x_2, \ldots, x_m \in S$ we must have $(x_1 P x_2 \wedge \cdots \wedge x_{m-1} P x_m \wedge x_m P x_1)$, where $3 \leq m \leq \#S$. Let $\{x_1, x_2, \ldots, x_m\} = S'$. Let $A = \{i \in N \mid (\forall a, b \in S')(a I_i b)\}$. Let $A_j = \{i \in N - A \mid x_j I_i x_{j+1}\}, j = 1, 2, \ldots, (m - 1); A_m = \{i \in N - A \mid x_m I_i x_1\}; V_j = \{i \in N - A \mid x_j P_i x_{j+1}\}, j = 1, 2, \ldots, (m - 1);$ and $V_m = \{i \in N - A \mid x_m P_i x_1\}$. By $\overline{P}$, for each $j \in 1, 2, \ldots, m$, $V_j$ is nonempty and consequently $A \neq N$. By monotonicity and neutrality we conclude that $V_j \cup A_j$ is $(N - A)$-decisive, $j = 1, 2, \ldots, m$. As individual weak preference relations are transitive, it follows that $(V_1 \cup A_1) \cap \cdots \cap (V_m \cup A_m) = \emptyset$.

Next suppose that for some $A \subset N$, there exists a collection $\{V_1, \ldots, V_m\}, 3 \leq m \leq \#S$, of $(N - A)$-decisive sets with empty intersection. Consider the following configuration of individual preferences:

$(\forall i \in A)(x_1 I_i \ldots I_i x_m)$
$(\forall i \in V_1)(x_1 P_i x_2)$
$(\forall i \in V_2)(x_2 P_i x_3)$
$\ldots\ldots\ldots\ldots$
$(\forall i \in V_{m-1})(x_{m-1} P_i x_m)$
$(\forall i \in V_m)(x_m P_i x_1)$.

It is possible to have the above configuration of preferences without violating transitivity of individual weak preference relations because it is given that $(V_1 \cap \cdots \cap V_m) = \emptyset$. As $V_j$ is $(N - A)$-decisive, $j = 1, 2, \ldots, m$, we conclude that $(x_1 P x_2 \wedge x_2 P x_3 \wedge \cdots \wedge x_{m-1} P x_m \wedge x_m P x_1)$ holds, which violates acyclicity. This establishes the theorem. □

## 10.2  Conditions for Transitivity

**Theorem 10.5** *Let $f : \mathscr{T}^n \mapsto \mathscr{C}$ satisfy conditions I, M and N. Let $\mathscr{D} \subseteq \mathscr{T}$. If $\mathscr{D}$ satisfies the condition of strict placement restriction then $f$ yields transitive social $R$, $R = f(R_1, \ldots, R_n)$, for every $(R_1, \ldots, R_n) \in \mathscr{D}^n$.*

*Proof* Suppose $f$ does not yield transitive $R$ for every $(R_1, \ldots, R_n) \in \mathscr{D}^n$. Then $(\exists (R_1, \ldots, R_n) \in \mathscr{D}^n)(\exists x, y, z \in S)[xRy \wedge yRz \wedge zPx]$.

In view of conditions I, M and N it follow that we must have:

$$\sim(\forall i \in N)[(z P_i x \to y P_i x) \wedge (z I_i x \to y R_i x)] \qquad (T10.5\text{-}1)$$

otherwise we will obtain $y P x$ in view of $z P x$, contradicting the hypothesis $x R y$. Similarly,

$$y R z \wedge z P x \to \sim (\forall i \in N)[(z P_i x \to z P_i y) \wedge (z I_i x \to z R_i y)]. \qquad (T10.5\text{-}2)$$

$$(T10.5\text{-}1) \to (\exists i \in N)[z P_i x R_i y \vee z I_i x P_i y]$$

$$\to (\exists i \in N)[R_i \text{ is concerned over } \{x, y, z\} \wedge z R_i x R_i y] \qquad (T10.5\text{-}3)$$

$$(T10.5\text{-}2) \to (\exists i \in N)[y R_i z P_i x \vee y P_i z I_i x]$$

$$\to (\exists i \in N)[R_i \text{ is concerned over } \{x, y, z\} \wedge y R_i z R_i x] \qquad (T10.5\text{-}4)$$

$$z P x \to (\exists i \in N)(z P_i x), \text{ in view of conditions I, M and N} \qquad (T10.5\text{-}5)$$

(T10.5-3)–(T10.5-5) imply that $\mathscr{D}$ violates SPR over $\{x, y, z\}$; which establishes the theorem. □

The next theorem shows that SPR is maximally sufficient for the class of social decision rules satisfying independence of irrelevant alternatives, neutrality and monotonicity.

Let $\mathscr{D}(SPR) = \{\mathscr{D} \subseteq \mathscr{T} \mid \mathscr{D} \text{ satisfies SPR}\}$.

**Theorem 10.6** *Let $\mathscr{D} \subseteq \mathscr{T}$ be such that every $f : \mathscr{T}^n \mapsto \mathscr{C}$ satisfying conditions I, M, and N yields transitive social R for every $(R_1, \ldots, R_n) \in \mathscr{D}^n$. Then we must have: $\mathscr{D} \in \mathscr{D}(SPR)$.*

*Proof* The proof of the theorem consists of showing the existence of an $f : \mathscr{T}^n \mapsto \mathscr{C}$ satisfying I, M and N which is such that it yields transitive social $R$ for every $(R_1, \ldots, R_n) \in \mathscr{D}^n$, $\mathscr{D} \subseteq \mathscr{T}$, iff $\mathscr{D} \in \mathscr{D}(SPR)$.

Let $N = \{1, 2, 3, 4\}$. Consider the social decision rule $f$ defined by:
$(\forall(R_1, \ldots, R_n) \in \mathscr{T}^n)(\forall x, y \in S)[xPy \leftrightarrow (\forall i \in N)(xR_iy) \wedge n(xP_iy) \geq 3]$.
From the definition of $f$ it is clear that it satisfies I, M and N. Therefore from Theorem 10.5 it follows that if $\mathscr{D} \subseteq \mathscr{T}$ satisfies SPR then social $R$ for every $(R_1, \ldots, R_n) \in \mathscr{D}^n$ would be transitive.

Now, let $\mathscr{D}$ violate SPR. Then $\mathscr{D}$ violates SPR over some triple, say, $\{x, y, z\} \subseteq S$. Then, by Lemma 5.1 $\mathscr{D}$ must contain, except for a formal interchange of alternatives, one of the eight sets listed in the statement of Lemma 5.1. If $\mathscr{D}$ contains set (i) or (iv) or (v) or (viii) then consider any $(R_1, \ldots, R_n) \in \mathscr{D}^n$ such that its restriction over $\{x, y, z\}$ is given by: $\#\{i \in N \mid R_i \mid \{x, y, z\} = R^s \mid \{x, y, z\}\} = \#\{i \in N \mid R_i \mid \{x, y, z\} = R^t \mid \{x, y, z\}\} = 2$; and if $\mathscr{D}$ contains set (ii) or (iii) or (vi) or (vii) then consider any $(R_1, \ldots, R_n) \in \mathscr{D}^n$ such that its restriction over $\{x, y, z\}$ is given by: $\#\{i \in N \mid R_i \mid \{x, y, z\} = R^s \mid \{x, y, z\}\} = 3 \wedge \#\{i \in N \mid R_i \mid \{x, y, z\} = R^t \mid \{x, y, z\}\} = 1$. This results in $(xIy \wedge yPz \wedge xIz)$ for sets (i), (ii), (iv) and (vii); in $(xPy \wedge yIz \wedge xIz)$ for sets (iii) and (v); and in $(xIy \wedge yIz \wedge xPz)$ for sets (vi) and (viii). Thus transitivity is violated in each case. This establishes the theorem. $\qquad\square$

**Theorem 10.7** *Let $f : \mathscr{T}^n \mapsto \mathscr{C}$ satisfy conditions I, M, N and $\overline{P}$. Let $\mathscr{D} \subseteq \mathscr{T}$. If $\mathscr{D}$ satisfies the condition of placement restriction (PR) then $f$ yields transitive social $R$, $R = f(R_1, \ldots, R_n)$, for every $(R_1, \ldots, R_n) \in \mathscr{D}^n$.*

*Proof* Suppose $f$ does not yield transitive $R$ for every $(R_1, \ldots, R_n) \in \mathscr{D}^n$. Then $(\exists(R_1, \ldots, R_n) \in \mathscr{D}^n)(\exists x, y, z \in S)[xRy \wedge yRz \wedge zPx]$.
As in Theorem 10.5, we can infer that (T10.7-1)–(T10.7-3) hold:

$(\exists i \in N)[R_i$ is concerned over $\{x, y, z\} \wedge zR_ixR_iy]$ \hfill (T10.7-1)

$(\exists i \in N)[R_i$ is concerned over $\{x, y, z\} \wedge yR_izR_ix]$ \hfill (T10.7-2)

$(\exists i \in N)[zP_ix]$ \hfill (T10.7-3)

$(T10.7\text{-}1) \rightarrow (\exists i \in N)[zP_iy]$ \hfill (T10.7-4)

$(T10.7\text{-}2) \rightarrow (\exists i \in N)[yP_ix]$ \hfill (T10.7-5)

$(T10.7\text{-}4) \wedge yRz \wedge \overline{P} \rightarrow (\exists i \in N)[yP_iz]$ \hfill (T10.7-6)

$(T10.7\text{-}5) \wedge xRy \wedge \overline{P} \rightarrow (\exists i \in N)[xP_iy]$ \hfill (T10.7-7)

(T10.7-1)–(T10.7-7) imply that $\mathscr{D}$ violates PR over $\{x, y, z\}$, which establishes the theorem. $\qquad\square$

The next theorem shows that PR is maximally sufficient for the class of social decision rules satisfying independence of irrelevant alternatives, neutrality, monotonicity and Pareto-criterion.

Let $\mathscr{D}(PR) = \{\mathscr{D} \subseteq \mathscr{T} \mid \mathscr{D} \text{ satisfies PR}\}$.

**Theorem 10.8** *Let $\mathscr{D} \subseteq \mathscr{T}$ be such that every $f : \mathscr{T}^n \mapsto \mathscr{C}$ satisfying conditions I, M, N and $\overline{P}$ yields transitive social R for every $(R_1, \ldots, R_n) \in \mathscr{D}^n$. Then we must have: $\mathscr{D} \in \mathscr{D}(PR)$.*

*Proof* The proof of the theorem consists of showing the existence of an $f : \mathscr{T}^n \mapsto \mathscr{C}$ satisfying I, M, N and $\overline{P}$ which is such that it yields transitive social $R$ for every $(R_1, \ldots, R_n) \in \mathscr{D}^n$, $\mathscr{D} \subseteq \mathscr{T}$, iff $\mathscr{D} \in \mathscr{D}(PR)$.

Let $\#N = 10$. Consider the social decision rule $f$ defined by:

$(\forall (R_1, \ldots, R_n) \in \mathscr{T}^n)(\forall x, y \in S)[xPy \leftrightarrow n(xP_iy) > \frac{3}{4}(n(xP_iy) + n(yP_ix))]$.

From the definition of $f$ it is clear that it satisfies I, M, N and $\overline{P}$. Therefore from Theorem 10.7 it follows that if $\mathscr{D} \subseteq \mathscr{T}$ satisfies PR then social $R$ for every $(R_1, \ldots, R_n) \in \mathscr{D}^n$ would be transitive.

Now, let $\mathscr{D}$ violate PR. Then $\mathscr{D}$ violates PR over some triple, say, $\{x, y, z\} \subseteq S$. Then, by Lemma 6.1 $\mathscr{D}$ must contain, except for a formal interchange of alternatives, one of the 10 sets listed in the statement of the lemma. If $\mathscr{D}$ contains set (i) or (ii) or (iii) then consider any $(R_1, \ldots, R_n) \in \mathscr{D}^n$ such that its restriction over $\{x, y, z\}$ is given by: $\#\{i \in N \mid R_i \mid \{x, y, z\} = R^s \mid \{x, y, z\}\} = \#\{i \in N \mid R_i \mid \{x, y, z\} = R^t \mid \{x, y, z\}\} = 5$; if $\mathscr{D}$ contains set (iv) or (v) then consider any $(R_1, \ldots, R_n) \in \mathscr{D}^n$ such that its restriction over $\{x, y, z\}$ is given by: $\#\{i \in N \mid R_i \mid \{x, y, z\} = R^s \mid \{x, y, z\}\} = 5 \wedge \#\{i \in N \mid R_i \mid \{x, y, z\} = R^t \mid \{x, y, z\}\} = 2 \wedge \#\{i \in N \mid R_i \mid \{x, y, z\} = R^u \mid \{x, y, z\}\} = 3$; if $\mathscr{D}$ contains set (vi) or (vii) or (viii) then consider any $(R_1, \ldots, R_n) \in \mathscr{D}^n$ such that its restriction over $\{x, y, z\}$ is given by: $\#\{i \in N \mid R_i \mid \{x, y, z\} = R^s \mid \{x, y, z\}\} = 4 \wedge \#\{i \in N \mid R_i \mid \{x, y, z\} = R^t \mid \{x, y, z\}\} = 3 \wedge \#\{i \in N \mid R_i \mid \{x, y, z\} = R^u \mid \{x, y, z\}\} = 3$; and if $\mathscr{D}$ contains set (ix) or (x) then consider any $(R_1, \ldots, R_n) \in \mathscr{D}^n$ such that its restriction over $\{x, y, z\}$ is given by: $\#\{i \in N \mid R_i \mid \{x, y, z\} = R^s \mid \{x, y, z\}\} = 6 \wedge \#\{i \in N \mid R_i \mid \{x, y, z\} = R^t \mid \{x, y, z\}\} = 1 \wedge \#\{i \in N \mid R_i \mid \{x, y, z\} = R^u \mid \{x, y, z\}\} = 3$. This results in $(xIy \wedge yPz \wedge xIz)$ for sets (i), (iii) and (iv); in $(xPy \wedge yIz \wedge xIz)$ for sets (ii), (v) and (ix); and in $(xIy \wedge yIz \wedge xPz)$ for sets (vi), (vii), (viii) and (x). Thus transitivity is violated in each case. This establishes the theorem.  □

## 10.3  Conditions for Quasi-transitivity

**Theorem 10.9** *Let $f : \mathscr{T}^n \mapsto \mathscr{C}$ satisfy conditions I, M and N. Let $\mathscr{D} \subseteq \mathscr{T}$. If $\mathscr{D}$ satisfies the condition of value restriction (2) then $f$ yields quasi-transitive social $R$, $R = f(R_1, \ldots, R_n)$, for every $(R_1, \ldots, R_n) \in \mathscr{D}^n$.*

*Proof* Suppose $f$ does not yield quasi-transitive $R$ for every $(R_1, \ldots, R_n) \in \mathcal{D}^n$. Then $(\exists (R_1, \ldots, R_n) \in \mathcal{D}^n)(\exists x, y, z \in S)[x P y \wedge y P z \wedge z R x]$.

In view of conditions I, M and N it follow that we must have:

$$\sim(\forall i \in N)[(x P_i y \to z P_i y) \wedge (x I_i y \to z R_i y)] \tag{T10.9-1}$$

otherwise we will obtain $z P y$ in view of $x P y$, contradicting the hypothesis $y P z$. Similarly,

$$y P z \wedge z R x \to \sim (\forall i \in N)[(y P_i z \to x P_i z) \wedge (y I_i z \to x R_i z)], \text{ and} \tag{T10.9-2}$$

$$x P y \wedge z R x \to \sim (\forall i \in N)[(x P_i y \to x P_i z) \wedge (x I_i y \to x R_i z)]. \tag{T10.9-3}$$

$$(\text{T10.9-1}) \to (\exists i \in N)[x P_i y R_i z \vee x I_i y P_i z]$$

$$\to (\exists i \in N)[R_i \text{ is concerned over } \{x, y, z\} \wedge x R_i y R_i z] \tag{T10.9-4}$$

$$(\text{T10.9-2}) \to (\exists i \in N)[y P_i z R_i x \vee y I_i z P_i x]$$

$$\to (\exists i \in N)[R_i \text{ is concerned over } \{x, y, z\} \wedge y R_i z R_i x] \tag{T10.9-5}$$

$$(\text{T10.9-3}) \to (\exists i \in N)[z R_i x P_i y \vee z P_i x I_i y]$$

$$\to (\exists i \in N)[R_i \text{ is concerned over } \{x, y, z\} \wedge z R_i x R_i y] \tag{T10.9-6}$$

(T10.9-4)–(T10.9-6) imply that $\mathcal{D}$ violates VR(2) over $\{x, y, z\}$. The theorem, therefore, stands established.  $\square$

**Theorem 10.10** *Let* $f : \mathcal{T}^n \mapsto \mathcal{C}$ *satisfy conditions I, M, N and* $\overline{P}$*. Let* $\mathcal{D} \subseteq \mathcal{T}$*. If* $\mathcal{D}$ *satisfies the condition of limited agreement then* $f$ *yields quasi-transitive social* $R, R = f(R_1, \ldots, R_n)$*, for every* $(R_1, \ldots, R_n) \in \mathcal{D}^n$*.*

*Proof* Suppose $f$ does not yield quasi-transitive $R$ for every $(R_1, \ldots, R_n) \in \mathcal{D}^n$. Then $(\exists (R_1, \ldots, R_n) \in \mathcal{D}^n)(\exists x, y, z \in S)[x P y \wedge y P z \wedge z R x]$.

From Theorem 10.9 it follows that (T10.10-1)–(T10.10-3) hold:

$$(\exists i \in N)[R_i \text{ is concerned over } \{x, y, z\} \wedge x R_i y R_i z] \tag{T10.10-1}$$

$$(\exists i \in N)[R_i \text{ is concerned over } \{x, y, z\} \wedge y R_i z R_i x] \tag{T10.10-2}$$

$$(\exists i \in N)[R_i \text{ is concerned over } \{x, y, z\} \wedge z R_i x R_i y] \tag{T10.10-3}$$

$$(\text{T10.10-1}) \to (\exists i \in N)[x P_i z] \tag{T10.10-4}$$

$$(\text{T10.10-2}) \to (\exists i \in N)[y P_i x] \tag{T10.10-5}$$

$$(\text{T10.10-3}) \to (\exists i \in N)[z P_i y] \tag{T10.10-6}$$

$$(\text{T10.10-4}) \wedge z R x \wedge \overline{P} \to (\exists i \in N)[z P_i x] \tag{T10.10-7}$$

$$(\text{T10.10-5}) \wedge x P y \wedge \overline{P} \to (\exists i \in N)[x P_i y] \tag{T10.10-8}$$

$$(\text{T10.10-6}) \wedge y P z \wedge \overline{P} \to (\exists i \in N)[y P_i z] \tag{T10.10-9}$$

(T10.10-4)–(T10.10-9) establish that $\mathcal{D}$ violates LA over $\{x, y, z\}$, establishing the theorem.  $\square$

## 10.4  Conditions for Transitivity and Quasi-transitivity When Individual Orderings Are Linear

**Theorem 10.11** *Let social decision rule $f : \mathscr{L}^n \mapsto \mathscr{C}$ satisfy independence of irrelevant alternatives, neutrality and monotonicity. Then, $f$ yields transitive social weak preference relation for every $(R_1, \ldots, R_n) \in \mathscr{L}^n$ iff it is null or dictatorial.*

*Proof* Let $f : \mathscr{L}^n \mapsto \mathscr{C}$ satisfy I, N and M. Suppose $f$ yields transitive social weak preference relation for every $(R_1, \ldots, R_n) \in \mathscr{L}^n$. Suppose there is a minimal decisive set $V$ consisting of more than one individual. Let $(V_1, V_2)$ be a partition of $V$ [*i.e.*, $V_1 \neq \emptyset$, $V_2 \neq \emptyset$, $V_1 \cap V_2 = \emptyset$, $V_1 \cup V_2 = V$]. Consider the following configuration of individual preferences:

$(\forall i \in V_1)[x P_i y P_i z]$
$(\forall i \in V_2)[y P_i z P_i x]$
$(\forall i \in N - V)[z P_i x P_i y]$,

where $x, y, z$ are all distinct and belong to $S$.
From $(\forall i \in V)(y P_i z)$ we obtain $y P z$, by the decisiveness of $V$.
$y P x \vee x R y$, as $R$ is connected.
$y P x \rightarrow V_2$ is almost decisive for $(y, x)$, by condition I
$\rightarrow V_2$ is a decisive set, by conditions I, M and N.
This contradicts the minimality of $V$.
$x R y \rightarrow x P z$, by transitivity of $R$
$\rightarrow V_1$ is almost decisive for $(x, z)$, by condition I
$\rightarrow V_1$ is a decisive set, by conditions I, M and N.
This contradicts the minimality of $V$.
As each of $y P x$ and $x R y$ leads to a contradiction, we conclude that no minimal decisive set can consist of more than one individual. Finiteness of $N$ implies that if there exists a decisive set then there must exist a nonempty minimal decisive set. Thus there are only two possibilities, either there is no decisive set implying $f$ is null or there is a minimal decisive set consisting of a single individual implying that $f$ is dictatorial.
Now, let $f$ be null or dictatorial. If $f$ is null then social $R$ is transitive for every $(R_1, \ldots, R_n) \in \mathscr{L}^n$. Let $f$ be dictatorial with individual $j$ as the dictator. Consider any $(R_1, \ldots, R_n) \in \mathscr{L}^n$ and any $x, y, z \in S$. Suppose $x R y \wedge y R z$. $x R y$ implies that we must have $x P_j y$, as $y P_j x$ would imply $y P x$ contradicting the hypothesis $x R y$. Similarly $y R z$ implies $y P_j z$. $x P_j y$ and $y P_j z$ imply $x P_j z$. Therefore we must have $x P z$ implying $x R z$ holds. Thus $R$ is transitive. This establishes the theorem.  $\square$

**Theorem 10.12** *Let social decision rule $f : \mathscr{L}^n \mapsto \mathscr{C}$ satisfy I, M, and N. Then, $f$ yields quasi-transitive social weak preference relation for every $(R_1, \ldots, R_n) \in \mathscr{L}^n$ iff there is at most one minimal decisive set.*

*Proof* Let social decision rule $f : \mathscr{L}^n \mapsto \mathscr{C}$ satisfy I, M and N. Suppose $f$ yields quasi-transitive social weak preference relation for every $(R_1, \ldots, R_n) \in \mathscr{L}^n$. Suppose there are two distinct minimal decisive sets $V$ and $V'$. Consider the following configuration of individual preferences:

$(\forall i \in V \cap V')(x P_i y P_i z)$

$(\forall i \in V - V')(z P_i x P_i y)$

$(\forall i \in N - V)(y P_i z P_i x)$.

As $(\forall i \in V)(x P_i y)$ and $(\forall i \in V')(y P_i z)$, we obtain $x P y$ and $y P z$, which imply $x P z$ as social $R$ is quasi-transitive. $x P z$, in view of condition I, M and N, implies that $V \cap V'$ is a decisive set. This, however, leads to a contradiction as the hypothesis of $V$ and $V'$ being distinct minimal decisive sets implies that $V \cap V'$ is not a decisive set. This establishes that there is at most one minimal decisive set.

Now, let $f$ be such that there is at most one minimal decisive set. Non-existence of a minimal decisive set implies non-existence of a decisive set, in view of the finiteness of $N$. Non-existence of a decisive set, in view of conditions I, M and N, implies that $f$ must be null. If $f$ is null then social $R$ is transitive, and hence quasi-transitive, for every $(R_1, \ldots, R_n) \in \mathscr{L}^n$. Next, let there be a unique minimal decisive set $V$. Take any $(R_1, \ldots, R_n) \in \mathscr{L}^n$ and any $x, y, z \in S$. Suppose $x P y \wedge y P z$. $x P y$ implies that we must have $(\forall i \in V)(x P_i y)$. Suppose not. Then $(\exists j \in V)(y P_i x)$. $x P y$ implies that $\{i \in N \mid x P_i y\}$ is a decisive set in view of conditions I, M and N. Therefore there exists a minimal decisive set $V'$ contained in $\{i \in N \mid x P_i y\}$. $V' \neq V$ as $(j \in V \wedge j \notin V')$. This, however, contradicts the supposition that there is a unique minimal decisive set. This establishes that we must have $(\forall i \in V)(x P_i y)$. Similarly, $y P z$ implies $(\forall i \in V)(y P_i z)$. From $(\forall i \in V)(x P_i y \wedge y P_i z)$, we conclude $(\forall i \in V)(x P_i z)$, which by the decisiveness of $V$ implies $x P z$. Thus social $R$ is quasi-transitive for every $(R_1, \ldots, R_n) \in \mathscr{L}^n$. This establishes the theorem.          □

**Theorem 10.13** *Let $f : \mathscr{L}^n \mapsto \mathscr{C}$ satisfy conditions I, M and N. Let $f$ be such that there are at least two minimal decisive sets. Let $\mathscr{D} \subseteq \mathscr{L}$. Then $f$ yields quasi-transitive social $R$, $R = f(R_1, \ldots, R_n)$, for every $(R_1, \ldots, R_n) \in \mathscr{D}^n$ iff $\mathscr{D}$ is such that for every triple $A$ contained in $S$, $\mathscr{D} \mid A$ does not contain a Latin Square.*

*Proof* Suppose $f$ does not yield quasi-transitive $R$ for every $(R_1, \ldots, R_n) \in \mathscr{D}^n$. Then $(\exists (R_1, \ldots, R_n) \in \mathscr{D}^n)(\exists x, y, z \in S)[x P y \wedge y P z \wedge z R x]$.

In view of conditions I, M and N it follow that we must have:

$\sim (\forall i \in N)(x P_i y \rightarrow z P_i y)$          (T10.13-1)

otherwise we will obtain $z P y$ in view of $x P y$, contradicting the hypothesis $y P z$. Similarly,

$y P z \wedge z R x \rightarrow \sim (\forall i \in N)(y P_i z \rightarrow x P_i z)$, and          (T10.13-2)

$x P y \wedge z R x \rightarrow \sim (\forall i \in N)(x P_i y \rightarrow x P_i z)$.          (T10.13-3)

$(\text{T10.13-1}) \rightarrow (\exists i \in N)(x P_i y P_i z)$          (T10.13-4)

$(\text{T10.13-2}) \rightarrow (\exists i \in N)(y P_i z P_i x)$          (T10.13-5)

$(\text{T10.13-3}) \rightarrow (\exists i \in N)(z P_i x P_i y)$          (T10.13-6)

(T10.13-4)–(T10.13-6) imply that $\mathscr{D} \mid \{x, y, z\}$ contains a Latin Square.

Next, let $\mathscr{D}$ contain a Latin Square over the triple $\{x, y, z\} \subseteq S$, say, $LS(xyzx)$. Let $V$ and $V'$ be two distinct minimal decisive sets. Consider any $(R_1, \ldots, R_n) \in \mathscr{D}^n$ such that $(R_1 \mid \{x, y, z\}, \ldots, R_n \mid \{x, y, z\})$ is given by: $(\forall i \in V \cap V')(x P_i y P_i z) \wedge$ $(\forall i \in V - V')(y P_i z P_i x) \wedge (\forall i \in N - V)(z P_i x P_i y)$. $(\forall i \in V')(x P_i y)$ and $(\forall i \in V)(y P_i z)$ imply $x P y$ and $y P z$, respectively. As $(\forall i \in V \cap V')(x P_i z) \wedge (\forall i \in N - (V \cap V'))(z P_i x)$, we must have $\sim x P z$ as $V \cap V'$ is a proper subset of both $V$ and $V'$ which are minimal decisive sets. Thus quasi-transitivity is violated. This establishes the theorem.  $\square$

**Theorem 10.14** *Let* $f : \mathscr{L}^n \mapsto \mathscr{C}$ *satisfy conditions I, M and N. Let $f$ be such that for every partition $(N_1, N_2)$ of $N$ either $N_1$ or $N_2$ is a decisive set. Then, social $R$ is quasi-transitive iff it is transitive.*

*Proof* Suppose $R$ violates transitivity. Then we must have for some $x, y, z \in S$, $x R y \wedge y R z \wedge z P x$. $x R y$ implies that $\{i \in N \mid y P_i x\}$ is not a decisive set, which in turn implies that $N - \{i \in N \mid y P_i x\} = \{i \in N \mid x P_i y\}$ is a decisive set. Therefore we must have $x P y$. Similarly, $y R z$ implies that $\{i \in N \mid z P_i y\}$ is not a decisive set, implying that $N - \{i \in N \mid z P_i y\} = \{i \in N \mid y P_i z\}$ is a decisive set. Consequently $y P z$ holds. $(x P y \wedge y P z \wedge z P x)$ constitutes a violation of quasi-transitivity, establishing the theorem.  $\square$

**Theorem 10.15** *Let* $f : \mathscr{L}^n \mapsto \mathscr{C}$ *be non-null, non-dictatorial, and satisfy conditions I, M and N. Let $f$ be such that for every partition $(N_1, N_2)$ of $N$ either $N_1$ or $N_2$ is a decisive set. Let $\mathscr{D} \subseteq \mathscr{L}$. Then $f$ yields transitive social $R$, $R = f(R_1, \ldots, R_n)$, for every $(R_1, \ldots, R_n) \in \mathscr{D}^n$ iff $\mathscr{D}$ is such that for every triple $A$ contained in $S$, $\mathscr{D} \mid A$ does not contain a Latin Square.*

*Proof* As $f$ is non-null, non-dictatorial, and satisfies conditions I, M and N, it follows, from Theorem 10.11, that $f$ does not yield transitive social $R$ for every $(R_1, \ldots, R_n) \in \mathscr{L}^n$. As $f$ is such that for every partition $(N_1, N_2)$ of $N$ either $N_1$ or $N_2$ is a decisive set, in view of Theorem 10.14, we conclude that $f$ does not yield quasi-transitive social $R$ for every $(R_1, \ldots, R_n) \in \mathscr{L}^n$. This, by Theorem 10.12, implies that there are at least two minimal decisive sets. Thus $f$ satisfies all the assumptions of Theorem 10.13. Consequently from Theorem 10.13 it follows that, for any $\mathscr{D} \subseteq \mathscr{L}$, $f$ yields quasi-transitive social $R$ for every $(R_1, \ldots, R_n) \in \mathscr{D}^n$ iff $\mathscr{D} \mid A$ does not contain a Latin Square. As social $R$ is quasi-transitive iff it is transitive, as established in Theorem 10.14, we conclude that for any $\mathscr{D} \subseteq \mathscr{L}$, $f$ yields transitive social $R$ for every $(R_1, \ldots, R_n) \in \mathscr{D}^n$ iff $\mathscr{D} \mid A$ does not contain a Latin Square.  $\square$

**Theorem 10.16** *Let* $f : \mathscr{L}^n \mapsto \mathscr{C}$ *be non-null, non-dictatorial, and satisfy conditions I, M and N. Let $f$ be such that there exists a partition $(N_1, N_2)$ of $N$ such that neither $N_1$ nor $N_2$ is a decisive set. Let $\mathscr{D} \subseteq \mathscr{L}$. Then $f$ yields transitive social $R$, $R = f(R_1, \ldots, R_n)$, for every $(R_1, \ldots, R_n) \in \mathscr{D}^n$ iff $\mathscr{D}$ is such that for every* *triple A contained in S, $\mathscr{D} \mid A$ does not contain more than one ordering belonging to the same Latin Square.*

*Proof* Suppose $f$ does not yield transitive $R$ for every $(R_1, \ldots, R_n) \in \mathscr{D}^n$. Then $(\exists(R_1, \ldots, R_n) \in \mathscr{D}^n)(\exists x, y, z \in S)[x R y \wedge y R z \wedge z P x]$.
In view of conditions I, M and N it follow that we must have:

$$\sim(\forall i \in N)(z P_i x \to y P_i x), \tag{T10.16-1}$$

otherwise we would have $y P x$ in view of $z P x$, contradicting $x R y$

Similarly,

$$\sim(\forall i \in N)(z P_i x \to z P_i y), \tag{T10.16-2}$$

otherwise we would have $z P y$ in view of $z P x$, contradicting $y R z$

$$(\text{T}10.16\text{-}1) \to (\exists i \in N)(z P_i x P_i y) \tag{T10.16-3}$$

$$(\text{T}10.16\text{-}2) \to (\exists j \in N)(y P_j z P_j x) \tag{T10.16-4}$$

$z P_i x P_i y$ and $y P_j z P_j x$ belong to the same Latin Square $LS(xyzx)$. This establishes that if $\mathscr{D} \subseteq \mathscr{L}$ is such that it does not contain more than one ordering belonging to the same Latin Square then every $(R_1, \ldots, R_n) \in \mathscr{D}^n$ yields transitive social $R$.

Next, let $\mathscr{D}$ be such that it contains more than one ordering belonging to the same Latin Square over the triple $\{x, y, z\} \subseteq S$, say, $(x P^i y P^i z \wedge y P^j z P^j x)$. Let $(N_1, N_2)$ be a partition of $N$ such that neither $N_1$ nor $N_2$ is a decisive set. As $f$ is non-null and satisfies I, N and M, it follows that the set of all individuals $N$ is a decisive set. Consider any $(R_1, \ldots, R_n) \in \mathscr{D}^n$ such that $(R_1 \mid \{x, y, z\}, \ldots, R_n \mid \{x, y, z\})$ is given by: $(\forall i \in N_1)(x P_i y P_i z) \wedge (\forall i \in N_2)(y P_i z P_i x)$. $(\forall i \in N)(y P_i z)$ implies $y P z$. As neither $N_1$ nor $N_2$ is a decisive set and $f$ satisfies I, N and M, it follows, from $(\forall i \in N_1)(x P_i y) \wedge (\forall i \in N_2)(y P_i x) \wedge (\forall i \in N_1)(x P_i z) \wedge (\forall i \in N_2)(z P_i x)$, that we must have $x I y \wedge x I z$. Thus transitivity is violated. This establishes the theorem. $\square$

# Appendix

## 10.5   Notes on Literature

The characterization of monotonicity and neutrality for the class of binary social decision rules with unrestricted domain and satisfying Pareto-indifference given in this chapter is from Jain (1988). Characterization of transitivity for the class of binary social decision rules with unrestricted domain and satisfying the Pareto-criterion was obtained in Jain (1977) and for quasi-transitivity in Guha (1972) and Blau (1976). The characterization of acyclicity for the class of social decision rules with unrestricted domain and satisfying the conditions of neutrality, monotonicity and the Pareto-criterion is from Jain (1977). The maximal sufficient conditions for transitivity for the class of binary, neutral and monotonic rules, and for its subclass satisfying the Pareto-criterion, were derived in Jain (1987). Sufficiency of value restriction (2) for quasi-transitivity under every neutral and monotonic binary social decision rule, and sufficiency of limited agreement for quasi-transitivity under every neutral and monotonic binary social decision rule satisfying the Pareto-criterion, were established in Sen and Pattanaik (1969). The proofs given here of these two theorems are, however, different from those given in Sen and Pattanaik (1969).

In this chapter the class of neutral and monotonic binary social decision rules has been discussed. A closely related class is that of Pareto-transitive binary social decision rules. A social decision rule $f : \mathbb{D} \mapsto \mathscr{C}$ is Pareto-transitive iff $(\forall (R_1, \ldots, R_n) \in \mathbb{D})(\forall x, y, z \in S)[(xRy \wedge (\forall i \in N)(yP_iz) \to xPz]$. In Salles (1975) it is claimed that for the class of Pareto-transitive binary social decision rules an Inada-type necessary and sufficient condition for transitivity is that over every triple of alternatives value restriction (1) or cyclical indifference (CI) holds. $\mathscr{D}$ satisfies CI over a triple of alternatives $A$ iff $(\exists$ distinct $x, y, z \in A)[(\mathscr{D} \mid A \subseteq \{xPyIz, yPzIx, zPxIy\}) \vee (\mathscr{D} \mid A \subseteq \{xIyPz, yIzPx, zIxPy\})$. We show in what follows that: (i) The satisfaction of (VR (1) or CI) over every triple of alternatives is not an Inada-type necessary condition for transitivity for the class of Pareto-transitive binary social decision rules. (ii) The satisfaction of (VR (1) or CI) over every triple of alternatives is not a sufficient condition for transitivity for the class of Pareto-transitive binary social decision rules. (iii) There does not exist an Inada-type necessary and sufficient condition for transitivity for the class of Pareto-transitive binary social decision rules.

*Example 10.1* Let the social decision rule $f$ be the MMD defined for an odd number of individuals, $n = 2k + 1, k \geq 1$. Suppose $xRy \wedge (\forall i \in N)(yP_iz)$. As $xRy \to n(xR_iy) \geq k + 1$, it follows that $n(xP_iz) \geq k + 1$. Thus we must have $xPz$. This establishes that MMD defined for an odd number of individuals is Pareto-transitive.

Now consider $\mathscr{D} = \{xPyPz, zPyPx, yPzIx, zIxPy\}$. $\mathscr{D}$ violates both VR(1) and CI. As every $(R_1, \ldots, R_n) \in \mathscr{D}^n$ yields transitive social $R$ under the MMD, it follows that the claim that the satisfaction of (VR (1) or CI) over every triple of alternatives is an Inada-type necessary condition for transitivity for every Pareto-transitive binary social decision rule is false.

*Example 10.2* Let $S = \{x, y, z\}$; $N = \{1, 2, \ldots, 9\}$. Let the social decision rule $f : \mathscr{T}^9 \mapsto \mathscr{C}$ be characterized as follows: $(\forall (a, b) \in S \times S)[(a, b) \neq (x, y), (y, x) \to (aRb \leftrightarrow n(aR_ib) \geq n(bR_ia))] \wedge [xRy \leftrightarrow \sim (n(yP_ix) \geq 5)] \wedge [yRx \leftrightarrow \sim (n(xP_iy) \geq 5)]$. It is immediate that $f$ is binary. Consider any permutation $(u, v, w)$ of $(x, y, z)$. Suppose $uRv \wedge (\forall i \in N)(vP_iw)$. As $uRv \to n(uR_iv) \geq 5$, it follows that we must have $n(uP_iw) \geq 5$. Thus $uPw$ must hold. This establishes that $f$ is Pareto-transitive. Now, let $\mathscr{D} = \{xIyPz, yIzPx, zIxPy\}$, and consider the following $(R_1, \ldots, R_n) \in \mathscr{D}^n$ : $(\forall i \in \{1, 2, 3\})(xI_iyP_iz) \wedge (\forall i \in \{4, 5, 6, 7\})(yI_izP_ix) \wedge (\forall i \in \{8, 9\})(zI_ixP_iy))$. As $[n(xP_iy) = 2 \wedge n(yP_ix) = 4 \wedge n(yP_iz) = 3 \wedge n(zP_iy) = 2 \wedge n(xP_iz) = 3 \wedge n(zP_ix) = 4]$, we obtain $(xIy \wedge yPz \wedge zPx)$ violating transitivity. This establishes that the claim that (VR (1) or CI) is sufficient for transitivity under every Pareto-transitive binary social decision rule is false.

An Inada-type necessary and sufficient condition for transitivity partitions the set of all nonempty subsets of $\mathcal{T}$ into two subsets $\mathcal{T}_1^T$ and $\mathcal{T}_2^T$ such that: for every $\mathcal{D} \in \mathcal{T}_1^T$, every $(R_1, \ldots, R_n) \in \mathcal{D}^n$ yields transitive social $R$; and for every $\mathcal{D} \in \mathcal{T}_2^T$, there exists a $(R_1, \ldots, R_n) \in \mathcal{D}^n$ which yields intransitive social $R$. As MMD defined for an odd number of individuals yields transitive social $R$ for every $(R_1, \ldots, R_n) \in \mathcal{D}^n$, $\mathcal{D} = \{xPyPz, zPyPx, yPzIx, zIxPy\}$, it follows that if there exists an Inada-type necessary and sufficient condition for transitivity for the class of Pareto-transitive binary social decision rules then it must be the case that $\{xPyPz, zPyPx, yPzIx, zIxPy\} \in \mathcal{T}_1^T$. Like the MMD defined for an odd number of individuals, the strict majority rule defined for an odd number of individuals is also a Pareto-transitive binary social decision rule. Consider the strict majority rule defined for $S = \{x, y, z\}$; $n = 2k + 1$, $k \geq 5$. Let $(R_1, \ldots, R_n) \in \mathcal{D}^n$, $\mathcal{D} = \{xPyPz, zPyPx, yPzIx, zIxPy\}$, be such that: $n(xP_iyP_iz) = k \wedge n(zP_iyP_ix) = k - 1 \wedge n(y_iPzI_ix) = 1 \wedge n(z_iIxP_iy) = 1$. We obtain: $[n(xP_iy) = k + 1 \wedge n(yP_ix) = k \wedge n(yP_iz) = k + 1 \wedge n(zP_iy) = k \wedge n(xP_iz) = k \wedge n(zP_ix) = k - 1]$, which results in $[xPy \wedge yPz \wedge xIz]$, violating transitivity. Therefore, if there exists an Inada-type necessary and sufficient condition for transitivity for the class of Pareto-transitive binary social decision rules then it must be the case that $\{xPyPz, zPyPx, yPzIx, zIxPy\} \in \mathcal{T}_2^T$. As $\{xPyPz, zPyPx, yPzIx, zIxPy\}$ cannot belong to both $\mathcal{T}_1^T$ and $\mathcal{T}_2^T$, it follows that for the class of Pareto-transitive binary social decision rules there does not exist any condition whatsoever that is an Inada-type necessary and sufficient condition for transitivity.

# References

Blau, J.H. 1976. Neutrality, monotonicity and the right of veto: A comment. *Econometrica* 44: 603.

Guha, A.S. 1972. Neutrality, monotonicity and the right of veto. *Econometrica* 40: 821–826.

Jain, Satish K. 1977. Characterization of rationality conditions in terms of minimal decisive sets. Discussion Paper No. 7715. Indian Statistical Institute, New Delhi.

Jain, Satish K. 1987. Maximal conditions for transitivity under neutral and monotonic binary social decision rules. *The Economic Studies Quarterly* 38: 124–130.

Jain, Satish K. 1988. Characterization of monotonicity and neutrality for binary Paretian social decision rules. *Mathematical Social Sciences* 15: 307–312.

Salles, Maurice. 1975. A general possibility theorem for group decision rules with Pareto-transitivity. *Journal of Economic Theory* 11: 110–118.

Sen, Amartya K., and Prasanta K. Pattanaik. 1969. Necessary and sufficient conditions for rational choice under majority decision. *Journal of Economic Theory* 1: 178–202.

# Quasi-transitive Individual Preferences

So far in this volume it has been assumed that individual weak preference relations are orderings, i.e., are reflexive, connected and transitive. This chapter explores how some of the results that have been derived in Chaps. 3–10 change when individuals are assumed to have reflexive, connected and quasi-transitive weak preference relations over the set of social alternatives, rather than orderings. Specifically, when individual weak preference relations are reflexive, connected and quasi-transitive, conditions for quasi-transitivity under the method of majority decision, under the class of special majority rules and under the class of social decision rules which are simple games are derived; and the structure of neutral and monotonic binary social decision rules is analysed.

In the context of the method of majority decision, when individual weak preference relations are reflexive, connected and quasi-transitive, it turns out that the absence of Latin Squares involving intransitive relations is crucial for quasi-transitivity. Let $\mathscr{D}$ be a set of reflexive, connected and quasi-transitive relations on the set of social alternatives. If the number of individuals is at least five then every profile belonging to $\mathscr{D}^n$ yields quasi-transitive social weak preference relation under the method of majority decision iff $\mathscr{D}$ is such that it does not contain a Latin Square involving an intransitive relation and it satisfies the condition of Latin Square partial agreement. Thus the only difference from the case of individual preferences being orderings is the addition of the requirement that there be no Latin Square involving an intransitive relation. The conjunction of these two requirements of Latin Square partial agreement and absence of a Latin Square involving an intransitive relation has been termed as Latin Square partial agreement-Q. Similar results are established for the cases of three and four individuals. If the number of individuals is four then every profile belonging to $\mathscr{D}^4$ yields quasi-transitive social weak preference relation iff $\mathscr{D}$ satisfies the conjunction of weak extremal restriction and the absence of a Latin Square involving an intransitive relation; and if the number of individuals is three then

S. K. Jain, *Domain Conditions and Social Rationality*,
https://doi.org/10.1007/978-981-13-9672-4_11

every profile belonging to $\mathscr{D}^3$ yields quasi-transitive social preference relation iff $\mathscr{D}$ satisfies the conjunction of Latin Square linear ordering restriction and the absence of a Latin Square involving an intransitive relation. Conjunctions of weak extremal restriction and Latin Square linear ordering restriction with the condition of the absence of a Latin Square involving an intransitive relation will be called weak extremal restriction-Q and Latin Square linear ordering restriction-Q, respectively. When there are only two individuals and individuals have orderings over the set of alternatives then violation of quasi-transitivity is not possible. However, when individual weak preference relations are reflexive, connected and quasi-transitive, even when there are only two individuals violation of quasi-transitivity is possible under the method of majority decision. In the case of two individuals, every profile belonging to $\mathscr{D}^2$ yields quasi-transitive social weak preference relation iff $\mathscr{D}$ satisfies the Latin Square intransitive relation restriction; which requires that there be no Latin Square consisting of two intransitive relations or one intransitive relation and a linear ordering.

In Chap. 6 we saw that when individuals have orderings of social alternatives, Latin Square partial agreement is a sufficient condition for quasi-transitivity under every special majority rule. A similar result holds when individual weak preference relations are reflexive, connected and quasi-transitive. In this case, Latin Square partial agreement-Q turns out to be sufficient for quasi-transitivity under every special majority rule. Also, in the case of every special majority rule, if the number of individuals $n$ is sufficiently large then Latin Square partial agreement-Q completely characterizes the sets $\mathscr{D}$ which are such that every profile belonging to $\mathscr{D}^n$ yields quasi-transitive social weak preference relation.

In Chap. 9 the condition of Latin Square unique value restriction had turned out to be the crucial condition in the context of quasi-transitivity under the social decision rules which are simple games. Here, we redefine the condition but in such a way that in the case of orderings the condition reduces to the same as defined earlier. Under the redefinition, a set of reflexive, connected and quasi-transitive weak preference relations over a triple of alternatives satisfies Latin Square unique value restriction iff it is not the case that there is an alternative which is uniquely proper best in a weak preference relation belonging to the set, and the same alternative is uniquely proper medium in a weak preference relation belonging to the set, and the same alternative is uniquely proper worst in a weak preference relation belonging to the set, and these three weak preference relations form a Latin Square. It is shown in this chapter that (i) a simple game social decision rule yields quasi-transitive social weak preference relation for every profile belonging to $\mathscr{D}^n$ iff it is null or there is a unique minimal winning coalition and (ii) a non-null simple game social decision rule which is such that there are at least two minimal winning coalitions yields quasi-transitive social weak preference relation for every profile belonging to $\mathscr{D}^n$, $\mathscr{D} \subseteq \mathscr{Q}$, iff $\mathscr{D}$ satisfies the condition of Latin Square unique value restriction. Thus, with the redefinition of Latin Square unique value restriction, the quasi-transitivity results for social decision rules that are simple games become identical for the cases of individuals having orderings and individuals having reflexive, connected and quasi-transitive weak preference relations.

With respect to characterization of neutrality and monotonicity for binary social decision rules, and characterizations of rationality conditions for the subclass of neutral and monotonic binary social decision rules, with domain $\mathscr{Q}^n$ one gets much sharper results than with domain $\mathscr{T}^n$. In this chapter we show that (i) a binary social decision rule with domain $\mathscr{Q}^n$ is neutral and monotonic iff it satisfies weak Pareto quasi-transitivity, (ii) a neutral and monotonic binary social decision rule with domain $\mathscr{Q}^n$ yields transitive social weak preference relation for every profile belonging to $\mathscr{Q}^n$ iff it is null, (iii) a neutral and monotonic binary social decision rule with domain $\mathscr{Q}^n$ yields a quasi-transitive social weak preference relation for every profile belonging to $\mathscr{Q}^n$ iff it is null or it is an oligarchic simple game and (iv) a neutral and monotonic binary social decision rule with domain $\mathscr{Q}^n$ yields acyclic social weak preference relation for every profile belonging to $\mathscr{Q}^n$ iff there do not exist $V_1, ..., V_m$, $3 \le m \le \#S$, such that $V_j$ is $(N - A_j)$-decisive for some $A_j \subset N$, $V_j \cap A_j = \emptyset$, $j = 1, 2, ..., m$, and intersection of all $V_i, i \in \{1, ..., m\} - \{j\}$ has empty intersection with $(V_j \cup A_j)$, $j = 1, 2, ..., m$.

In Chap. 10 we saw that, when individuals have orderings over the set of social alternatives, value restriction (2) is a sufficient condition for quasi-transitivity under every neutral and monotonic binary social decision rule. This result continues to hold even when individuals have reflexive, connected and quasi-transitive weak preference relations.

The chapter is divided into seven sections. Section 11.1 contains some definitions involving quasi-transitive binary relations. Sections 11.2, 11.3 and 11.4 are concerned with the conditions for quasi-transitivity under the method of majority decision, under the special majority rules and under the simple game social decision rules respectively. Section 11.5 contains the characterization of monotonicity and neutrality for binary social decision rules with domain $\mathscr{Q}^n$, characterization of rationality conditions for the class of neutral and monotonic binary social decision rules with domain $\mathscr{Q}^n$ and related theorems. The last two sections, which constitute the appendix to the chapter, contain notes on the literature and a brief discussion on domain conditions for acyclicity.

## 11.1 Some Definitions Involving Quasi-transitive Binary Relations

Over a triple of alternatives $A = \{x, y, z\}$ there are 19 logically possible reflexive, connected and quasi-transitive binary relations. These are: 1. $xPyPz$, 2. $yPzPx$, 3. $zPxPy$, 4. $xPzPy$, 5. $zPyPx$, 6. $yPxPz$, 7. $xPyIz$, 8. $yPzIx$, 9. $zPxIy$, 10. $xIyPz$, 11. $yIzPx$, 12. $zIxPy$, 13. $xIyIz$, 14. $xPy, yIz, xIz$, 15. $yPz, zIx, yIx$, 16. $zPx, xIy, zIy$, 17. $xPz, zIy, xIy$, 18. $zPy, yIx, zIx$ and 19. $yPx, xIz, yIz$. The first 13 of these are transitive and the remaining six intransitive.

Let $A = \{x, y, z\} \subseteq S$ be a triple of alternatives. For any distinct $a, b, c \in A$, we define:
$Q[LS(abca)] = \{R \in \mathscr{Q}|A \mid (a$ is best and $b$ is proper medium and $c$ is worst in $R)$

$\vee$ ($b$ is best and $c$ is proper medium and $a$ is worst in $R$) $\vee$ ($c$ is best and $a$ is proper medium and $b$ is worst in $R$)}.

Thus we have:

$Q[LS(xyzx)] = Q[LS(yzxy)] = Q[LS(zxyz)] = \{xPyPz, xPyIz, xIyPz, yPzPx,$
$yPzIx,\ yIzPx, zPxPy, zPxIy, zIxPy, (xPy, yIz, xIz), (yPz, zIx, yIx),$
$(zPx, xIy, zIy)\}$
$Q[LS(xzyx)] = Q[LS(zyxz)] = Q[LS(yxzy)] = \{xPzPy, xPzIy,$
$xIzPy, zPyPx,\ zPyIx, zIyPx, yPxPz, yPxIz, yIxPz, (yPx, xIz, yIz),$
$(xPz, zIy, xIy), (zPy, yIx, zIx)\}$.

In previous chapters, in defining the notions of social decisions rules and their various subclasses, and in defining properties thereof, the domain has been taken to be $\mathscr{T}^n$ or a subset of it. In this chapter the same definitions will be used with domain $\mathscr{Q}^n$ or a subset of it.

## 11.2   The Method of Majority Decision

We generalize the three restrictions on preferences defined in the context of quasi-transitivity under the method of majority decision for sets of orderings to sets of binary relations that are reflexive, connected and quasi-transitive; and introduce one new restriction on preferences which turns out to be relevant for quasi-transitivity under the MMD.

Latin Square Partial Agreement-Q (LSPA-Q): $\mathscr{D} \subseteq \mathscr{Q}$ satisfies LSPA-Q over the triple $A \subseteq S$ iff $[[(\forall R^s, R^t, R^u \in \mathscr{D}|A)(R^s, R^t, R^u$ form a Latin Square over $A \rightarrow R^s, R^t, R^u$ are orderings over $A)] \wedge [(\forall$ distinct $a, b, c \in A)[(\exists R^s, R^t, R^u \in \mathscr{D}|A \cap \mathscr{T}|A)(R^s, R^t, R^u$ are concerned over $A \wedge aP^sbP^sc \wedge bR^tcR^ta \wedge cR^u aR^ub) \rightarrow (\forall R \in \mathscr{D}|A \cap T[LS(abca)])(aRc)]]]$. $\mathscr{D}$ satisfies LSPA-Q iff it satisfies LSPA-Q over every triple contained in $S$.

The satisfaction of Latin Square partial agreement-Q by $\mathscr{D}$ over the triple $A$ requires that there be no Latin Square contained in $\mathscr{D}|A$ involving an intransitive binary relation; and that $\mathscr{D}|A \cap \mathscr{T}|A$ satisfy Latin Square partial agreement.

Weak Extremal Restriction-Q (WER-Q): $\mathscr{D} \subseteq \mathscr{Q}$ satisfies WER-Q over the triple $A \subseteq S$ iff $[(\forall R^s, R^t, R^u \in \mathscr{D}|A)(R^s, R^t, R^u$ form a Latin Square over $A \rightarrow R^s, R^t, R^u$ are orderings over $A) \wedge \sim (\exists$ distinct $a, b, c \in A)(\exists R^s, R^t, R^u \in \mathscr{D}|A \cap \mathscr{T}|A)(aP^sbP^sc \wedge bR^tcP^ta \wedge cP^u aR^ub)]$. $\mathscr{D}$ satisfies WER-Q iff it satisfies WER-Q over every triple contained in $S$.

The satisfaction of weak Extremal Restriction-Q by $\mathscr{D}$ over the triple $A$ requires that there be no Latin Square contained in $\mathscr{D}|A$ involving an intransitive binary relation; and that $\mathscr{D}|A \cap \mathscr{T}|A$ satisfy weak extremal restriction.

Latin Square Linear Ordering Restriction-Q (LSLOR-Q): $\mathscr{D} \subseteq \mathscr{Q}$ satisfies LSLOR-Q over the triple $A \subseteq S$ iff $[(\forall R^s, R^t, R^u \in \mathscr{D}|A)(R^s, R^t, R^u$ form a Latin Square over $A \rightarrow R^s, R^t, R^u$ are orderings over $A) \wedge \sim (\exists$ distinct $a, b, c \in A)(\exists R^s, R^t, R^u \in \mathscr{D}|A \cap \mathscr{T}|A)(R^s, R^t, R^u$ are concerned over $A \wedge aP^sbP^sc \wedge bP^tcP^ta \wedge cR^u aR^ub)]$. $\mathscr{D}$ satisfies LSLOR-Q iff it satisfies LSLOR-Q over every triple contained in $S$.

The satisfaction of Latin Square linear ordering restriction-Q by $\mathcal{D}$ over the triple $A$ requires that there be no Latin Square contained in $\mathcal{D}|A$ involving an intransitive binary relation; and that $\mathcal{D}|A \cap \mathcal{T}|A$ satisfy Latin Square linear ordering restriction.

Latin Square Intransitive Relation Restriction-Q (LSIRR-Q): $\mathcal{D} \subseteq \mathcal{Q}$ satisfies LSIRR-Q over the triple $A \subseteq S$ iff $\sim (\exists R^s, R^t \in \mathcal{D}|A)[(R^s, R^t$ form a Latin Square over $A) \wedge (R^s$ is intransitive$) \wedge (R^t$ is intransitive $\vee R^t$ is a linear ordering$)]$. $\mathcal{D}$ satisfies LSIRR-Q iff it satisfies LSIRR-Q over every triple contained in $S$.

The satisfaction of Latin Square intransitive relation restriction-Q by $\mathcal{D}$ over the triple $A$ requires that there be no Latin Square contained in $\mathcal{D}|A$ consisting of two intransitive relations or consisting of one intransitive relation and one linear ordering.

*Remark 11.1* It is clear from the definitions of the above four conditions that LSPA-Q implies WER-Q, WER-Q implies LSLOR-Q and LSLOR-Q implies LSIRR-Q.   $\Diamond$

*Remark 11.2* If every preference relation in $\mathcal{D}$, $\mathcal{D} \subseteq \mathcal{Q}$, is an ordering then the definitions of Latin Square partial agreement-Q, weak extremal restriction-Q, and Latin Square linear ordering restriction-Q reduce to definitions of Latin Square partial agreement, weak extremal restriction and Latin Square linear ordering restriction, respectively.                                                                                         $\Diamond$

**Lemma 11.1** *Let* $f : \mathcal{Q}^n \mapsto \mathscr{C}$ *be the method of majority decision. Let* $(R_1, \ldots, R_n)$ $\in \mathcal{Q}^n$ *and* $R = f(R_1, \ldots, R_n)$. *Let* $A = \{x, y, z\} \subseteq S$ *be a triple of alternatives and suppose* $xPy$, $yPz$ *and* $zRx$. *Then we have:*
*(i)* $(\exists i, j, k \in N)[R_i|A, R_j|A, R_k|A$ *form Latin Square LS(xyzx) over* $A]$.
*(ii)* $(\exists i \in N)[xP_iyP_iz \vee yP_izP_ix \vee zP_ixP_iy \vee (xP_iy \wedge yI_iz \wedge xI_iz) \vee (yP_iz \wedge zI_ix \wedge yI_ix) \vee (zP_ix \wedge xI_iy \wedge zI_iy)]$.
*(iii)* $(\exists i, j, k \in N)[R_i|A, R_j|A, R_k|A \in Q[LS(xyzx)] \wedge (xP_iy \vee R_i|A$ *is intransi-* $tive) \wedge (yP_jz \vee R_j|A$ *is intransitive*$) \wedge (zP_kx \vee R_k|A$ *is intransitive*$)]$.

*Proof* $xPy \rightarrow N(xP_iy) > N(yP_ix)$                                                   (L11.1-1)

$yPz \rightarrow N(yP_iz) > N(zP_iy)$                                                              (L11.1-2)

$zRx \rightarrow N(zP_ix) \geq N(xP_iz)$                                                             (L11.1-3)

(L11.1-1), (L11.1-2) and (L11.1-3) imply, respectively:

$N(zP_ixP_iy) + N(zI_ixP_iy) + N(xP_izP_iy) + N(xP_izI_iy) + N(xP_iyP_iz) + N(x$
$P_iy \wedge yI_iz \wedge xI_iz) > N(zP_iyP_ix) + N(zI_iyP_ix) + N(yP_izP_ix) + N(yP_izI_ix)$
$+ N(yP_ixP_iz) + N(yP_ix \wedge xI_iz \wedge yI_iz)$                                            (L11.1-4)

$N(xP_iyP_iz) + N(xI_iyP_iz) + N(yP_ixP_iz) + N(yP_ixI_iz) + N(yP_izP_ix) + N(y$
$P_iz \wedge zI_ix \wedge yI_ix) > N(xP_izP_iy) + N(xI_izP_iy) + N(zP_ixP_iy) + N(zP_ixI_iy) +$
$N(zP_iyP_ix) + N(zP_iy \wedge yI_ix \wedge zI_ix)$                                            (L11.1-5)

$N(yP_izP_ix) + N(yI_izP_ix) + N(zP_iyP_ix) + N(zP_iyI_ix) + N(zP_ixP_iy) + N(z$
$P_ix \wedge xI_iy \wedge zI_iy) \geq N(yP_ixP_iz) + N(yI_ixP_iz) + N(xP_iyP_iz) + N(xP_iyI_iz) +$
$N(xP_izP_iy) + N(xP_iz \wedge zI_iy \wedge xI_iy)$      (L11.1-6)

By adding (L11.1-4) and (L11.1-5) we obtain:

$2N(xP_iyP_iz) + N(xP_iyI_iz) + N(xI_iyP_iz) + N(xP_iy \wedge yI_iz \wedge xI_iz) + N(yP_iz$
$\wedge zI_ix \wedge yI_ix) > 2N(zP_iyP_ix) + N(zP_iyI_ix) + N(zI_iyP_ix) + N(zP_iy \wedge yI_ix \wedge$
$zI_ix) + N(yP_ix \wedge xI_iz \wedge yI_iz)$      (L11.1-7)

(L11.1-7) $\rightarrow$ $N(xP_iyP_iz) + N(xP_iyI_iz) + N(xI_iyP_iz) + N(xP_iy \wedge yI_iz \wedge xI_iz) +$
$N(yP_iz \wedge zI_ix \wedge yI_ix) > 0$      (L11.1-8)

(L11.1-8) $\rightarrow$ $(\exists i \in N)[\text{in } R_i | A \text{ } x \text{ is best} \wedge y \text{ is proper medium } \wedge z \text{ is worst}]$
     (L11.1-9)

Analogously we can show that:

(L11.1-5) $\wedge$ (L11.1-6) $\rightarrow$ $(\exists j \in N)[\text{in } R_j | A \text{ } y \text{ is best} \wedge z \text{ is proper medium } \wedge$
$x \text{ is worst}]$      (L11.1-10)

(L11.1-6) $\wedge$ (L11.1-4) $\rightarrow$ $(\exists k \in N)[\text{in } R_k | A \text{ } z \text{ is best} \wedge x \text{ is proper medium } \wedge$
$y \text{ is worst}]$      (L11.1-11)

(L11.1-9), (L11.1-10) and (L11.1-11) imply:

$(\exists i, j, k \in N)[R_i | A, R_j | A, R_k | A \text{ form Latin Square } LS(xyzx) \text{ over } A]$
     (L11.1-12)

Adding (L11.1-4), (L11.1-5) and (L11.1-6), we obtain:

$N(xP_iyP_iz) + N(yP_izP_ix) + N(zP_ixP_iy) + N(xP_iy \wedge yI_iz \wedge xI_iz) + N(yP_iz \wedge$
$zI_ix \wedge yI_ix) + N(zP_ix \wedge xI_iy \wedge zI_iy) > N(zP_iyP_ix) + N(yP_ixP_iz) +$
$N(xP_izP_iy) + N(zP_iy \wedge yI_ix \wedge zI_ix) + N(yP_ix \wedge xI_iz \wedge yI_iz) + N(xP_iz$
$\wedge zI_iy \wedge xI_iy)$      (L11.1-13)

(L11.1-13) $\rightarrow$ $N(xP_iyP_iz) + N(yP_izP_ix) + N(zP_ixP_iy) + N(xP_iy \wedge yI_iz \wedge$
$xI_iz) + N(yP_iz \wedge zI_ix \wedge yI_ix) + N(zP_ix \wedge xI_iy \wedge zI_iy) > 0$    (L11.1-14)

(L11.1-14) $\rightarrow$ $(\exists i \in N)[xP_iyP_iz \vee yP_izP_ix \vee zP_ixP_iy \vee (xP_iy \wedge yI_iz \wedge xI_iz) \vee$
$(yP_iz \wedge zI_ix \wedge yI_ix) \vee (zP_ix \wedge xI_iy \wedge zI_iy)]$      (L11.1-15)

Adding (L11.1-4) and (L11.1-13) we obtain:

$2N(zP_ixP_iy) + N(zI_ixP_iy) + 2N(xP_iyP_iz) + N(xP_iyI_iz) + 2N(xP_iy \wedge yI_iz \wedge$
$xI_iz) + N(yP_iz \wedge zI_ix \wedge yI_ix) + N(zP_ix \wedge xI_iy \wedge zI_iy) > 2N(zP_iyP_ix) +$
$N(zI_iyP_ix) + 2N(yP_ixP_iz) + N(yP_ixI_iz) + 2N(yP_ix \wedge xI_iz \wedge yI_iz) + N(xP_iz \wedge$
$zI_iy \wedge xI_iy) + N(zP_iy \wedge yI_ix \wedge zI_ix)$      (L11.1-16)

(L11.1-16) $\rightarrow (\exists i \in N)[R_i|A \in Q[LS(xyzx)] \wedge (xP_iy \vee R_i|A$ is intransitive)]

(L11.1-17)

Analogously we can show that:

(L11.1-5) $\wedge$ (L11.1-13) $\rightarrow (\exists j \in N)[R_j|A \in Q[LS(xyzx)] \wedge (yP_jz \vee R_j|A$ is intransitive)]

(L11.1-18)

(L11.1-6) $\wedge$ (L11.1-13) $\rightarrow (\exists k \in N)[R_k|A \in Q[LS(xyzx)] \wedge (zP_kx \vee R_k|A$ is intransitive)]

(L11.1-19)

(L11.1-17)-(L11.1-19) imply:

$(\exists i, j, k \in N)[R_i|A, R_j|A, R_k|A \in Q[LS(xyzx)] \wedge (xP_iy \vee R_i|A$ is intransitive) $\wedge$ $(yP_jz \vee R_j|A$ is intransitive) $\wedge (zP_kx \vee R_k|A$ is intransitive)]

(L11.1-20)

(L11.1-12), (L11.1-15) and (L11.1-20) establish the lemma.                    □

**Theorem 11.1** *Let* $\#N = n \geq 5$. *Let* $\mathscr{D} \subseteq \mathscr{Q}$. *Then the method of majority decision* $f$ *yields quasi-transitive social* $R$, $R = f(R_1, \ldots, R_n)$, *for every* $(R_1, \ldots, R_n) \in$ $\mathscr{D}^n$ *iff* $\mathscr{D}$ *satisfies the condition of Latin Square partial agreement-Q.*

*Proof* Suppose $f$ does not yield quasi-transitive social $R$ for every $(R_1, \ldots, R_n) \in$ $\mathscr{D}^n, \mathscr{D} \subseteq \mathscr{Q}$. Then $(\exists (R_1, \ldots, R_n) \in \mathscr{D}^n)(\exists x, y, z \in S)(xPy \wedge yPz \wedge zRx)$. Denote $\{x, y, z\}$ by $A$. By Lemma 11.1 we obtain:

$(i)(\exists i, j, k \in N)[R_i|A, R_j|A, R_k|A$ form Latin Square $LS(xyzx)$ over $A]$.

$(ii)(\exists i \in N)[xP_iyP_iz \vee yP_izP_ix \vee zP_ixP_iy \vee (xP_iy \wedge yI_iz \wedge xI_iz) \vee (yP_iz \wedge zI_ix \wedge yI_ix) \vee (zP_ix \wedge xI_iy \wedge zI_iy)]$.

$(iii)(\exists i, j, k \in N)[R_i|A, R_j|A, R_k|A \in Q[LS(xyzx)] \wedge (xP_iy \vee R_i|A$ is intransitive) $\wedge (yP_jz \vee R_j|A$ is intransitive) $\wedge (zP_kx \vee R_k|A$ is intransitive)].

If $(\exists i \in N)[R_i|A \in Q[LS(xyzx)] \wedge R_i|A$ is intransitive] then in view of (i) there is a Latin Square involving an intransitive binary relation; which would imply violation of LSPA-Q.

(T11.1-1)

If $\sim (\exists i \in N)[R_i|A \in Q[LS(xyzx)] \wedge R_i|A$ is intransitive] then (i)-(iii) imply that: there exist $R_i|A, R_j|A, R_k|A \in T[LS(xyzx)]$, of which at least one is a linear ordering over $A$, which form a Latin Square; and furthermore $(\exists i, j, k \in N)[R_i|A, R_j|A, R_k|A \in T[LS(xyzx)] \wedge xP_iy \wedge yP_jz \wedge zP_kx]$. This implies that LSPA-Q is violated.

(T11.1-2)

(T11.1-1) and (T11.1-2) establish the sufficiency of LSPA-Q.

Suppose $\mathscr{D} \subseteq \mathscr{Q}$ violates LSPA-Q. Then there is some triple $A = \{x, y, z\}$ over which LSPA-Q is violated. Violation of LSPA-Q over the triple $A$ implies that there exists a Latin Square over $A$ involving an intransitive binary relation or LSPA is violated over $A$. If LSPA is violated over $A$ then by Theorem 3.4 there exists $(R_1, \ldots, R_n) \in \mathscr{D}^n$ for which $R = f(R_1, \ldots, R_n)$ violates quasi-transitivity. If there exists a Latin Square over $A$ involving an intransitive binary relation then we must have: ($\exists$ distinct $a, b, c \in \{x, y, z\}$)($\exists R^s, R^t \in \mathscr{D}$)$[[(aP^sb \wedge bI^sc \wedge aI^sc) \wedge$ $(bP^tc \wedge cI^ta \wedge bI^ta)] \vee [(aP^sb \wedge bI^sc \wedge aI^sc) \wedge (bR^tcR^ta \wedge R^t|A$ is concerned)]]. Consider any $(R_1, \ldots, R_n) \in \mathscr{D}^n$ such that $\#\{i \in N \mid R_i|A = R^s|A\} = n - 1, \#\{i \in N \mid R_i|A = R^t|A\} = 1$. Then in each case MMD yields an $R$ which violates quasi-transitivity.                                                                 $\square$

**Theorem 11.2** *Let $\#N = n = 4$. Let $\mathscr{D} \subseteq \mathscr{Q}$. Then the method of majority decision $f$ yields quasi-transitive social $R$, $R = f(R_1, \ldots, R_4)$, for every $(R_1, \ldots, R_4) \in \mathscr{D}^4$ iff $\mathscr{D}$ satisfies the condition of Weak Extremal Restriction-Q.*

*Proof* Suppose $f$ does not yield quasi-transitive social $R$ for every $(R_1, \ldots, R_4) \in \mathscr{D}^4$, $\mathscr{D} \subseteq \mathscr{Q}$. Then $(\exists (R_1, \ldots, R_4) \in \mathscr{D}^4)(\exists x, y, z \in S)(xPy \wedge yPz \wedge zRx)$. By Lemma 11.1 we obtain:

$(\exists i, j, k \in N)[R_i|A, R_j|A, R_k|A$ form Latin Square $LS(xyzx)$ over $A]$

(T11.2-1)

$(\exists i \in N)[xP_iyP_iz \vee yP_izP_ix \vee zP_ixP_iy \vee (xP_iy \wedge yI_iz \wedge xI_iz) \vee (yP_iz \wedge zI_i$ $x \wedge yI_ix) \vee (zP_ix \wedge xI_iy \wedge zI_iy)]$                            (T11.2-2)

If $(\exists i \in N) [R_i|A \in Q[LS(xyzx)] \wedge R_i|A$ is intransitive] then in view of (T11.2-1) there is a Latin Square involving an intransitive binary relation; which would imply violation of WER-Q.                            (T11.2-3)

If $\sim (\exists i \in N)[R_i|A \in Q[LS(xyzx)] \wedge R_i|A$ is intransitive] then (T11.2-1) and (T11.2-2) imply that: there exist $R_i|A, R_j|A, R_k|A \in T[LS(xyzx)]$ which form a Latin Square over $A$, with at least one of them being a linear ordering.    (T11.2-4)

(T11.2-4) $\to (\exists i, j, k \in N)(xP_iz \wedge yP_jx \wedge zP_ky)$                (T11.2-5)

$(\exists j \in N)(yP_jx) \wedge xPy \to N(xP_iy) \geq 2 \wedge N(xR_iy) = 3 \wedge N(yP_ix) = 1$

(T11.2-6)

(T11.2-6) $\to N(R_i$ concerned over $\{x, y, z\} \wedge yR_izR_ix)=1 \wedge N(zP_iyP_ix) = 0 \wedge$ $N(yP_ixP_iz)=0 \wedge N(yP_ix \wedge xI_iz \wedge yI_iz) = 0$                (T11.2-7)

$(\exists k \in N)(zP_ky) \wedge yPz \to N(yP_iz) \geq 2 \wedge N(yR_iz) = 3 \wedge N(zP_iy) = 1$

(T11.2-8)

(T11.2-8)$\to N(R_i$ concerned over $\{x, y, z\} \wedge zR_ixR_iy) = 1 \wedge N(xP_izP_iy) = 0 \wedge$ $N(zP_iy \wedge yI_ix \wedge zI_ix) = 0$                            (T11.2-9)

(T11.2-4) $\wedge$ (T11.2-7) $\wedge$ (T11.2-9) $\to N(xR_iy \wedge yR_iz \wedge xR_iz) = 2 \wedge N(R_i$ concerned over $\{x, y, z\} \wedge xR_iyR_iz) \geq 1 \wedge N(R_i$ concerned over $\{x, y, z\} \wedge$ $yR_izR_ix) = 1 \wedge N(R_i$ concerned over $\{x, y, z\} \wedge zR_ixR_iy) = 1$        (T11.2-10)

$zRx \wedge N(xP_iz) = 1 \wedge (\text{T11.2-6}) \wedge (\text{T11.2-8}) \wedge (\text{T11.2-10}) \rightarrow (\exists i, j, k \in N)[(xP_iy$
$P_iz \wedge yP_izP_ix \wedge zR_ixP_iy) \vee (xP_iyP_iz \wedge yP_izR_ix \wedge zP_ixP_iy)]$
$\rightarrow$ WER-Q is violated. (T11.2-11)

$zRx \wedge N(xP_iz) = 2 \wedge (\exists i \in N)(xP_iz \wedge zI_iy \wedge xI_iy) \wedge (\text{T11.2-6}) \wedge (\text{T11.2-8}) \wedge$
$(\text{T11.2-10}) \rightarrow (\exists i, j, k \in N)(xP_iyP_iz \wedge yP_izP_ix \wedge zP_ixP_iy)$
$\rightarrow$ WER-Q is violated. (T11.2-12)

$zRx \wedge N(xP_iz) = 2 \wedge \sim (\exists i \in N)(xP_iz \wedge zI_iy \wedge xI_iy) \wedge (\text{T11.2-6}) \wedge (\text{T11.2-8}) \wedge$
$(\text{T11.2-10}) \rightarrow (\exists i, j, k \in N)(xP_iyP_iz \wedge yR_jzP_jx \wedge zP_kxR_ky) \vee (\exists i, j, k, l \in$
$L)(xP_iyI_iz \wedge xI_jyP_jz \wedge yP_kzP_kx \wedge zP_lxP_ly)$
$\rightarrow$ WER-Q is violated. (T11.2-13)

(T11.2-3), (T11.2-11), (T11.2-12) and (T11.2-13) establish the sufficiency of WER.

Suppose $\mathscr{D} \subseteq \mathscr{Q}$ violates WER-Q. Then there is some triple $A = \{x, y, z\}$ over which WER-Q is violated. Violation of WER-Q over the triple $A$ implies that there exists a Latin Square over $A$ involving an intransitive binary relation or WER is violated over $A$. If WER is violated over $A$ then by Theorem 3.5 there exists $(R_1, \ldots, R_4) \in \mathscr{D}^4$ for which $R = f(R_1, \ldots, R_4)$ violates quasi-transitivity. If there exists a Latin Square over $A$ involving an intransitive binary relation then we must have: $(\exists$ distinct $a, b, c \in \{x, y, z\})(\exists R^s, R^t \in \mathscr{D})[[(aP^sb \wedge bI^sc \wedge aI^sc) \wedge (bP^tc \wedge cI^ta \wedge bI^ta)] \vee [(aP^sb \wedge bI^sc \wedge aI^sc) \wedge (bR^tcR^ta \wedge R^t|A$ is concerned$)]]$. Consider any $(R_1, \ldots, R_4) \in \mathscr{D}^4$ such that $\#\{i \in N \mid R_i|A = R^s|A\} = 3$ and $\#\{i \in N \mid R_i|A = R^t|A\} = 1$. Then in each case MMD yields an $R$ which violates quasi-transitivity. $\square$

**Theorem 11.3** *Let $\#N = n = 3$. Let $\mathscr{D} \subseteq \mathscr{Q}$. Then the method of majority decision $f$ yields quasi-transitive social $R$, $R = f(R_1, R_2, R_3)$, for every $(R_1, R_2, R_3) \in \mathscr{D}^3$ iff $\mathscr{D}$ satisfies the condition of Latin Square linear ordering restriction-Q.*

*Proof* Suppose $f$ does not yield quasi-transitive social $R$ for every $(R_1, R_2, R_3) \in \mathscr{D}^3$. Then $(\exists(R_1, R_2, R_3) \in \mathscr{D}^3)(\exists x, y, z \in S)(xPy \wedge yPz \wedge zRx)$. By Lemma 11.1 we have: $(\exists i, j, k \in N)[R_i|A, R_j|A, R_k|A$ form Latin Square $LS(xyzx)$ over $A]$
(T11.3-1)

$(\exists i \in N)[xP_iyP_iz \vee yP_izP_ix \vee zP_ixP_iy \vee (xP_iy \wedge yI_iz \wedge xI_iz) \vee (yP_iz \wedge zI_ix \wedge yI_ix) \vee (zP_ix \wedge xI_iy \wedge zI_iy)]$
(T11.3-2)

If $(\exists i \in N)$ $[R_i|A \in Q[LS(xyzx)] \wedge R_i|A$ is intransitive$]$ then in view of (T11.3-1) there is a Latin Square involving an intransitive binary relation; which would imply violation of LSLOR-Q.
(T11.3-3)

$\sim (\exists i \in N)$ $[R_i|A \in Q[LS(xyzx)] \wedge R_i|A$ is intransitive$] \wedge$ (T11.3-1)
$\wedge n = 3 \rightarrow (\forall i \in N)(R_i|A$ is transitive$)$
(T11.3-4)

(T11.3-4) $\wedge$ (T11.3-1) $\wedge$ (T11.3-2) $\rightarrow (\forall i \in N)(R_i|A$ is transitive$) \wedge (\exists i, j, k \in N)(R_i|A, R_j|A, R_k|A$ form Latin Square $LS(xyzx)$ over $A) \wedge (\exists i \in N)(xP_iyP_iz \vee yP_izP_ix \vee zP_ixP_iy)$
(T11.3-5)

Now, by proceeding as in Theorem 3.6 one can show that (T11.3-5) implies that LSLOR is violated; which coupled with (T11.3-3) establishes the sufficiency of LSLOR-Q for quasi-transitivity.

Suppose $\mathscr{D} \subseteq \mathscr{Q}$ violates LSLOR-Q. Then there is some triple $A = \{x, y, z\}$ over which LSLOR-Q is violated. Violation of LSLOR-Q over the triple $A$ implies that there exists a Latin Square over $A$ involving an intransitive binary relation or LSLOR is violated over $A$. If LSLOR is violated over $A$ then by Theorem 3.6 there exists $(R_1, R_2, R_3) \in \mathscr{D}^3$ for which $R = f(R_1, R_2, R_3)$ violates quasi-transitivity. If there exists a Latin Square over $A$ involving an intransitive binary relation then we must have: $(\exists$ distinct $a, b, c \in \{x, y, z\})(\exists R^s, R^t \in \mathscr{D})[[(aP^sb \wedge bI^sc \wedge aI^sc) \wedge (bP^tc \wedge cI^ta \wedge bI^ta)] \vee [(aP^sb \wedge bI^sc \wedge aI^sc) \wedge (bR^tcR^ta \wedge R^t|A$ is concerned)]]. Consider any $(R_1, R_2, R_3) \in \mathscr{D}^3$ such that $\#\{i \in N \mid R_i \mid A = R^s \mid A\} = 2$ and $\#\{i \in N \mid R_i \mid A = R^t \mid A\} = 1$. Then in each case MMD yields an $R$ which violates quasi-transitivity.                                                                                      $\square$

**Theorem 11.4** *Let $\#N = n = 2$. Let $\mathscr{D} \subseteq \mathscr{Q}$. Then the method of majority decision $f$ yields quasi-transitive social $R$, $R = f(R_1, R_2)$, for every $(R_1, R_2) \in \mathscr{D}^2$ iff D satisfies the condition of Latin Square intransitive relation restriction-Q.*

*Proof* Suppose $f$ does not yield quasi-transitive social $R$ for every $(R_1, R_2) \in \mathscr{D}^2$. Then $(\exists(R_1, R_2) \in \mathscr{D}^2)(\exists x, y, z \in S)(xPy \wedge yPz \wedge zRx)$.

$$xPy \rightarrow N(xP_iy) \geq 1 \wedge N(xR_iy) = 2 \tag{T11.4-1}$$

$$yPz \rightarrow N(yP_iz) \geq 1 \wedge N(yR_iz) = 2 \tag{T11.4-2}$$

$(\text{T11.4-1}) \wedge (\text{T11.4-2}) \wedge zRx \rightarrow (\exists i, j \in N)[[xP_iyP_iz \wedge (zP_jx \wedge xI_jy \wedge zI_jy)] \vee [(xP_iy \wedge yI_iz \wedge xI_iz) \wedge (yP_jz \wedge zI_jx \wedge yI_jx)]]$
$\rightarrow$ LSIRR-Q is violated.

Suppose $\mathscr{D} \subseteq \mathscr{Q}$ violates LSIRR-Q. Then there is some triple $A = \{x, y, z\}$ over which LSIRR-Q is violated. Violation of LSIRR-Q over the triple $A$ implies that: $(\exists$ distinct $a, b, c \in \{x, y, z\})(\exists R^s, R^t \in \mathscr{D})[[(aP^sb \wedge bI^sc \wedge aI^sc) \wedge (bP^tc \wedge cI^ta \wedge bI^ta)] \vee [(aP^sb \wedge bI^sc \wedge aI^sc) \wedge bP^tcP^ta]]$. Consider any $(R_1, R_2) \in \mathscr{D}^2$ such that $\#\{i \in N \mid R_i \mid A = R^s \mid A\} = \#\{i \in N \mid R_i \mid A = R^t \mid A\} = 1$. Then in each case MMD yields an $R$ which violates quasi-transitivity.                              $\square$

## 11.3  The Special Majority Rules

**Theorem 11.5** *Let $f : \mathscr{Q}^n \mapsto \mathscr{C}$ be a p-majority rule, $p \in (\frac{1}{2}, 1)$. Let $\mathscr{D} \subseteq \mathscr{Q}$. Then $f$ yields quasi-transitive social $R$, $R = f(R_1, \ldots, R_n)$, for every $(R_1, \ldots, R_n) \in \mathscr{D}^n$ if $\mathscr{D}$ satisfies the condition of Latin Square partial agreement-Q.*

*Proof* Suppose $f$ does not yield quasi-transitive social $R$ for every $(R_1, \ldots, R_n) \in \mathscr{D}^n$. Then $(\exists (R_1, \ldots, R_n) \in \mathscr{D}^n)(\exists x, y, z \in S)(xPy \wedge yPz \wedge zRx)$.

$$xPy \to n(xP_iy) > \tfrac{p}{1-p}n(yP_ix) \tag{T11.5-1}$$

$$yPz \to n(yP_iz) > \tfrac{p}{1-p}n(zP_iy) \tag{T11.5-2}$$

$$zRx \to n(zP_ix) \geq \tfrac{1-p}{p}n(xP_iz) \tag{T11.5-3}$$

(T11.5-1), (T11.5-2) and (T11.5-3) imply, respectively:

$$n(zP_ixP_iy)+n(zI_ixP_iy)+n(xP_izP_iy) + n(xP_izI_iy) + n(xP_iyP_iz) + n(xP_iy \wedge yI_iz \wedge xI_iz) > \tfrac{p}{1-p}[n(zP_iyP_ix) + n(zI_iyP_ix) + n(yP_izP_ix) + n(yP_izI_ix) + n(yP_ixP_iz) + n(yP_ix \wedge xI_iz \wedge yI_iz)] \tag{T11.5-4}$$

$$n(xP_iyP_iz) + n(xI_iyP_iz) + n(yP_ixP_iz) + n(yP_ixI_iz) + n(yP_izP_ix) + n(yP_iz \wedge zI_ix \wedge yI_ix) > \tfrac{p}{1-p}[n(xP_izP_iy) + n(xI_izP_iy) + n(zP_ixP_iy) + n(zP_ixI_iy) + n(zP_iyP_ix) + n(zP_iy \wedge yI_ix \wedge zI_ix)] \tag{T11.5-5}$$

$$n(yP_izP_ix) + n(yI_izP_ix) + n(zP_iyP_ix) + n(zP_iyI_ix) + n(zP_ixP_iy) + n(zP_ix \wedge xI_iy \wedge zI_iy) \geq \tfrac{1-p}{p}[n(yP_ixP_iz) + n(yI_ixP_iz) + n(xP_iyP_iz) + n(xP_iyI_iz) + n(xP_izP_iy) + n(xP_iz \wedge zI_iy \wedge xI_iy)] \tag{T11.5-6}$$

Multiplying (T11.5-6) by $\tfrac{p}{1-p}$ we obtain:

$$\tfrac{p}{1-p}[n(yP_izP_ix) + n(yI_izP_ix) + n(zP_iyP_ix) + n(zP_iyI_ix) + n(zP_ixP_iy) + n(zP_ix \wedge xI_iy \wedge zI_iy)] \geq n(yP_ixP_iz) + n(yI_ixP_iz) + n(xP_iyP_iz) + n(xP_iyI_iz) + n(xP_izP_iy) + n(xP_iz \wedge zI_iy \wedge xI_iy) \tag{T11.5-7}$$

Adding (T11.5-4) and (T11.5-5), adding (T11.5-5) and (T11.5-7), and adding (T11.5-7) and (T11.5-4), we obtain, respectively:

$$2n(xP_iyP_iz) + n(xP_iyI_iz) + n(xI_iyP_iz) + n(xP_iy \wedge yI_iz \wedge xI_iz) + n(yP_iz \wedge zI_ix \wedge yI_ix) > \tfrac{2p-1}{1-p}[n(yP_izP_ix) + n(zP_ixP_iy) + n(yP_izI_ix) + n(zI_ixP_iy) + n(xP_izP_iy) + n(yP_ixP_iz)] + \tfrac{2p}{1-p}n(zP_iyP_ix) + \tfrac{p}{1-p}[n(zP_iyI_ix) + n(zI_iyP_ix) + n(yP_ix \wedge xI_iz \wedge yI_iz) + n(zP_iy \wedge yI_ix \wedge zI_ix)] \tag{T11.5-8}$$

$$\tfrac{1}{1-p}n(yP_izP_ix) + n(yP_izI_ix) + \tfrac{p}{1-p}n(yI_izP_ix) + n(yP_iz \wedge zI_ix \wedge yI_ix) + \tfrac{p}{1-p}n(zP_ix \wedge xI_iy \wedge zI_iy) > \tfrac{1}{1-p}n(xP_izP_iy) + n(xP_izI_iy) + \tfrac{p}{1-p}n(xI_izP_iy) + n(xP_iz \wedge zI_iy \wedge xI_iy) + \tfrac{p}{1-p}n(zP_iy \wedge yI_ix \wedge zI_ix) \tag{T11.5-9}$$

$$\tfrac{1}{1-p}n(zP_ixP_iy) + \tfrac{p}{1-p}n(zP_ixI_iy) + n(zI_ixP_iy) + n(xP_iy \wedge yI_iz \wedge xI_iz) + \tfrac{p}{1-p}n(zP_ix \wedge xI_iy \wedge zI_iy) > \tfrac{1}{1-p}n(yP_ixP_iz) + \tfrac{p}{1-p}n(yP_ixI_iz) + n(yI_ixP_iz) + n(xP_iz \wedge zI_iy \wedge xI_iy) + \tfrac{p}{1-p}n(yP_ix \wedge xI_iz \wedge yI_iz) \tag{T11.5-10}$$

We have $\frac{2p-1}{1-p} > 0$ as $\frac{1}{2} < p < 1$. Therefore, (T11.5-8)–(T11.5-10) imply, respectively:

$2n(xP_iyP_iz) + n(xP_iyI_iz) + n(xI_iyP_iz) + n(xP_iy \wedge yI_iz \wedge xI_iz) + n(yP_iz \wedge$

$zI_ix \wedge yI_ix) > 0$ \hfill (T11.5-11)

$\frac{1}{1-p}n(yP_izP_ix) + n(yP_izI_ix) + \frac{p}{1-p}n(yI_izP_ix) + n(yP_iz \wedge zI_ix \wedge yI_ix) +$

$\frac{p}{1-p}n(zP_ix \wedge xI_iy \wedge zI_iy) > 0$ \hfill (T11.5-12)

$\frac{1}{1-p}n(zP_ixP_iy) + \frac{p}{1-p}n(zP_ixI_iy) + n(zI_ixP_iy) + n(xP_iy \wedge yI_iz \wedge xI_iz) +$

$\frac{p}{1-p}n(zP_ix \wedge xI_iy \wedge zI_iy) > 0$ \hfill (T11.5-13)

(T11.5-11)–(T11.5-13) imply, respectively:

$n(xP_iyP_iz) + n(xP_iyI_iz) + n(xI_iyP_iz) + n(xP_iy \wedge yI_iz \wedge xI_iz) + n(yP_iz \wedge zI_i$

$x \wedge yI_ix) > 0$ \hfill (T11.5-14)

$n(yP_izP_ix) + n(yP_izI_ix) + n(yI_izP_ix) + n(yP_iz \wedge zI_ix \wedge yI_ix) + n(zP_ix \wedge xI_i$

$y \wedge zI_iy) > 0$ \hfill (T11.5-15)

$n(zP_ixP_iy) + n(zP_ixI_iy) + n(zI_ixP_iy) + n(xP_iy \wedge yI_iz \wedge xI_iz) + n(zP_ix \wedge xI_i$

$y \wedge zI_iy) > 0$ \hfill (T11.5-16)

(T11.5-14) $\rightarrow$ $(\exists i \in N)[(R_i$ is concerned over $\{x, y, z\} \wedge xR_iyR_iz) \vee (xP_iy \wedge$

$yI_iz \wedge xI_iz) \vee (yP_iz \wedge zI_ix \wedge yI_ix)]$

$\rightarrow (\exists i \in N)[x$ is best in $R_i|\{x, y, z\} \wedge y$ is proper medium in $R_i|\{x, y, z\} \wedge z$ is

worst in $R_i|\{x, y, z\}]$ \hfill (T11.5-17)

(T11.5-15) $\rightarrow$ $(\exists j \in N)[(R_j$ is concerned over $\{x, y, z\} \wedge yR_jzR_jx) \vee (yP_jz \wedge$

$zI_jx \wedge yI_jx) \vee (zP_jx \wedge xI_jy \wedge zI_jy)]$

$\rightarrow (\exists j \in N)[y$ is best in $R_j|\{x, y, z\} \wedge z$ is proper medium in $R_j|\{x, y, z\} \wedge x$ is

worst in $R_j|\{x, y, z\}]$ \hfill (T11.5-18)

(T11.5-16) $\rightarrow$ $(\exists k \in N)[(R_k$ is concerned over $\{x, y, z\} \wedge zR_kxR_ky) \vee (xP_ky \wedge$

$yI_kz \wedge xI_kz) \vee (zP_kx \wedge xI_ky \wedge zI_ky)]$

$\rightarrow (\exists k \in N)[z$ is best in $R_k|\{x, y, z\} \wedge x$ is proper medium in $R_k|\{x, y, z\} \wedge y$ is

worst in $R_k|\{x, y, z\}]$ \hfill (T11.5-19)

(T11.5-17)–(T11.5-19) imply that $R_i|\{x, y, z\}, R_j|\{x, y, z\}, R_k|\{x, y, z\}$ form $LS$

$(xyzx)$ and consequently it follows that $\mathscr{D} \mid \{x, y, z\}$ contains $LS(xyzx)$

\hfill (T11.5-20)

Adding (T11.5-4), (T11.5-5) and (T11.5-7) we obtain:

$n(xP_iyP_iz) + n(yP_izP_ix) + n(zP_ixP_iy) + n(xP_iy \wedge yI_iz \wedge xI_iz) + n(yP_iz \wedge$

$zI_ix \wedge yI_ix) + \frac{p}{1-p}n(zP_ix \wedge xI_iy \wedge zI_iy) > \frac{p}{1-p}[n(xP_izP_iy) + n(zP_iyP_ix) +$

$n(yP_ixP_iz)] + \frac{2p-1}{1-p}[n(xI_izP_iy) + n(yP_ixI_iz)] + n(xP_iz \wedge zI_iy \wedge xI_iy) +$

$\frac{p}{1-p}[n(zP_iy \wedge yI_ix \wedge zI_ix) + n(yP_ix \wedge xI_iz \wedge yI_iz)]$ 　　　　(T11.5-21)

(T11.5-21) $\rightarrow (\exists l \in N)[xP_lyP_lz \vee yP_lzP_lx \vee zP_lxP_ly \vee (xP_ly \wedge yI_lz \wedge xI_lz) \vee$

$(yP_lz \wedge zI_lx \wedge yI_lx) \vee (zP_lx \wedge xI_ly \wedge zI_ly)]$ 　　　　(T11.5-22)

(T11.5-22) $\rightarrow (\exists l \in N)[R_l \mid \{x, y, z\} \in Q[LS(xyzx)] \wedge R_l \mid \{x, y, z\}$ is linear or intransitive] 　　　　(T11.5-23)

Adding (T11.5-21) to (T11.5-4), (T11.5-5) and (T11.5-7) we obtain, respectively:

$2n(xP_iyP_iz) + 2n(zP_ixP_iy) + n(xP_iyI_iz) + \frac{2-3p}{1-p}n(zI_ixP_iy) + 2n(xP_iy \wedge yI_iz \wedge xI_iz) + n(yP_iz \wedge zI_ix \wedge yI_ix) + \frac{p}{1-p}n(zP_ix \wedge xI_iy \wedge zI_iy) > \frac{2p}{1-p}[n(zP_iyP_ix) + n(yP_ixP_iz)] + \frac{3p-1}{1-p}n(yP_ixI_iz) + \frac{p}{1-p}n(zI_iyP_ix) + \frac{2p-1}{1-p}[n(xP_izP_iy) + n(yP_izP_ix)] + n(xP_iz \wedge zI_iy \wedge xI_iy) + \frac{p}{1-p}n(zP_iy \wedge yI_ix \wedge zI_ix) + \frac{2p}{1-p}n(yP_ix \wedge xI_iz \wedge yI_iz)$ 　　　　(T11.5-24)

$2n(xP_iyP_iz) + 2n(yP_izP_ix) + n(xI_iyP_iz) + \frac{2-3p}{1-p}n(yP_izI_ix) + n(xP_iy \wedge yI_iz \wedge xI_iz) + 2n(yP_iz \wedge zI_ix \wedge yI_ix) + \frac{p}{1-p}n(zP_ix \wedge xI_iy \wedge zI_iy) > \frac{2p}{1-p}[n(xP_izP_iy) + n(zP_iyP_ix)] + \frac{p}{1-p}n(zP_iyI_ix) + \frac{3p-1}{1-p}n(xI_izP_iy) + n(xP_iz \wedge zI_iy \wedge xI_iy) + \frac{2p}{1-p}n(zP_iy \wedge yI_ix \wedge zI_ix) + \frac{p}{1-p}n(yP_ix \wedge xI_iz \wedge yI_iz) + \frac{2p-1}{1-p}[n(zP_ixP_iy) + n(yP_ixP_iz)]$ 　　　　(T11.5-25)

$\frac{1}{1-p}[n(yP_izP_ix) + n(zP_ixP_iy)] + \frac{p}{1-p}[n(yI_izP_ix) + n(zP_ixI_iy)] + n(xP_iy \wedge yI_iz \wedge xI_iz) + n(yP_iz \wedge zI_ix \wedge yI_ix) + \frac{2p}{1-p}n(zP_ix \wedge xI_iy \wedge zI_iy) > \frac{1}{1-p}[n(xP_izP_iy) + n(yP_ixP_iz)] + n(xP_izI_iy) + n(yI_ixP_iz) + \frac{2p-1}{1-p}[n(xI_izP_iy) + n(yP_ixI_iz)] + 2n(xP_iz \wedge zI_iy \wedge xI_iy) + \frac{p}{1-p}[n(zP_iy \wedge yI_ix \wedge zI_ix) + n(yP_ix \wedge xI_iz \wedge yI_iz)]$ 　　　　(T11.5-26)

(T11.5-24)–(T11.5-26) imply, respectively:

$n(xP_iyP_iz) + n(zP_ixP_iy) + n(xP_iyI_iz) + n(zI_ixP_iy) + n(xP_iy \wedge yI_iz \wedge xI_iz) + n(yP_iz \wedge zI_ix \wedge yI_ix) + n(zP_ix \wedge xI_iy \wedge zI_iy) > 0$ 　　　　(T11.5-27)

$n(xP_iyP_iz) + n(yP_izP_ix) + n(xI_iyP_iz) + n(yP_izI_ix) + n(xP_iy \wedge yI_iz \wedge xI_iz) + n(yP_iz \wedge zI_ix \wedge yI_ix) + n(zP_ix \wedge xI_iy \wedge zI_iy) > 0$ 　　　　(T11.5-28)

$n(yP_izP_ix) + n(zP_ixP_iy) + n(yI_izP_ix) + n(zP_ixI_iy) + n(xP_iy \wedge yI_iz \wedge xI_iz) + n(yP_iz \wedge zI_ix \wedge yI_ix) + n(zP_ix \wedge xI_iy \wedge zI_iy) > 0$ 　　　　(T11.5-29)

(T11.5-27)–(T11.5-29) imply, respectively:

$(\exists i \in N)[R_i | \{x, y, z\} \in Q[LS(xyzx) \wedge (xP_iy \vee R_i | \{x, y, z\}$ is intransitive)]

　　　　(T11.5-30)

$(\exists j \in N)[R_j | \{x, y, z\} \in Q[LS(xyzx) \wedge (yP_iz \vee R_j | \{x, y, z\}$ is intransitive)]

(T11.5-31)

$(\exists k \in N)[R_k|\{x, y, z\} \in Q[LS(xyzx) \wedge (z P_i x \vee R_k|\{x, y, z\} \text{ is intransitive})]$

(T11.5-32)

If $(\exists l \in N)[R_l|\{x, y, z\} \in Q[LS(xyzx)] \wedge R_l|\{x, y, z\}$ is intransitive] then in view of (T11.5-20) there is a Latin Square involving an intransitive binary relation; which would imply violation of LSPA-Q.                                            (T11.5-33)

If $\sim (\exists l \in N)[R_l|\{x, y, z\} \in Q[LS(xyzx)] \wedge R_l|\{x, y, z\}$ is intransitive] then (T11.5-20), (T11.5-22), (T11.5-30), (T11.5-31) and (T11.5-32) imply that: there exist $R_i|\{x, y, z\}, R_j|\{x, y, z\}, R_k|\{x, y, z\} \in T[LS(xyzx)]$, of which at least one is a linear ordering over $\{x, y, z\}$, which form a Latin Square; and furthermore $(\exists i', j', k' \in N)[R_{i'}|\{x, y, z\}, R_{j'}|\{x, y, z\}, R_{k'}|\{x, y, z\} \in T[LS(xyzx)] \wedge x P_{i'} y \wedge y P_{j'} z \wedge z P_{k'} x]$. This implies that LSPA-Q is violated.                               (T11.5-34)

(T11.5-33) and (T11.5-34) establish the theorem.                                      □

**Theorem 11.6** *Let $f : \mathcal{Q}^n \mapsto \mathcal{C}$ be a p-majority rule, $p \in (\frac{1}{2}, 1)$. Let $n$ be a positive integer such that $n > max\{\frac{m+2}{(1-p)^2}, \frac{1}{2p-1}\}$, where $m$ is a positive integer greater than $\frac{p}{1-p}$. Let $\mathcal{D} \subseteq \mathcal{Q}$. Then $f$ yields quasi-transitive social $R$, $R = f(R_1, \ldots, R_n)$, for every $(R_1, \ldots, R_n) \in \mathcal{D}^n$ iff $\mathcal{D}$ satisfies the condition of Latin Square partial agreement-Q.*

*Proof* Let $f : \mathcal{Q}^n \mapsto \mathcal{C}$ be a $p$-majority rule, $p \in (\frac{1}{2}, 1)$; and let $\mathcal{D} \subseteq \mathcal{Q}$. If $\mathcal{D}$ satisfies LSPA-Q then it follows from Theorem 11.5 that $f$ yields quasi-transitive social $R$ for every $(R_1, \ldots, R_n) \in \mathcal{D}^n$.

Suppose $\mathcal{D} \subseteq \mathcal{Q}$ violates LSPA-Q. Then there is some triple $A = \{x, y, z\}$ over which LSPA-Q is violated. Violation of LSPA-Q over the triple $A$ implies that there exists a Latin Square over $A$ involving an intransitive binary relation or LSPA is violated over $A$. If LSPA is violated over $A$ then by Theorem 6.6 there exists $(R_1, \ldots, R_n) \in \mathcal{D}^n$ for which $R = f(R_1, \ldots, R_n)$ violates quasi-transitivity. If there exists a Latin Square over $A$ involving an intransitive binary relation then we must have: $(\exists$ distinct $a, b, c \in \{x, y, z\})(\exists R^s, R^t \in \mathcal{D})[[(a P^s b \wedge b I^s c \wedge a I^s c) \wedge (b P^t c \wedge c I^t a \wedge b I^t a)] \vee [(a P^s b \wedge b I^s c \wedge a I^s c) \wedge (b R^t c R^t a \wedge R^t | A$ is concerned)]]. Consider any $(R_1, \ldots, R_n) \in \mathcal{D}^n$ such that $\#\{i \in N \mid R_i|A = R^s|A\} = n - 1, \#\{i \in N \mid R_i|A = R^t|A\} = 1$. Then in each case $p$-majority rule yields an $R$ which violates quasi-transitivity.                                      □

## 11.4 Social Decision Rules Which Are Simple Games

In Theorem 9.1 it was established that a social decision rule with domain $\mathscr{T}^n$ is a simple game iff it satisfies the conditions of (i) independence of irrelevant alternatives, (ii) neutrality and (iii) monotonicity, and (iv) its structure is such that a coalition is blocking iff it is strictly blocking. It can be easily checked that this characterization is valid with domain $\mathscr{Q}^n$ as well. The proof of Theorem 9.1, with the replacement of $\mathscr{T}^n$ by $\mathscr{Q}^n$, constitutes a proof of the corresponding theorem stated below.

**Theorem 11.7** *A social decision rule $f : \mathscr{Q}^n \mapsto \mathscr{C}$ is a simple game iff it satisfies the conditions of (i) independence of irrelevant alternatives, (ii) neutrality and (iii) monotonicity, and (iv) its structure is such that a coalition is blocking iff it is strictly blocking.*

Proposition 9.3 stating that a simple game SDR with domain $\mathscr{T}^n$ yields quasi-transitive social preferences for every profile of individual preferences iff it null or oligarchic holds with domain $\mathscr{Q}^n$ as well. The proof of Proposition 9.3, with the replacement of $\mathscr{T}^n$ by $\mathscr{Q}^n$, constitutes a proof of the corresponding proposition stated below.

**Proposition 11.1** *Let social decision rule $f : \mathscr{Q}^n \mapsto \mathscr{C}$ be a simple game. Then, $f$ yields quasi-transitive social weak preference relation for every $(R_1, \ldots, R_n) \in \mathscr{Q}^n$ iff it is null or there is a unique minimal winning coalition.*

In Chap. 4 the condition of Latin Square unique value restriction was defined for sets of orderings. The condition can be generalized for sets of binary weak preference relations that are reflexive, connected and quasi-transitive.

Latin Square Unique Value Restriction (LSUVR): Let $\mathscr{D} \subseteq \mathscr{Q}$ be a set of reflexive, connected and quasi-transitive weak preference relations of $S$. Let $A = \{x, y, z\} \subseteq S$ be a triple of alternatives. $\mathscr{D}$ satisfies LSUVR over the triple $A$ iff there do not exist distinct $a, b, c \in A$ and $R^s, R^t, R^u \in \mathscr{D}|A \cap Q[LS(abca)]$ such that (i) alternative $b$ is uniquely proper medium in $R^s$, uniquely proper best in $R^t$, uniquely proper worst in $R^u$ and (ii) $R^s, R^t, R^u$ form $LS(abca)$. More formally, $\mathscr{D} \subseteq \mathscr{Q}$ satisfies LSUVR over the triple $A$ iff $\sim [(\exists \text{ distinct } a, b, c \in A)(\exists R^s, R^t, R^u \in \mathscr{D}|A \cap Q[LS(abca)])[aP^sbP^sc \wedge [bP^tcR^ta \vee (bP^tc \wedge cI^ta \wedge bI^ta)] \wedge [cR^uaP^ub \vee (aP^ub \wedge bI^uc \wedge aI^uc)]]]$. $\mathscr{D}$ satisfies LSUVR iff it satisfies LSUVR over every triple of alternatives contained in $S$.

If $\mathscr{D} \subseteq \mathscr{T}$ then the above definition reduces to that of LSUVR as given in Chap. 4.

**Theorem 11.8** *Let social decision rule $f : \mathscr{Q}^n \mapsto \mathscr{C}$ be a non-null simple game such that there are at least two minimal winning coalitions. Let $\mathscr{D} \subseteq \mathscr{Q}$. Then $f$ yields quasi-transitive social $R$, $R = f(R_1, \ldots, R_n)$, for every $(R_1, \ldots, R_n) \in \mathscr{D}^n$ iff $\mathscr{D}$ satisfies the condition of Latin Square unique value restriction.*

*Proof* Let $f : \mathscr{Q}^n \mapsto \mathscr{C}$ be a non-null simple game such that there are at least two minimal winning coalitions; and let $\mathscr{D} \subseteq \mathscr{Q}$. Suppose $f$ does not yield quasi-transitive social $R$ for every $(R_1, \ldots, R_n) \in \mathscr{D}^n$. Then:

$$(\exists (R_1, \ldots, R_n) \in \mathscr{D}^n)(\exists x, y, z \in S)(x P y \wedge y P z \wedge z R x). \tag{T11.8-1}$$

$$x P y \rightarrow (\exists V_1 \in W)(\forall i \in V_1)(x P_i y), \tag{T11.8-2}$$

by the definition of a simple game

$$y P z \rightarrow (\exists V_2 \in W)(\forall i \in V_2)(y P_i z), \tag{T11.8-3}$$

by the definition of a simple game.

(T11.8-2) $\wedge$ (T11.8-3) $\rightarrow (\exists i \in V_1 \cap V_2)(x P_i y \wedge y P_i z)$, as $V_1 \cap V_2 \neq \emptyset$, by Remark 4.1[1]

$\rightarrow (\exists i \in V_1 \cap V_2)(x P_i y P_i z)$, as individual weak preference relations are reflexive, connected and quasi-transitive.

$z R x \rightarrow (\exists j \in V_2)(y P_j z \wedge z R_j x)$, as $(\forall i \in V_2)(x P_i z)$ would imply $x P z$

$\rightarrow (\exists j \in V_2)(y P_j z R_j x \vee (y P_j z \wedge z I_j x \wedge y I_j x))$, as individual weak preference relations are reflexive, connected and quasi-transitive

$z R x \rightarrow (\exists k \in V_1)(z R_k x \wedge x P_k y)$, as $(\forall i \in V_1)(x P_i z)$ would imply $x P z$

$\rightarrow (\exists k \in V_1)(z R_k x P_k y \vee (z I_k x \wedge x P_k y \wedge z I_k y))$, as individual weak preference relations are reflexive, connected and quasi-transitive.

(T11.8-1) implies that $x, y, z$ are distinct alternatives. $R_i|\{x, y, z\}$, $R_j|\{x, y, z\}$, $R_k|\{x, y, z\}$ belong to $Q[LS(xyzx)]$, and form $LS(xyzx)$. In the triple $\{x, y, z\}$, $y$ is uniquely proper medium according to $x P_i y P_i z$; is uniquely proper best according to both $y P_j z R_j x$ and $(y P_j z \wedge z I_j x \wedge y I_j x)$ and is uniquely proper worst according to both $z R_k x P_k y$ and $(z I_k x \wedge x P_k y \wedge z I_k y)$. Therefore LSUVR is violated over the triple $\{x, y, z\}$. Thus $\mathscr{D}$ violates LSUVR. We have shown that violation of quasi-transitivity by $R = f(R_1, \ldots, R_n)$, $(R_1, \ldots, R_n) \in \mathscr{D}^n$, implies violation of LSUVR by $\mathscr{D}$, i.e., if $\mathscr{D}$ satisfies LSUVR then every $(R_1, \ldots, R_n) \in \mathscr{D}^n$ yields quasi-transitive social $R$.

Suppose $\mathscr{D} \subseteq \mathscr{Q}$ violates LSUVR. Then there exist distinct $x, y, z \in S$ such that $\mathscr{D}$ violates LSUVR over the triple $\{x, y, z\}$. Violation of LSUVR by $\mathscr{D}$ over $\{x, y, z\}$ implies ($\exists$ distinct $a, b, c \in \{x, y, z\}$)($\exists R^s, R^t, R^u \in \mathscr{D}$)$[a P^s b P^s c \wedge [b P^t c R^t a \vee (b P^t c \wedge c I^t a \wedge b I^t a)] \wedge [c R^u a P^u b \vee (c I^u a \wedge a P^u b \wedge c I^u b)]]$. It is given that $f$ is non-null and that there are at least two minimal winning coalitions. Let $V_1$ and $V_2$ be distinct minimal winning coalitions. $V_1 \cap V_2 \neq \emptyset$ in view of Remark 4.1; and $[V_1 - V_2 \neq \emptyset \wedge N - V_1 \neq \emptyset]$ as $V_1$ and $V_2$ are distinct minimal winning coalitions. From the fact that $V_1$ and $V_2$ are distinct minimal winning coalitions we conclude that $V_1 \cap V_2 \notin W$. Now consider any $(R_1, \ldots, R_n) \in \mathscr{D}^n$ such that the restriction of $(R_1, \ldots, R_n)$ to $\{x, y, z\} = \{a, b, c\} = A$, $(R_1|\{x, y, z\}, \ldots, R_n|\{x, y, z\})$, is given by: $[(\forall i \in V_1 \cap V_2)(R_i|A = R^s|A) \wedge (\forall i \in V_1 - V_2)(R_i|A = R^t|A) \wedge (\forall i \in N - V_1)(R_i|A = R^u|A)]$. $[V_2 \in W \wedge (\forall i \in V_2)(a P_i b) \rightarrow a P b]$ and $[V_1 \in W \wedge (\forall i \in$

---

[1] Remark 4.1 was made in the context of domain $\mathscr{T}^n$. It is, however, clear that the assertion made there holds for domain $\mathscr{Q}^n$ as well.

$V_1)(bP_ic) \rightarrow bPc]$. $\{i \in N \mid aP_ic\} = V_1 \cap V_2$ and $V_1 \cap V_2 \notin W$ imply $cRa$ as $f$ is a simple game. $(aPb \wedge bPc \wedge cRa)$ implies that $R$ violates quasi-transitivity. We have shown that if $\mathscr{D} \subseteq \mathscr{Q}$ violates LSUVR then there exists $(R_1, \ldots, R_n) \in \mathscr{D}^n$ such that $R = f(R_1, \ldots, R_n)$ violates quasi-transitivity, i.e., if $f$ yields quasi-transitive $R$ for every $(R_1, \ldots, R_n) \in \mathscr{D}^n$ then $\mathscr{D}$ must satisfy LSUVR. This establishes the theorem. □

## 11.5 Neutral and Monotonic Social Decision Rules

### 11.5.1 Characterization of Neutral and Monotonic Binary Social Decision Rules

**Lemma 11.2** *If a social decision rule $f : \mathscr{Q}^n \mapsto \mathscr{C}$ satisfies independence of irrelevant alternatives and weak Pareto quasi-transitivity, then it satisfies the condition of Pareto-indifference.*

*Proof* Suppose $f : \mathscr{Q}^n \mapsto \mathscr{C}$ satisfies condition I and WPQT, but violates the condition of Pareto-indifference. Then,

$(\exists (R_1, \ldots, R_n) \in \mathscr{Q}^n)(\exists x, y \in S)[(\forall i \in N)(xI_iy) \wedge xPy]$

By condition I we conclude,

$(\forall (R_1, \ldots, R_n) \in \mathscr{Q}^n)[(\forall i \in N)(xI_iy) \rightarrow xPy]$  (L11.2-1)

Let $z$ be an alternative distinct from $x$ and $y$, and consider the configuration $(\forall i \in N)(xI_iy \wedge yP_iz \wedge xI_iz)$. We conclude by (L11.2-1) and WPQT $xPz$ and by Condition I,

$(\forall (R_1, \ldots, R_n) \in \mathscr{Q}^n)[(\forall i \in N)(xI_iz) \rightarrow xPz]$  (L11.2-2)

Next we consider the configuration $(\forall i \in N)(zP_ix \wedge xI_iy \wedge zI_iy)$. We conclude by (L11.2-1) and WPQT $zPy$ and by Condition I,

$(\forall (R_1, \ldots, R_n) \in \mathscr{Q}^n)[(\forall i \in N)(zI_iy) \rightarrow zPy]$  (L11.2-3)

Finally consider the configuration $(\forall i \in N)(zI_iy \wedge yP_ix \wedge zI_ix)$. (L11.2-3) and WPQT imply $zPx$; and $zPx$ and Condition I imply,

$(\forall (R_1, \ldots, R_n) \in \mathscr{Q}^n)[(\forall i \in N)(xI_iz) \rightarrow zPx]$.  (L11.2-4)

As (L11.2-2) and (L11.2-4) contradict each other, the lemma is established. □

**Lemma 11.3** *Let social decision rule $f : \mathscr{Q}^n \mapsto \mathscr{C}$ satisfy independence of irrelevant alternatives and weak Pareto quasi-transitivity. Then: $(\forall (R_1, \ldots, R_n) \in \mathscr{Q}^n)$ $(\forall x, y \in S)[(\forall i \in N)(xR_iy) \wedge (\exists i \in N)(xP_iy) \rightarrow xRy]$.*

*Proof* Let $f : \mathscr{Q}^n \mapsto \mathscr{C}$ satisfy condition I and WPQT. Suppose $(\exists (R_1, \ldots, R_n) \in \mathscr{Q}^n)(\exists x, y \in S)(\exists$ nonempty $N_1 \subseteq N)[(\forall i \in N_1)(xP_iy) \wedge (\forall i \in N-N_1)(xI_iy) \wedge yPx]$.

Let $z$ be an alternative distinct from $x$ and $y$, and consider the following configuration of individual preferences:

$(\forall i \in N_1)(x P_i y \wedge x P_i z \wedge y I_i z)$

$(\forall i \in N - N_1)(x I_i y \wedge x P_i z \wedge y I_i z)$

$y P x$ and $(\forall i \in N)(x P_i z)$ imply $y P z$ by WPQT. $(\forall i \in N)(y I_i z) \wedge y P z$, however, contradicts the result of the previous lemma that Pareto-indifference holds. This contradiction establishes the lemma.  $\square$

**Lemma 11.4** *Let social decision rule $f : \mathcal{Q}^n \mapsto \mathcal{C}$ satisfy independence of irrelevant alternatives and weak Pareto quasi-transitivity. Then, whenever a group of individuals $V$ is almost $(N - A)$−decisive for some ordered pair of distinct alternatives, it is $(N - A)$-decisive for every ordered pair of distinct alternatives, where $A \subset N$ and $V \subseteq N - A$.*

*Proof* Let $f$ satisfy I and WPQT. Let $V$ be almost $(N - A)$−decisive for $(x, y)$; $x \neq y, x, y \in S, A \subset N, V \subseteq N - A$. Let $z$ be an alternative distinct from $x$ and $y$, and consider the following configuration of individual preferences:

$(\forall i \in A)[x I_i y \wedge y P_i z \wedge x I_i z]$

$(\forall i \in V)[x P_i y \wedge y P_i z \wedge x P_i z]$

$(\forall i \in N - (A \cup V))[y P_i x \wedge y P_i z].$

In view of the almost $(N - A)$−decisiveness of $V$ for $(x, y)$ and the fact that $[(\forall i \in A)(x I_i y) \wedge (\forall i \in V)(x P_i y) \wedge (\forall i \in N - (A \cup V))(y P_i x)]$, we obtain $x P y$. From $x P y$ and $(\forall i \in N)(y P_i z)$ we conclude $x P z$ by WPQT. As $(\forall i \in A)(x I_i z)$, $(\forall i \in V)(x P_i z)$, and the preferences of individuals in $N - (A \cup V)$ have not been specified over $\{x, z\}$, it follows, in view of condition I, that $V$ is $(N - A)$−decisive for $(x, z)$. Similarly, by considering the configuration $[(\forall i \in A)(z P_i x \wedge x I_i y \wedge z I_i y) \wedge (\forall i \in V)(z P_i x \wedge x P_i y \wedge z P_i y) \wedge (\forall i \in N - (A \cup V))(z P_i x \wedge y P_i x)]$, we can show $[D_{N-A}(x, y) \rightarrow \overline{D}_{N-A}(z, y)]$. By appropriate interchanges of alternatives it follows that $D_{N-A}(x, y) \rightarrow \overline{D}_{N-A}(a, b)$, for all $(a, b) \in \{x, y, z\} \times \{x, y, z\}$, where $a \neq b$. To prove the assertion for any $(a, b) \in S \times S, a \neq b$, first we note that if $[(a = x \vee a = y) \vee (b = x \vee b = y)]$, the desired conclusion $\overline{D}_{N-A}(a, b)$ can be obtained by considering a triple which includes all of $x$, $y$, $a$ and $b$. If both $a$ and $b$ are different from $x$ and $y$, then one first considers the triple $\{x, y, a\}$ and deduces $\overline{D}_{N-A}(x, a)$ and hence $D_{N-A}(x, a)$, and then considers the triple $\{x, a, b\}$ and obtains $\overline{D}_{N-A}(a, b)$.  $\square$

**Lemma 11.5** *Let social decision rule $f : \mathcal{Q}^n \mapsto \mathcal{C}$ satisfy independence of irrelevant alternatives and weak Pareto quasi-transitivity. Then, whenever a group of individuals $V$ is almost $(N - A)$−semidecisive for some ordered pair of distinct alternatives, it is $(N - A)$−semidecisive for every ordered pair of distinct alternatives, where $A \subset N$ and $V \subseteq N - A$.*

*Proof* Let $f$ satisfy I and WPQT. Let $V$ be almost $(N - A)$−semidecisive for $(x, y)$; $x \neq y, x, y \in S, A \subset N, V \subseteq N - A$. Let $z$ be an alternative distinct from $x$ and $y$, and consider the following configuration of individual preferences:

$(\forall i \in A)[x I_i y \wedge y P_i z \wedge x I_i z]$

$(\forall i \in V)[x P_i y \wedge y P_i z \wedge x P_i z]$

$(\forall i \in N - (A \cup V))[y P_i x \wedge y P_i z].$

From the almost $(N - A)$−semidecisiveness of $V$ for $(x, y)$ and $[(\forall i \in A)(x I_i y) \wedge (\forall i \in V)(x P_i y) \wedge (\forall i \in N - (A \cup V))(y P_i x)]$, we obtain $x R y$. Suppose $z P x$. $(\forall i \in N)(y P_i z)$ and $z P x$ imply $y P x$ by WPQT, which contradicts $x R y$. Therefore $z P x$ cannot be true, which by connectedness of $R$ implies $x R z$. As $(\forall i \in A)(x I_i z)$, $(\forall i \in V)(x P_i z)$ and the preferences of individuals belonging to $N - (A \cup V)$ have not been specified over $\{x, z\}$, $x R z$ implies that $\overline{S}_{N-A}(x, z)$ holds in view of condition I. Thus $S_{N-A}(x, y) \rightarrow \overline{S}_{N-A}(x, z)$. Similarly, by considering the configuration $[(\forall i \in A)(z P_i x \wedge x I_i y \wedge z I_i y) \wedge (\forall i \in V)(z P_i x \wedge x P_i y \wedge z P_i y) \wedge (\forall i \in N - (A \cup V))(z P_i x \wedge y P_i x)]$, we can show that $S_{N-A}(x, y) \rightarrow \overline{S}_{N-A}(z, y)$. By appropriate interchanges of alternatives it follows that $S_{N-A}(x, y)$ implies $\overline{S}_{N-A}(a, b)$ for all $(a, b) \in \{x, y, z\} \times \{x, y, z\}, a \neq b$. To prove the assertion for any $(a, b) \in S \times S, a \neq b$, first we note that if $[(a = x \vee a = y) \vee (b = x \vee b = y)]$, the desired conclusion $\overline{S}_{N-A}(a, b)$ can be obtained by considering a triple which includes all of $x$, $y$, $a$ and $b$. If both $a$ and $b$ are different from $x$ and $y$, then one first considers the triple $\{x, y, a\}$ and deduces $\overline{S}_{N-A}(x, a)$ and hence $S_{N-A}(x, a)$, and then considers the triple $\{x, a, b\}$ and obtains $\overline{S}_{N-A}(a, b)$. □

**Proposition 11.2** *If social decision rule $f : \mathscr{Q}^n \mapsto \mathscr{C}$ satisfies independence of irrelevant alternatives and weak Pareto quasi-transitivity then it is neutral.*

*Proof* Consider any $(R_1, \ldots, R_n), (R'_1, \ldots, R'_n) \in \mathscr{Q}^n$ such that $(\forall i \in N)[(x R_i y \leftrightarrow z R'_i w) \wedge (y R_i x \leftrightarrow w R'_i z)]$, $x, y, z, w \in S$. Designate by $N_1, N_2$ and $N_3$ the sets $\{i \in N \mid x P_i y \wedge z P'_i w\}$, $\{i \in N \mid x I_i y \wedge z I'_i w\}$ and $\{i \in N \mid y P_i x \wedge w P'_i z\}$, respectively. If $N_1 \cup N_3 = \emptyset$, then $x I y$ and $z I' w$ follow from the condition of Pareto-indifference which hold in view of Lemma 11.2.

Now assume that $N_1 \cup N_3 \neq \emptyset$. Nonemptiness of $N_1 \cup N_3$ implies that $x \neq y$ and $z \neq w$. Suppose $x P y$. Then $N_1$ is nonempty (Lemma 11.3) and almost $(N - N_2)$−decisive for $(x, y)$ by condition I. In view of Lemma 11.4 it follows that $N_1$ is $(N - N_2)$−decisive for every ordered pair of distinct alternatives. Therefore $z P' w$ must hold as $[(\forall i \in N_2)(z I'_i w) \wedge (\forall i \in N_1)(z P'_i w)]$. We have shown that $(x P y \rightarrow z P' w)$. By an analogous argument it can be shown that $(z P' w \rightarrow x P y)$. So we have $(x P y \leftrightarrow z P' w)$. Next suppose $y P x$. Then $N_3$ is nonempty (Lemma 11.3) and an almost $(N - N_2)$−decisive set for $(y, x)$ by condition I; and hence an $(N - N_2)$-decisive set by Lemma 11.4. Therefore, we must have $w P' z$ as $[(\forall i \in N_2)(w I'_i z) \wedge (\forall i \in N_3)(w P'_i z)]$. So $(y P x \rightarrow w P' z)$. By a similar argument one obtains $(w P' z \rightarrow y P x)$. Therefore we have $(y P x \leftrightarrow w P' z)$. As $(x P y \leftrightarrow z P' w)$ and $(y P x \leftrightarrow w P' z)$, by the connectedness of $R$ and $R'$ it follows that $(x I y \leftrightarrow z I' w)$. This establishes that the SDR is neutral. □

*Remark 11.3* In Lemma 10.4 it was shown that an SDR with domain $\mathscr{T}^n$ is monotonic iff it is weakly monotonic. From the proof of the lemma it is clear that it holds for the domain $\mathscr{Q}^n$ as well. ◇

**Proposition 11.3** *If social decision rule* $f : \mathcal{Q}^n \mapsto \mathscr{C}$ *satisfies independence of irrelevant alternatives and the weak Pareto quasi-transitivity then it is monotonic.*

*Proof* Let SDR $f : \mathcal{Q}^n \mapsto \mathscr{C}$ satisfy condition I and WPQT. In view of Remark 11.3 it suffices to show that the SDR is weakly monotonic. Consider any $(R_1, \ldots, R_n)$, $(R'_1, \ldots, R'_n) \in \mathcal{Q}^n$, any $x, y \in S$ and any individual $k \in N$ such that $(\forall i \in N - \{k\})[(x R_i y \leftrightarrow x R'_i y) \wedge (y R_i x \leftrightarrow y R'_i x)]$ and $[(y P_k x \wedge x R'_k y) \vee (x I_k y \wedge x P'_k y)]$. Designate by $N_1$, $N_2$ and $N_3$ the sets $\{i \in N \mid x P_i y\}, \{i \in N \mid x I_i y\}, \{i \in N \mid y P_i x\}$, respectively.
If $N_1 \cup N_3 = \emptyset$ then $x I y$ and $x R' y$ follow in view of Lemmas 11.2 and 11.3.
Now let $N_1 \cup N_3 \neq \emptyset$.
Suppose $x P y$. Then $N_1$ is nonempty by Lemma 11.3 and almost $(N - N_2)$−decisive for $(x, y)$ as a consequence of condition I, and hence an $(N - N_2)$−decisive set in view of Lemma 11.4. If $k \in N_3$ then it follows that we must have $x P' y$. Now suppose $k \in N_2$. If $y R' x$ then $N_3$ is almost $(N - (N_2 - \{k\}))$−semidecisive for $(y, x)$ and hence an $(N - (N_2 - \{k\}))$−semidecisive set by Lemma 11.5. As $[(\forall i \in N_3)(y P_i x) \wedge (\forall i \in N_2 - \{k\})(x I_i y)]$, it follows that we must have $y R x$. This, however, contradicts the hypothesis that $x P y$ holds. So $y R' x$ is impossible and therefore $x P' y$ must obtain.

Next suppose $x I y$. Then $N_1$ is an $(N - N_2)$−semidecisive set in view of Lemma 11.5. If $k \in N_3$ then it follows that we must have $x R' y$ as $[(\forall i \in N_1)(x P'_i y) \wedge (\forall i \in N_2)(x I'_i y)]$. Suppose $k \in N_2$ and $y P' x$. $y P' x$ implies that $N_3$ is nonempty (Lemma 11.3) and an $(N - (N_2 - \{k\}))$−decisive set, which in turn implies that $y P x$ must obtain as $[(\forall i \in N_3)(y P_i x) \wedge (\forall i \in N_2 - \{k\})(x I_i y)]$, contradicting the hypothesis of $x I y$. So $y P' x$ is impossible and by the connectedness of $R'$ we conclude that $x R' y$ must hold. Thus we have shown that $[(x P y \rightarrow x P' y) \wedge (x I y \rightarrow x R' y)]$, which establishes that the SDR is weakly monotonic.                                    $\square$

**Proposition 11.4** *Let social decision rule* $f : \mathcal{Q}^n \mapsto \mathscr{C}$ *satisfy independence of irrelevant alternatives, neutrality and monotonicity. Then* $f$ *satisfies weak Pareto quasi-transitivity.*

*Proof* Consider any $x, y, z \in S$ and any $(R_1, \ldots, R_n) \in \mathcal{Q}^n$ such that $[x P y \wedge (\forall i \in N)(y P_i z)]$. Designate by $N_1, N_2, N_3$ the sets $\{i \in N \mid x P_i y\}, \{i \in N \mid x I_i y\}, \{i \in N \mid y P_i x\}$, respectively, and by $N'_1, N'_2, N'_3$ the sets $\{i \in N \mid x P_i z\}, \{i \in N \mid x I_i z\}, \{i \in N \mid z P_i x\}$, respectively. As individual weak preference relations are quasi-transitive, from $(\forall i \in N)(y P_i z)$ we conclude that $N_1 \subseteq N'_1$ and $N'_3 \subseteq N_3$. Let $(R'_1, \ldots, R'_n) \in \mathcal{Q}^n$ be any configuration such that $[(\forall i \in N_1)(x P'_i z) \wedge (\forall i \in N_2)(x I'_i z) \wedge (\forall i \in N'_3)(z P'_i x)]$. As $x P y$, we conclude $x P' z$ by conditions I and $N$. $x P' z$ in turn implies $x P z$ in view of $N_1 \subseteq N'_1$ and $N'_3 \subseteq N_3$, as a consequence of conditions I and $M$. Thus we have shown that $x P y$ and $(\forall i \in N)(y P_i z)$ imply $x P z$. By an analogous argument it can be shown that $(\forall i \in N)(x P_i y)$ and $y P z$ imply $x P z$. This establishes that WPQT holds.                                    $\square$

Combining Propositions 11.2, 11.3 and 11.4 we obtain:

**Theorem 11.9** *A binary social decision rule* $f : \mathcal{Q}^n \mapsto \mathscr{C}$ *is neutral and monotonic iff weak Pareto quasi-transitivity holds.*

### 11.5.2 Characterization of Transitivity

**Theorem 11.10** *A neutral and monotonic binary social decision rule* $f : \mathcal{Q}^n \mapsto \mathscr{C}$ *yields a transitive social weak preference relation $R$ for every* $(R_1, \ldots, R_n) \in \mathcal{Q}^n$ *iff it is null.*

*Proof* If $f$ is null then social weak preference relation for every $(R_1, \ldots, R_n) \in \mathcal{Q}^n$ is transitive.

Let $f$ yield a transitive social weak preference relation for every $(R_1, \ldots, R_n) \in \mathcal{Q}^n$. Suppose for some $(R_1, \ldots, R_n) \in \mathcal{Q}^n$ and some $x, y \in S$, $xPy$ obtains. Designate by $N_1, N_2, N_3$ the sets $\{i \in N \mid x P_i y\}, \{i \in N \mid x I_i y\}, \{i \in N \mid y P_i x\}$, respectively. Let $z \in S$ be an alternative distinct from $x$ and $y$. Consider any $(R'_1, \ldots, R'_n) \in \mathcal{Q}^n$ such that $(\forall i \in N_1)(x P'_i y) \wedge (\forall i \in N_2)(x I'_i y) \wedge (\forall i \in N_3)(y P'_i x) \wedge (\forall i \in N)(y I'_i z \wedge x I'_i z)$. We obtain $x P' y$ by hypothesis and condition I, and $y I' z$ by conditions I and N. $x P' y$ and $y I' z$ imply $x P' z$ by transitivity. But this is a contradiction as $x I' z$ holds in view of $(\forall i \in N)(x I' z)$, by conditions I and N. Therefore we conclude that there do not exist any $(R_1, \ldots, R_n) \in \mathcal{Q}^n$ and $x, y \in S$ such that $xPy$ obtains, i.e., f is null. ∎

In the proof of Theorem 11.10, monotonicity has not been used, and neutrality has been used only to infer Pareto-indifference. Therefore, the following theorem holds:

**Theorem 11.11** *Let* $f : \mathcal{Q}^n \mapsto \mathscr{C}$ *be a binary social decision rule satisfying the condition of Pareto-indifference. Then* $f$ *yields a transitive social weak preference relation for every* $(R_1, \ldots, R_n) \in \mathcal{Q}^n$ *iff it is null.*

### 11.5.3 Alternative Characterization of the Null Social Decision Rule

Weak Pareto quasi-transitivity characterizes neutrality and monotonicity for the class of binary SDRs $f : \mathcal{Q}^n \mapsto \mathscr{C}$. The stronger condition of Pareto quasi-transitivity characterizes neutrality and monotonicity for the class of Paretian binary SDRs $f : \mathcal{T}^n \mapsto \mathscr{C}$. In the context of binary SDRs $f : \mathcal{Q}^n \mapsto \mathscr{C}$, however, Pareto quasi-transitivity characterizes the null social decision rule as is shown by the following theorem.

**Theorem 11.12** *Let the social decision rule* $f : \mathcal{Q}^n \mapsto \mathscr{C}$ *satisfy independence of irrelevant alternatives. Then* $f$ *is null iff Pareto quasi-transitivity holds.*

*Proof* If $f$ is null then $(\forall (R_1, \ldots, R_n) \in \mathscr{Q}^n)(\forall x, y \in S)(\sim x P y)$, and therefore PQT is trivially satisfied.

Let PQT hold. Suppose for some $(R_1, \ldots, R_n) \in \mathscr{Q}^n$ and some $x, y \in S, x P y$ holds. First we show that this implies that there must exist an $(R_1', \ldots, R_n') \in \mathscr{Q}^n$ such that $x P' y$ obtains and $\{i \in N \mid x P_i' y\} \neq N$. If $\{i \in N \mid x P_i y\} \neq N$ then there is nothing to prove. Suppose $\{i \in N \mid x P_i y\} = N$ and let $(N', N - N')$ be a partition of $N$ such that both $N'$ and $N - N'$ are nonempty. Let $z$ be an alternative distinct from $x$ and $y$, and consider the following configuration of individual preferences:

$(\forall i \in N')(x P_i^1 y \wedge y P_i^1 z \wedge x P_i^1 z)$
$(\forall i \in N - N')(x P_i^1 y \wedge y I_i^1 z \wedge x I_i^1 z)$

We have $x P^1 y$ by hypothesis and condition I, and $[(\forall i \in N)(y R_i^1 z) \wedge (\exists i \in N)$ $(y P_i^1 z)]$ by construction. So by PQT we must have $x P^1 z$. Next consider the following configuration:

$(\forall i \in N')(x P_i^2 y \wedge z P_i^2 y \wedge x P_i^2 z)$
$(\forall i \in N - N')(x I_i^2 y \wedge y I_i^2 z \wedge x I_i^2 z)$

We have $x P^2 z$ by our demonstration and condition I, and $[(\forall i \in N)(z R_i^2 y) \wedge (\exists i \in N)(z P_i^2 y)]$ by construction. Therefore by PQT we must have $x P^2 y$. This establishes the claim.

Let $(R_1', \ldots, R_n') \in \mathscr{Q}^n$ be any configuration such that $x P' y$ and $\{i \in N \mid x P_i' y\} \neq N$. Designate by $N_1, N_2, N_3$ the sets $\{i \in N \mid x P_i' y\}, \{i \in N \mid x I_i' y\}, \{i \in N \mid y P_i' x\}$, respectively. Let $z$ be an alternative distinct from $x$ and $y$, and consider the following configuration of individual preferences:

$(\forall i \in N_1)(x P_i'' y \wedge y I_i'' z \wedge x I_i'' z)$
$(\forall i \in N_2)(x I_i'' y \wedge y P_i'' z \wedge x I_i'' z)$
$(\forall i \in N_3)(y P_i'' x \wedge y P_i'' z \wedge x I_i'' z)$

We have $x P'' y$ by hypothesis and condition I, and $[(\forall i \in N)(y R_i'' z) \wedge (\exists i \in N)(y P_i'' z)]$ in view of the fact that $N_1 \neq N$. Therefore we must have $x P'' z$ by PQT. As PQT implies WPQT, it contradicts the result of Lemma 11.2 that Pareto-indifference holds. Therefore it cannot be the case that $(\exists (R_1, \ldots, R_n) \in \mathscr{Q}^n)(\exists x, y \in S)(x P y)$, i.e., $f$ must be null.                                                                                □

### 11.5.4 Characterization of Quasi-transitivity

**Theorem 11.13** *A neutral and monotonic binary social decision rule* $f : \mathscr{Q}^n \mapsto \mathscr{C}$ *yields a quasi-transitive social weak preference relation for every* $(R_1, \ldots, R_n) \in \mathscr{Q}^n$ *iff it is null or it is an oligarchic simple game.*

*Proof* If $f$ is null then social weak preference relation is transitive for every $(R_1, \ldots, R_n) \in \mathscr{Q}^n$. Let $f$ be an oligarchic simple game. Consider any $(R_1, \ldots, R_n) \in \mathscr{Q}^n$ and any $x, y, z \in S$ such that $x P y$ and $y P z$. Let $V$ be the oligarchy. Let $V_1 = \{i \in N \mid x P_i y\}$ and $V_2 = \{i \in N \mid y P_i z\}$. Then it follows that $V_1$ and $V_2$ are decisive sets as $f$ is a simple game. In view of the fact that $f$ is oligarchic it follows that both $V_1$ and $V_2$ contain $V$. Thus $(\forall i \in V)(x P_i y \wedge y P_i z)$. As individual weak preference relations are quasi-transitive, it follows that $(\forall i \in V)(x P_i z)$. So $x P z$ must hold. This establishes that social weak preference relation is quasi-transitive for every $(R_1, \ldots, R_n) \in \mathscr{Q}^n$.

If $f$ yields a transitive social weak preference relation for every $(R_1, \ldots, R_n) \in \mathscr{Q}^n$ then $f$ is null by Theorem 11.10. Suppose that $f$ yields a quasi-transitive social weak preference relation for every $(R_1, \ldots, R_n) \in \mathscr{Q}^n$ but does not yield a transitive social weak preference relation for every $(R_1, \ldots, R_n) \in \mathscr{Q}^n$. Then for some $(R_1, \ldots, R_n) \in \mathscr{Q}^n$ and some $x, y, z \in S$ we must have $(x P y \wedge y I z \wedge x I z)$. $x P y$ implies by conditions I, M and N that the set of all individuals $N$ is a decisive set and therefore WP holds. As a consequence of WP there exists a nonempty set $V \in W$ such that $V$ is minimally decisive. Suppose $V'$ is a minimal decisive set and $V' \neq V$. Consider the following configuration of individual preferences: $[(\forall i \in V \cap V')(x P_i y \wedge y P_i z \wedge x P_i z) \wedge (\forall i \in V - V')(z P_i x \wedge x P_i y \wedge z P_i y) \wedge (\forall i \in N-V)(y P_i z \wedge z P_i x \wedge y P_i x)]$. As $[(\forall i \in V)(x P_i y) \wedge (\forall i \in V')(y P_i z)]$, we obtain $x P y$ and $y P z$, which imply $x P z$ as social $R$ is quasi-transitive. $x P z$, in view of conditions I, M and N, implies that $V \cap V'$ is a decisive set. This, however, leads to a contradiction as the hypothesis of $V$ and $V'$ being distinct minimal decisive sets implies that $V \cap V'$ is not a decisive set. This establishes that $V$ is the unique minimal decisive set.

Let $j \in V$. We will show that $(\forall (R_1, \ldots, R_n) \in \mathscr{Q}^n)(\forall x, y \in S)[x R_j y \rightarrow x R y]$. Suppose not. Then $(\exists (R_1, \ldots, R_n) \in \mathscr{Q}^n)(\exists x, y \in S)[x R_j y \wedge y P x]$. Then by conditions I, M and N we conclude that:
$$(\forall (R_1, \ldots, R_n) \in \mathscr{Q}^n)(\forall x, y \in S)[x I_j y \wedge (\forall i \in N - \{j\})(y P_i x) \rightarrow y P x].$$
$$\text{(T11.13-1)}$$
Now consider the following configuration of individual preferences: $[(\forall i \in N - \{j\})(z P_i y \wedge y P_i x \wedge z P_i x) \wedge (z I_j y \wedge y I_j x \wedge x P_j z)]$, where $x, y, z \in S$ are all distinct. We obtain $z P y$ and $y P x$ by (T11.13-1), which in turn imply $z P x$ by quasi-transitivity. $z P x$ implies that $N - \{j\}$ is a decisive set, by conditions I, M and N. Therefore there exists a nonempty $V' \subseteq N - \{j\}$ such that $V'$ is minimally decisive. As $V' \neq V$, it contradicts the fact that $V$ is the unique minimal decisive set. Therefore we conclude that it is impossible that for some $(R_1, \ldots, R_n) \in \mathscr{Q}^n$ and some $x, y \in S$, $x R_j y$ and $y P x$ hold, i.e., $V$ is a strict oligarchy.

Consider any $(R_1, \ldots, R_n) \in \mathscr{Q}^n$ and any $x, y \in S$. Suppose $x P y$ holds. As $y R_j x$ for some $i \in V$ would imply $y R x$ as shown above, we conclude that $(\forall i \in V)(x P_i y)$. Consequently $\{i \in N \mid x P_i y\}$ is a decisive set. Thus we have shown that $(\forall (R_1, \ldots, R_n) \in \mathscr{Q}^n)(\forall x, y \in S)[x P y \rightarrow (\exists V'' \in W)(\forall i \in V'')(x P_i y)]$, which proves that $f$ is a simple game. $\qquad \square$

The conjunction of WP and quasi-transitivity implies WPQT. WPQT, given condition I, in turn implies neutrality and monotonicity. As in the proof of the first part of Theorem 11.13, neutrality and monotonicity have not been used, we conclude that the following theorem holds:

**Theorem 11.14** *Let $f : \mathcal{Q}^n \mapsto \mathcal{C}$ be a binary social decision rule satisfying the weak Pareto-criterion. Then, $f$ yields a quasi-transitive social weak preference relation for every $(R_1, \dots, R_n) \in \mathcal{Q}^n$ iff $f$ is an oligarchic simple game.*

*Remark 11.4* An SDR $f : \mathcal{Q}^n \mapsto \mathcal{C}$ is an oligarchic simple game iff it is strictly oligarchic.

*Proof* Suppose $f : \mathcal{Q}^n \mapsto \mathcal{C}$ is an oligarchic simple game. Let $V$ be the oligarchy. Then $V$ is the unique minimal decisive set. Consequently $(\forall V' \in W)(V \subseteq V')$. As $f$ is a simple game it follows that $(\forall (R_1, \dots, R_n) \in \mathcal{Q}^n)(\forall x, y \in S)[xPy \leftrightarrow (\exists V' \in W)(\forall i \in V')(xP_i y)]$. This implies that $(\forall (R_1, \dots, R_n) \in \mathcal{Q}^n)(\forall x, y \in S)[xPy \leftrightarrow (\forall i \in V)(xP_i y)]$, as $V \in W$ and $(\forall V' \in W)(V \subseteq V')$. This in turn implies $(\forall (R_1, \dots, R_n) \in \mathcal{Q}^n)(\forall x, y \in S)[yRx \leftrightarrow (\exists i \in V)(yR_i x)]$, which establishes that $f$ is strictly oligarchic.

Now suppose that $f$ is strictly oligarchic. Then $f$ is obviously oligarchic. Let $V$ be the strict oligarchy. Then by the definition of strict oligarchy, we obtain $(\forall (R_1, \dots, R_n) \in \mathcal{Q}^n)(\forall x, y \in S)[(\exists i \in V)(yR_i x) \to yRx]$, which is equivalent to $(\forall (R_1, \dots, R_n) \in \mathcal{Q}^n)(\forall x, y \in S)[xPy \to (\forall i \in V)(xP_i y)]$. As $V \in W$ it follows that $(\forall (R_1, \dots, R_n) \in \mathcal{Q}^n)(\forall x, y \in S)[xPy \to (\exists V' \in W)(\forall i \in V')(xP_i y)]$. From the definition of a decisive set then it follows that $(\forall (R_1, \dots, R_n) \in \mathcal{Q}^n)(\forall x, y \in S)[xPy \leftrightarrow (\exists V' \in W)(\forall i \in V')(xP_i y)]$. This establishes that $f$ is a simple game.                                                                              $\Diamond$

   In view of the above remark, the expression 'oligarchic simple game' in the statements of Theorems 11.13 and 11.14 can be replaced by the expression 'strict oligarchy'.

## 11.5.5  Characterization of Acyclicity

**Theorem 11.15** *Let $f : \mathcal{Q}^n \mapsto \mathcal{C}$ be a neutral and monotonic binary social decision rule. Then, $f$ yields an acyclic social weak preference relation for every $(R_1, \dots, R_n) \in \mathcal{Q}^n$ iff there does not exist a collection $\{V_1, \dots, V_m\}$ of nonempty subsets of the set of individuals $N$ such that:*
*(a) for each $j \in \{1, 2, \dots, m\}$, $V_j$ is $(N - A_j)$-decisive for some $A_j \subset N$, $V_j \cap A_j = \emptyset$*
*(b) for each $j \in \{1, 2, \dots, m\}$, $(V_j \cup A_j) \cap V_{\sim j} = \emptyset$; where $V_{\sim j}$, $j = 1, 2, \dots, m$, is defined by: $V_{\sim j} = \bigcap_k V_k$, $k \in \{1, 2, \dots, m\} - \{j\}$*
*(c) $3 \le m \le \#S$.*

*Proof* Suppose acyclicity is violated. Then for some $(R_1, \dots, R_n) \in \mathcal{Q}^n$ and some distinct $x_1, x_2, \dots, x_m \in S$ we must have $(x_1 P x_2 \wedge \dots \wedge x_{m-1} P x_m \wedge x_m P x_1)$, where $3 \le m \le \#S$. Let $A_j = \{i \in N \mid x_j I_i x_{j+1}\}$, $j = 1, 2, \dots, (m-1)$; $A_m = \{i \in N \mid x_m I_i x_1\}$; $V_j = \{i \in N \mid x_j P_i x_{j+1}\}$, $j = 1, 2, \dots, (m-1)$ and $V_m = \{i \in N \mid x_m P_i x_1\}$. Thus we have $V_j \cap A_j = \emptyset$, $j = 1, 2, \dots, m$. By neutrality and monotonicity, it follows that for each $j \in \{1, 2, \dots, m\}$, $V_j$ is nonempty and consequently $A_j \neq N$. By a further appeal to monotonicity and neutrality we conclude that

$V_j$ is $(N - A_j)$-decisive, $j = 1, 2, ..., m$. As individual weak preference relations are quasi-transitive, it follows that $(\forall i \in V_{\sim j})(x_{j+1} P_i x_j)$, $j = 1, 2, ..., (m - 1)$, and $(\forall i \in V_{\sim m})(x_1 P x_m)$. As $[(\forall i \in V_j)(x_j P_i x_{j+1}) \wedge (\forall i \in A_j)(x_j I_i x_{j+1})]$, $j = 1, 2, ..., (m - 1)$, and $[(\forall i \in V_m)(x_m P_i x_1) \wedge (\forall i \in A_m)(x_m I_i x_1)]$, it follows that $(A_j \cup V_j) \cap V_{\sim j} = \emptyset$, $j = 1, 2, ..., m$. This proves that the violation of acyclicity implies the existence of a collection $\{V_1, ..., V_m\}$ of nonempty subsets of the set of individuals $N$ satisfying (a), (b) and (c) mentioned in the statement of the theorem.

Next suppose that there exists a collection $\{V_1, ..., V_m\}$ of nonempty subsets of $N$ such that (a), (b) and (c) hold. Consider the following configuration of individual preferences:

$(\forall i \in A_1)(x_1 I_i x_2) \wedge (\forall i \in V_1)(x_1 P_i x_2)$

$(\forall i \in A_2)(x_2 I_i x_3) \wedge (\forall i \in V_2)(x_2 P_i x_3)$

$\cdots$

$\cdots$

$(\forall i \in A_{m-1})(x_{m-1} I_i x_m) \wedge (\forall i \in V_{m-1})(x_{m-1} P_i x_m)$

$(\forall i \in A_m)(x_m I_i x_1) \wedge (\forall i \in V_m)(x_m P_i x_1)$.

It is possible to have the above configuration of preferences without violating quasi-transitivity of individual weak preference relations because for each $j \in 1, 2, ..., m$, it is given that $(V_j \cup A_j) \cap V_{\sim j} = \emptyset$. As $V_j$ is $(N - A_j)$-decisive, $j = 1, 2, ..., m$, we conclude that $(x_1 P x_2 \wedge ... \wedge x_{m-1} P x_m \wedge x_m P x_1)$ holds, which violates acyclicity. This establishes the theorem. $\qquad\square$

## 11.5.6 Condition for Quasi-transitivity

**Theorem 11.16** *Let $f : \mathscr{D}^n \mapsto \mathscr{C}$ satisfy conditions I, M and N. Let $\mathscr{D} \subseteq \mathscr{Q}$. If $\mathscr{D}$ satisfies the condition of value restriction (2) then $f$ yields quasi-transitive social $R$, $R = f(R_1, ..., R_n)$, for every $(R_1, ..., R_n) \in \mathscr{D}^n$.*

*Proof* Suppose $f$ does not yield quasi-transitive $R$ for every $(R_1, ..., R_n) \in \mathscr{D}^n$. Then $(\exists (R_1, ..., R_n) \in \mathscr{D}^n)(\exists x, y, z \in S)[x P y \wedge y P z \wedge z R x]$.

In view of conditions I, M and N it follow that we must have:

$$\sim (\forall i \in N)[(x P_i y \to z P_i y) \wedge (x I_i y \to z R_i y)] \qquad (T11.16\text{-}1)$$

otherwise we will obtain $z P y$ in view of $x P y$, contradicting the hypothesis $y P z$. Similarly,

$$y P z \wedge z R x \to \sim (\forall i \in N)[(y P_i z \to x P_i z) \wedge (y I_i z \to x R_i z)], \text{ and} \qquad (T11.16\text{-}2)$$

$$x P y \wedge z R x \to \sim (\forall i \in N)[(x P_i y \to x P_i z) \wedge (x I_i y \to x R_i z)]. \qquad (T11.16\text{-}3)$$

$(T11.16\text{-}1) \to (\exists i \in N)[(x P_i y \wedge y R_i z) \vee (x I_i y \wedge y P_i z)]$

$\to (\exists i \in N)[(x P_i y \wedge y P_i z \wedge x P_i z) \vee (x P_i y \wedge y I_i z \wedge x P_i z) \vee (x P_i y \wedge y I_i z \wedge x I_i z) \vee (x I_i y \wedge y P_i z \wedge x P_i z) \vee (x I_i y \wedge y P_i z \wedge x I_i z)]$

$\to (\exists i \in N)[[R_i \text{ is concerned over } \{x, y, z\} \wedge x R_i y R_i z] \vee (x P_i y \wedge y I_i z \wedge x I_i z) \vee (x I_i y \wedge y P_i z \wedge x I_i z)]$

$\rightarrow (\exists i \in N)[x$ is best in $R_i \wedge y$ is proper medium in $R_i \wedge z$ is worst in $R_i]$

(T11.16-4)

(T11.16-2) $\rightarrow (\exists i \in N)[(yP_iz \wedge zR_ix) \vee (yI_iz \wedge zP_ix)]$

$\rightarrow (\exists i \in N)[(yP_iz \wedge zP_ix \wedge yP_ix) \vee (yP_iz \wedge zI_ix \wedge yP_ix) \vee (yP_iz \wedge zI_ix \wedge yI_ix) \vee (yI_iz \wedge zP_ix \wedge yP_ix) \vee (yI_iz \wedge zP_ix \wedge yI_ix)]$

$\rightarrow (\exists i \in N)[[R_i$ is concerned over $\{x, y, z\} \wedge yR_izR_ix] \vee (yP_iz \wedge zI_ix \wedge yI_ix) \vee (yI_iz \wedge zP_ix \wedge yI_ix)]$

$\rightarrow (\exists i \in N)[y$ is best in $R_i \wedge z$ is proper medium in $R_i \wedge x$ is worst in $R_i]$

(T11.16-5)

(T11.16-3) $\rightarrow (\exists i \in N)[(zR_ix \wedge xP_iy) \vee (zP_ix \wedge xI_iy)]$

$\rightarrow (\exists i \in N)[(zP_ix \wedge xP_iy \wedge zP_iy) \vee (zI_ix \wedge xP_iy \wedge zP_iy) \vee (zI_ix \wedge xP_iy \wedge zI_iy) \vee (zP_ix \wedge xI_iy \wedge zP_iy) \vee (zP_ix \wedge xI_iy \wedge zI_iy)]$

$\rightarrow (\exists i \in N)[[R_i$ is concerned over $\{x, y, z\} \wedge zR_ixR_iy] \vee (zI_ix \wedge xP_iy \wedge zI_iy) \vee (zP_ix \wedge xI_iy \wedge zI_iy)]$

$\rightarrow (\exists i \in N)[z$ is best in $R_i \wedge x$ is proper medium in $R_i \wedge y$ is worst in $R_i]$

(T11.16-6)

(T11.16-4)–(T11.16-6) imply that $\mathscr{D}$ violates VR(2) over $\{x, y, z\}$. The theorem, therefore, stands established.                                              $\square$

## Appendix

## 11.6   Notes on Literature

The most important contribution in the context of conditions for quasi-transitivity under the method of majority decision, when individuals have reflexive, connected and quasi-transitive weak preference relations over the set of social alternatives, is that of Inada (1970). In this paper, Inada generalized the conditions of value restriction (2) and of limited agreement for the case of individual weak preference relations being reflexive, connected and quasi-transitive; and showed that the satisfaction over every triple of alternatives of the generalized version of value restriction (2) or of the generalized version of limited agreement or of dichotomous preferences or of antagonistic preferences is sufficient for quasi-transitivity under the method of majority decision, when individual weak preference relations are reflexive, connected and quasi-transitive. These four conditions were combined into a single condition in Jain (1986a). This single condition has been called in this volume as Latin Square partial agreement-Q. Given that individual weak preference relations are reflexive, connected and quasi-transitive, Inada (1970) also showed that satisfaction of at least one of the above four conditions was necessary for quasi-transitivity under the method of majority decision, the term 'necessary' having the sense as in Inada (1969). In Jain (2009) it is shown that Latin Square partial agreement-Q completely characterizes the sets $\mathscr{D}$ of reflexive, connected and quasi-transitive weak preference relations over the set of social alternatives such that every profile belonging to $\mathscr{D}^n$ yields quasi-transitive social weak preference relation under the method of majority decision, provided the number of individuals is at least five. Complete characterizations of sets $\mathscr{D}$ of reflexive, connected and quasi-transitive weak preference relations over the set of social alternatives such that every profile belonging to $\mathscr{D}^n$ yields quasi-transitive social weak preference relation under the method of majority decision, for the cases of 2, 3 and 4 individuals were obtained in Jain (2009).

When individual weak preference relations are reflexive, connected and quasi-transitive, the sufficiency of Latin Square partial agreement-Q for quasi-transitivity under every special majority rule was established in Jain (1986a). Complete characterization of the sets $\mathscr{D}$ of reflexive, connected and quasi-transitive weak preference relations over the set of social alternatives such that every profile belonging to $\mathscr{D}^n$ yields quasi-transitive social weak preference relation under social decision rules that are simple games was obtained in Jain (1989).

The characterization of monotonicity and neutrality for the class of binary social decision rules with domain $\mathscr{Q}^n$ given in this chapter is from Jain (1996). Characterization of rationality conditions for the class of binary social decision rules with domain $\mathscr{Q}^n$ and satisfying the conditions of neutrality and monotonicity is also from Jain (1996).

The sufficiency of value restriction (2) for quasi-transitivity under every neutral and monotonic binary social decision rule, when individual weak preference relations are reflexive, connected and quasi-transitive, was shown in Pattanaik (1970). The proof of the theorem given in this chapter, however, is different.

## 11.7  Domain Conditions for Acyclicity

Let $f$ be a social decision rule. Unlike transitivity and quasi-transitivity, condition of acyclicity is not defined over triples. Consequently, there is no reason to expect existence of conditions defined only over triples which can completely characterize all $\mathscr{D}$, $\mathscr{D} \subseteq \mathscr{B}$, which are such that all logically possible $(R_1, \ldots, R_n) \in \mathscr{D}^n$ give rise to acyclic social $R$, $R = (R_1, \ldots, R_n)$, under $f$. In fact, if $\mathscr{B} = \mathscr{Q}$, then the subsets $\mathscr{D} \subseteq \mathscr{B}$ which are such that all logically possible $(R_1, \ldots, R_n) \in \mathscr{D}^n$ give rise to acyclic social $R$ under the MMD cannot be characterized by a condition defined only over triples as the following theorem shows.

**Theorem 11.17** *Let $f$ be the method of majority decision; and let $\#S = s \geq 4$ and $\#N = n \geq 2$. Let $\mathscr{D}_{\mathscr{Q}} = \{\mathscr{D} \subseteq \mathscr{Q} \mid (\forall (R_1, \ldots, R_n) \in \mathscr{D}^n)(R = f(R_1, \ldots, R_n)$ is acyclic)\}. Then, there does not exist any condition $\alpha$ defined only over triples such that $\mathscr{D}$, $\mathscr{D} \in 2^{\mathscr{Q}} - \{\emptyset\}$, belongs to $\mathscr{D}_{\mathscr{Q}}$ iff it satisfies condition $\alpha$.*

*Proof* Let condition $\alpha$ defined only over triples be such that $\mathscr{D}, \mathscr{D} \in 2^{\mathscr{D}} - \{\emptyset\}$, belongs to $\mathscr{D}_{\mathscr{Q}}$ iff it satisfies condition $\alpha$. Let $S = \{x, y, z, w, t_1, \ldots, t_{s-4}\}$. Consider $\mathscr{D} = \{(xPy, yIz, xIz; x, y, zPwPt_1P \ldots Pt_{s-4}), (yPz, zIx, yIx; x, y, zPwPt_1 P \ldots Pt_{s-4})\}$. It is immediate that the MMD yields acyclic $R$ for every $(R_1, \ldots, R_n)$ $\in \mathscr{D}^n$; and consequently it follows that $\mathscr{D} \in \mathscr{D}_{\mathscr{Q}}$. As condition $\alpha$ is defined only over triples, it follows that $\mathscr{D}$ must be satisfying $\alpha$ over every triple of alternatives. Therefore it follows that if $A \subseteq S$ is a triple and $(\exists$ distinct $a, b, c \in A)[\mathscr{D}|\{a, b, c\} = \{(aPb \wedge bIc \wedge aIc), (aIb \wedge bPc \wedge aIc)\} \vee \mathscr{D}|\{a, b, c\} = \{aPbPc, aIbPc\}]$ then $\mathscr{D}$ would satisfy $\alpha$ over $A$.

Now consider the following $(R_1, \ldots, R_n)$.

$(xP_1y, yI_1z, zP_1w, wI_1x, xI_1z, yI_1w; x, y, z, wP_1t_1P_1 \ldots P_1t_{s-4})$

$(\forall i \in N - \{1\})(xI_iy, yP_iz, zI_iw, wP_ix, xI_iz, yI_iw; x, y, z, wP_it_1P_i \ldots P_it_{s-4})$.

The $R$ yielded by the MMD for the above configuration is: $(xPy, yPz, zPw, wPx, xIz, yIw; x, y, z, wPt_1P \ldots Pt_{s-4})$, which violates acyclicity. Now, for every triple of alternatives $A \subseteq S$ we have $(\exists$ distinct $a, b, c \in A)[\{R_i|A|i \in N\} \subseteq \{(aPb, bIc, aIc), (bPc, cIa, bIa)\}] \vee (\exists$ distinct $a, b, c \in A)[\{R_i|A \mid i \in N\} \subseteq \{aPbPc, aIb Pc\}]$. Therefore, it follows that either $\{(aPb \wedge bIc \wedge aIc), (aIb \wedge bPc \wedge aIc)\}$ or $\{aPbPc, aIbPc\}$ must be violating $\alpha$, contradicting the earlier conclusion that both of these sets satisfy $\alpha$. This contradiction establishes the theorem.          $\square$

From the above theorem the following corollary follows immediately.

**Corollary 11.1** *Let $f$ be the method of majority decision; and let $\#S = s \geq 4$ and $\#N = n \geq 2$. Let $\mathscr{D}_{\mathscr{A}} = \{\mathscr{D} \subseteq \mathscr{A} \mid (\forall(R_1, \ldots, R_n) \in \mathscr{D}^n)(R = f(R_1, \ldots, R_n)$ is acyclic)\}. Then, there does not exist any condition $\alpha$ defined only over triples such that $\mathscr{D}, \mathscr{D} \in 2^{\mathscr{A}} - \{\emptyset\}$, belongs to $\mathscr{D}_{\mathscr{A}}$ iff it satisfies condition $\alpha$.*

If the number of individuals is greater than or equal to 11 then the sets $\mathscr{D} \subseteq \mathscr{T}$ which are such that all logically possible $(R_1, \ldots, R_n) \in \mathscr{D}^n$ give rise to acyclic social $R$ under the MMD, however, can be characterized by a condition defined only over triples. The following theorem can easily be proved.

**Theorem 11.18** *Let $\#S \geq 3$ and $\#N = n \geq 11$. Let $\mathscr{D} \subseteq \mathscr{T}$. Then the method of majority decision $f$ yields acyclic social $R$, $R = f(R_1, \ldots, R_n)$, for every $(R_1, \ldots, R_n) \in \mathscr{D}^n$ iff $\mathscr{D}$ satisfies the condition of Latin Square partial agreement.*[2]

As far as the remaining cases are concerned, there is no uniformity among them. In some cases it can be shown that no condition defined only over triples can be a characterizing condition while in some other cases it is possible to formulate a characterizing condition defined only over triples. For instance, it can be shown that if the number of individuals is four then there does not exist any condition defined

---

[2]See Sen and Pattanaik (1969) and Kelly (1974).

only over triples which can characterize the sets of orderings which invariably give rise to acyclic social $R$. On the other hand the validity of Theorem 11 can be shown for $n = 9$ as well.[3]

## References

Inada, Ken-ichi. 1969. The simple majority decision rule. *Econometrica* 37: 490–506.

Inada, Ken-ichi. 1970. Majority rule and rationality. *Journal of Economic Theory* 2: 27–40.

Jain, Satish K. 1989. Characterization theorems for social decision rules which are simple games. Paper presented at the IX World Congress of the International Economic Association, Economics Research Center, Athens School of Economics & Business, Athens, Greece, held on 28 August–September 1, 1989.

Jain, Satish K. 1986a. Special majority rules: Necessary and sufficient condition for quasi-transitivity with quasi-transitive individual preferences. *Social Choice and Welfare* 3: 99–106.

Jain, Satish K. 1996b. Structure of neutral and monotonic binary social decision rules with quasi-transitive individual preferences. *Journal of Economics (Zeitschrift für Nationalökonomie)* 64: 195–212.

Jain, Satish K. 2009. The method of majority decision and rationality conditions. In *Ethics, welfare, and measurement, Volume 1 of Arguments for a better world: Essays in honor of Amartya Sen*, ed. Kaushik Basu, and Ravi Kanbur, 167–192. New York: Oxford University Press.

Kelly, J.S. 1974. Necessity conditions in voting theory. *Journal of Economic Theory* 8: 149–160.

Pattanaik, Prasanta K. 1970. On social choice with quasitransitive individual preferences. *Journal of Economic Theory* 2: 267–275.

Sen, Amartya K., and Prasanta K. Pattanaik. 1969. Necessary and sufficient conditions for rational choice under majority decision. *Journal of Economic Theory* 1: 178–202.

---

[3] See Kelly (1974).

This monograph has been mainly concerned with the Inada-type necessary and sufficient conditions for transitivity and quasi-transitivity under a variety of social decision rules. The social decision rules and the classes of social decision rules that have been analysed from this perspective are: the method of majority decision, the class of special majority rules, the class of simple game social decision rules and its subclass of strict majority rules, the class of semi-strict majority rules, the class of Pareto-inclusive strict majority rules and the class of neutral and monotonic binary social decision rules.

For the method of majority decision, complete results exist on the domain conditions for transitivity and quasi-transitivity. We have the following: (i) When the number of individuals is even and greater than or equal to two, an Inada-type necessary and sufficient condition for transitivity is that the extremal restriction holds over every triple of alternatives. (ii) When the number of individuals is odd and greater than or equal to three, an Inada-type necessary and sufficient condition for transitivity is that the weak Latin Square partial agreement holds over every triple of alternatives. (iii) When the number of individuals is greater than or equal to five, an Inada-type necessary and sufficient condition for quasi-transitivity is that the Latin Square partial agreement holds over every triple of alternatives. (iv) When the number of individuals is four, an Inada-type necessary and sufficient condition for quasi-transitivity is that the weak extremal restriction holds over every triple of alternatives. (v) When the number of individuals is three, an Inada-type necessary and sufficient condition for quasi-transitivity is that the Latin Square linear ordering restriction holds over every triple of alternatives.

For the class of social decision rules that are simple games also complete results exist on the domain conditions for transitivity and quasi-transitivity. We have the following: (i) A simple game social decision rule yields transitive social weak preference relation for every logically possible profile of individual orderings iff it is

© Springer Nature Singapore Pte Ltd. 2019
S. K. Jain, *Domain Conditions and Social Rationality*,
https://doi.org/10.1007/978-981-13-9672-4_12

null or dictatorial. (ii) For a non-null non-strong simple game social decision rule an Inada-type necessary and sufficient condition for transitivity is that the condition of Latin Square extremal value restriction holds over every triple of alternatives. (iii) For a non-dictatorial strong simple game social decision rule an Inada-type necessary and sufficient condition for transitivity is that the condition of weak Latin Square extremal value restriction holds over every triple of alternatives. (iv) A simple game social decision rule yields quasi-transitive social weak preference relation for every logically possible profile of individual orderings iff it is null or is such that there is a unique minimal winning coalition. (v) For a non-null simple game social decision rule with at least two minimal winning coalitions an Inada-type necessary and sufficient condition for quasi-transitivity is that the condition of Latin Square unique value restriction holds over every triple of alternatives.

The class of strict majority rules is a subclass of the class of social decision rules that are simple games; consequently, we have complete results on the domain conditions for transitivity and quasi-transitivity for it. These are: (i) For a $p$-strict majority rule such that there exists a partition $(N_1, N_2)$ of the set of individuals $N$ such that $\#N_1 \le pn \wedge \#N_2 \le pn$ an Inada-type necessary and sufficient condition for transitivity is that the condition of Latin Square extremal value restriction holds over every triple of alternatives. (ii) For a $p$-strict majority rule such that there does not exist a partition $(N_1, N_2)$ of the set of individuals $N$, $\#N \ge 3$, such that $\#N_1 \le pn \wedge \#N_2 \le pn$ an Inada-type necessary and sufficient condition for transitivity is that the condition of weak Latin Square extremal value restriction holds over every triple of alternatives. (iii) If $n = \lfloor pn \rfloor + 1$, then $p$-strict majority rule yields quasi-transitive social weak preference relation for every logically possible profile of individual orderings. (iv) For a $p$-strict majority rule, if $n > \lfloor pn \rfloor + 1$, then an Inada-type necessary and sufficient condition for quasi-transitivity is that the condition of Latin Square unique value restriction holds over every triple of alternatives. In view of these results it follows that for the strict majority rule we have: (i) When the number of individuals is even and greater than or equal to two, an Inada-type necessary and sufficient condition for transitivity is that the condition of Latin Square extremal value restriction holds over every triple of alternatives. (ii) When the number of individuals is odd and greater than or equal to three, an Inada-type necessary and sufficient condition for transitivity is that the weak Latin Square extremal value restriction holds over every triple of alternatives. (iii) When the number of individuals is greater than or equal to three, an Inada-type necessary and sufficient condition for quasi-transitivity is that the condition of Latin Square unique value restriction holds over every triple of alternatives.

The results on the domain conditions for transitivity and quasi-transitivity for other classes of social decision rules considered in this monograph are less complete. For the class of special majority rules we have the following: (i) Satisfaction of placement restriction over every triple of alternatives is a sufficient condition for transitivity under every special majority rule. (ii) For every special majority rule, satisfaction of placement restriction over every triple of alternatives is an Inada-type necessary and sufficient condition for transitivity for infinitely many values of $n$. (iii) Satisfaction of Latin Square partial agreement over every triple of alternatives is a sufficient condition for quasi-transitivity under every special majority rule. (iv) For every special majority rule, there exists a positive integer $n'$ such that the satisfaction of Latin Square partial agreement over every triple of alternatives is an Inada-type necessary and sufficient condition for all $n \geq n'$. Two-thirds majority rule belongs to the class of special majority rules. For the two-thirds majority rule we have: (i) If the number of individuals $n \geq 10$, then the satisfaction of placement restriction over every triple of alternatives is an Inada-type necessary and sufficient condition for transitivity. (ii) If the number of individuals $n \geq 10$, then the satisfaction of Latin Square partial agreement over every triple of alternatives is an Inada-type necessary and sufficient condition for quasi-transitivity. From the transitivity result for the two-thirds majority rule it seems that it might be the case that for every special majority rule there exists a positive integer $n'$ such that the satisfaction of placement restriction over every triple of alternatives is an Inada-type necessary and sufficient condition for transitivity for all $n \geq n'$.

The satisfaction over every triple of alternatives of at least one of the three conditions of strict placement restriction, partial agreement and strongly antagonistic preferences (1) is a sufficient condition for transitivity under every $p$-semi-strict majority rule. It is also the case that for every $p$-semi-strict majority rule there exists a positive even integer $n'$ such that for every even integer $n \geq n'$ the satisfaction over every triple of alternatives of at least one of the three conditions of strict placement restriction, partial agreement and strongly antagonistic preferences (1) is an Inada-type necessary and sufficient condition for transitivity. From the proofs of the concerned theorems it is clear that the transitivity conditions for an odd number of individuals are likely to be weaker than the above ones, as indeed is the case with both the method of majority decision and the strict majority rule. The satisfaction over every triple of alternatives of value restriction (2) or absence of unique extremal value is a sufficient condition for quasi-transitivity under every $p$-semi-strict majority rule. It is also the case that for every $p$-semi-strict majority rule there exist infinitely many values of $n$ for which the satisfaction over every triple of alternatives of value restriction (2) or absence of unique extremal value is an Inada-type necessary and sufficient condition for quasi-transitivity.

For the class of Pareto-inclusive strict majority rules we have the following results: (i) If $n = 2$ then the satisfaction over every triple of alternatives of extremal restriction is an Inada-type necessary and sufficient condition for transitivity under every Pareto-inclusive strict majority rule. (ii) For Pareto-inclusive $p$-strict majority rule, if $n \geq 3$ and $n = \lfloor pn \rfloor + 1$, then the satisfaction over every triple of alternatives of at least one of the three conditions of placement restriction, absence of unique extremal value and strongly antagonistic preferences (2) is an Inada-type necessary and sufficient condition for transitivity. (iii) The satisfaction over every triple of alternatives of placement restriction or absence of unique extremal value is a sufficient condition for transitivity under every Pareto-inclusive strict majority rule. (iv) For the class of Pareto-inclusive strict majority rules, the satisfaction over every triple of alternatives of placement restriction or absence of unique extremal value is maximally sufficient for transitivity. (v) If $n = \lfloor pn \rfloor + 1$, then Pareto-inclusive $p$-strict majority rule yields quasi-transitive weak preference relation for every logically possible profile of individual orderings. (vi) If $n = \lfloor pn \rfloor + 2$, then the satisfaction over every triple of alternatives of Latin Square linear ordering restriction is an Inada-type necessary and sufficient condition for quasi-transitivity under Pareto-inclusive $p$-strict majority rule. (vii) If $n > \lfloor pn \rfloor + 2$, then the satisfaction over every triple of alternatives of Latin Square unique value restriction or limited agreement is an Inada-type necessary and sufficient condition for quasi-transitivity under Pareto-inclusive $p$-strict majority rule. Thus for the class of Pareto-inclusive strict majority rules, while the results for quasi-transitivity are complete, for transitivity they are not.

For the class of neutral and monotonic binary social decision rules we have the following results: (i) The satisfaction over every triple of alternatives of strict placement restriction is a sufficient condition for transitivity under every neutral and monotonic binary social decision rule. (ii) For the class of neutral and monotonic binary social decision rules, the satisfaction over every triple of alternatives of strict placement restriction is maximally sufficient for transitivity. (iii) The satisfaction over every triple of alternatives of value restriction (2) is a sufficient condition for quasi-transitivity under every neutral and monotonic binary social decision rule. Whether for the class of neutral and monotonic binary social decision rules satisfaction of value restriction (2) over every triple of alternatives is maximally sufficient is not known.

For the class of neutral and monotonic binary social decision rules satisfying the Pareto-criterion we have the following results: (i) The satisfaction over every triple of alternatives of placement restriction is a sufficient condition for transitivity under every neutral and monotonic binary social decision rule satisfying the Pareto-criterion. (ii) For the class of neutral and monotonic binary social decision rules satisfying the Pareto-criterion, the satisfaction over every triple of alternatives of placement restriction is maximally sufficient for transitivity. (iii) The satisfaction over every triple of alternatives of value restriction (2) or limited agreement is a sufficient condition for quasi-transitivity under every neutral and monotonic binary social decision rule satisfying the Pareto-criterion. Whether for the class of neutral and monotonic binary social decision rules satisfying the Pareto-criterion satisfaction of value restriction (2) or limited agreement over every triple of alternatives is maximally sufficient is not known.

All the above results have been derived under the assumption that every individual has an ordering over the set of social alternatives. Statements of domain conditions for the class of neutral and monotonic binary social decision rules become extremely simple if it is assumed that every individual has a linear ordering of social alternatives. We have the following results when individuals have linear orderings over the set of social alternatives: (i) A neutral and monotonic binary social decision rule yields transitive social weak preference relation for every logically possible profile of individual linear orderings iff it is null or dictatorial. (ii) For a non-dictatorial neutral and monotonic binary social decision rule which is such that for every partition $(N_1, N_2)$ of the set of individuals $N$, $N_1$ or $N_2$ is a decisive set, an Inada-type necessary and sufficient condition for transitivity is that there be no Latin Square over any triple of alternatives. (iii) For a non-null neutral and monotonic binary social decision rule which is such that there exists a partition $(N_1, N_2)$ of the set of individuals $N$ such that neither $N_1$ nor $N_2$ is a decisive set, an Inada-type necessary and sufficient condition for transitivity is that over every triple of alternatives there be at most one ordering belonging to any particular Latin Square. (iv) A neutral and monotonic binary social decision rule yields quasi-transitive social weak preference relation for every logically possible profile of individual linear orderings iff there is at most one minimal decisive set. (v) For a neutral and monotonic binary social decision rule which is such that there are at least two minimal decisive sets, an Inada-type necessary and sufficient condition for quasi-transitivity is that there be no Latin Square over any triple of alternatives.

Some results are available for the case of when individual weak preference relations are reflexive, connected and quasi-transitive. For the method of majority decision we have: (i) When the number of individuals is greater than or equal to five, an Inada-type necessary and sufficient condition for quasi-transitivity is that the Latin Square partial agreement—Q holds over every triple of alternatives. (ii) When the number of individuals is four, an Inada-type necessary and sufficient condition for quasi-transitivity is that the weak extremal restriction—Q holds over every triple of alternatives. (iii) When the number of individuals is three, an Inada-type necessary and sufficient condition for quasi-transitivity is that the Latin Square linear ordering restriction—Q holds over every triple of alternatives. (iv) When the number of individuals is two, an Inada-type necessary and sufficient condition for quasi-transitivity is that the Latin Square intransitive relation restriction—Q holds over every triple of alternatives. For quasi-transitivity under the class of social decision rules that are simple games we have the same results as in the case of individuals having orderings over social alternatives. For the class of neutral and monotonic binary social decision rules, satisfaction of VR (2) over every triple of alternatives continues to be sufficient for quasi-transitivity.

For acyclicity, excepting for the method of majority decision, there are no results. For the method of majority decision, for $n \geq 11$ and for $n = 9$, Inada-type necessary and sufficient condition for acyclicity is the same as for quasi-transitivity. It is of considerable interest, in view of the importance of the method of majority decision, to find out whether it is possible to formulate Inada-type necessary and sufficient conditions for acyclicity for any of the remaining cases defined only in terms of triples of alternatives.